POLYMER ADDITIVES

POLYMER SCIENCE AND TECHNOLOGY

Recent volumes in the series:

A Continuation Order Plan is available for this series. A continuation order will bring delivery of each new volume immediately upon publication. Volumes are billed only upon actual shipment. For further information please contact the publisher.

POLYMER ADDITIVES

Edited by
Jiri E. Kresta
Polymer Institute
University of Detroit
Detroit, Michigan

Plenum Press • New York and London

Library of Congress Cataloging in Publication Data

International Symposium on Polymer Additives
  (1982: Las Vegas, Nev.)
  Polymer additives.

  (Polymer science and technology; v. 26)
  "Proceedings of the International Symposium on Polymer Additives, held March 30–
April 1, 1982, at the American Chemical Society meeting in Las Vegas, Nevada"—
  "Sponsored by the American Chemical Society, Division of Polymeric Materials
Science and Engineering"—Pref.
  Includes bibliographical references and index.
  1. Polymers and polymerization—Additives—Congresses. I. Kresta, Jiri E., date— .
II. American Chemical Society. Division of Polymeric Materials: Science and Engineer-
ing. III. Title. IV. Series.
TP1142.I58 1982   668.4′11   84-13264

ISBN-13: 978-1-4612-9724-6   e-ISBN-13: 978-1-4613-2797-4
DOI: 10.1007/978-1-4613-2797-4

Proceedings of the International Symposium on Polymer Additives,
held March 30–April 1, 1982, at the American Chemical Society
Meeting, in Las Vegas, Nevada

©1984 Plenum Press, New York
Softcover reprint of the hardcover 1st edition 1984

A Division of Plenum Publishing Corporation
233 Spring Street, New York, N.Y. 10013

PREFACE

    Ever since the beginning of the plastics and rubber industry, it was realized that useful products could be produced only if certain additives were incorporated into polymers. With the help of these additives, when physically dispersed in a polymer matrix, it has been possible to improve stability against thermal, oxidative, UV, hydrolytic and biological degradation, mechanical properties, flammability, cost, and processibility of plastics. The enormous growth of the volume of plastics consumed by modern society, and new application areas for plastics, have created a demand for new, better additives and better understanding of their functions in polymer systems. As a result of these trends there is a need for sharing of information on progress achieved in the area of polymer additives among engineers and scientists of the plastics industry and academia.

    This book is based on expanded and updated papers originally presented at the International Symposium on Polymer Additives, which was held in Las Vegas, Nevada, and was sponsored by the American Chemical Society, Division of Polymeric Materials Science and Engineering. The book is divided into five parts which cover advances in various areas of polymer additives. The first part is devoted to the progress in understanding of UV degradation and stabilization of various polymers. Oxidation degradation and stabilization of plastic materials is covered in the second part. New developments in the stabilization of PVC are presented in the third part. Problems associated with flame retardance and smoke suppression are discussed in the fourth part. The last section covers plasticizers, fillers, fiber reinforcement and special additives.

I would like to thank Mrs. Iris Glebe for her efforts in typing this book and for help with editing, and Dianne Lewis for her assistance in preparation of the index. Thanks is also due to the editors of Plenum Press for their patience and for helpfulness with the manuscript.

Jiri E. Kresta
Polymer Institute
University of Detroit
April, 1984

CONTENTS

RECENT DEVELOPMENTS IN PHOTODEGRADATION

AND PHOTOSTABILIZATION OF POLYDIENES

Jan F. Rabek

Department of Polymer Technology
Royal Institute of Technology
Stockholm, Sweden

INTRODUCTION

Photooxidation reactions belong to the most important processes operating in the atmospheric aging of polydienes. Photooxidative degradation and/or crosslinking of polydienes is generally considered to proceed by reactions of free polymer alkyl-, alkylperoxy-, alkyloxy- and other free-radicals formed by photolysis or photoexcitation of internal (e.g., chromophoric groups such as carbonyl groups) and/or external impurities (e.g., traces of catalysts, initiators and other additives.[1-4]

Some recent publications on the photooxidative degradation and photostabilization of polydienes[5-20] have postulated the role of singlet oxygen ($^1O_2$) in the initial step of photooxidation in which allylic hydroperoxide groups are formed.

It is important to consider the fact that polymer hydroperoxides formed during photooxidation and/or singlet oxygen oxidation are not simple hydroperoxides but are most likely to have neighboring hydroxy, carbonyl, hydroperoxy groups and unsaturated bonds. These groups may be important in the photosensitized decomposition of hydroperoxide groups, formation of hydrogen bonds, charge transfer complexes and reaction with free radicals, etc.

The polymer alkyl radicals (P·) arising by the thermal and/or photoinduced initiation participate in a chain autooxidation reaction. In the propagation steps hydroperoxy groups are formed:

$$P\cdot + O_2 \longrightarrow POO\cdot \qquad (1)$$

1

$$POO\cdot + PH \longrightarrow POOH + P\cdot \tag{2}$$

The resulting hydroperoxide is decomposed in a monomolecular (3) or bimolecular (4) reaction:

$$POOH \longrightarrow PO\cdot + HO\cdot \tag{3}$$

$$2\ POOH \longrightarrow PO\cdot + POO\cdot + H_2O \tag{4}$$

The perfect knowledge of individual degradation mechanisms and of their specific roles under given conditions and the optimum utilization of all aspects of the mechanism of the stabilizers' action are the conditions for the effective stabilization of polydienes.

In a pure photochemical process, the formation of polymer alkyl radicals ($P\cdot$) can be considered as a result of a change in bond strength upon the redistribution of electrons in the excited state of a molecule and/or chromophoric groups in a macromolecule.

A new mechanism resulting from electronic-vibrational coupling should also be taken into consideration. In this mechanism reactions resulting from electronic-vibrational coupling leads to the excitation of specific molecular vibrations. Such coupling is strongest with high-energy stretching vibrations, and these vibrations are known to be the preferred energy acceptors in a radiationless relaxation process.[21-23]

The abundance of photochemical hydrogen abstraction reactions:

$$PH + h\nu \longrightarrow P\cdot + H\cdot \tag{5}$$

can be explained by preferential electronic energy transfer to stretching vibrations involving hydrogen, which are of relatively high frequency. The quantum yield of this reaction is high when the rate of reaction of the vibrationally excited molecule exceeds the rate of vibrational relaxation.

In polydienes the most reactive atoms in a macromolecule are hydrogens in methylene groups (CH_2) which are involved in the high frequency molecular vibrations. When vibrations can be assigned to stretching modes of bonds and for bonds of similar vibrational frequencies, the lower the heat of formation of the bond or the more anharmonic the vibration, the more reactive the bond.[21]

The sigma electrons of carbon-hydrogen bonds in the methylene groups (CH_2) cannot be excited at wavelengths exceeding 2000 Å. Electron redistribution through polarization following excitation cannot be responsible for the reactivity of the sigma bonds of methylene groups, since sigma bonds involving hydrogen are among the least polarizable.

In the radiationless relaxation process (in which electronic energy is converted into vibrational energy[23-27]), the best energy accepting vibrations are the more anharmonic vibrations that approach most closely the electronic energies. These are usually anharmonic stretching vibrations with high vibrational energies that involve the stretching of the weaker bonds of carbon with hydrogen, the lightest atom.[22,23] In the electronic-vibrational coupling mechanism electronic energy flows preferentially to stretching vibrations involving hydrogen atoms whenever a radiationless relaxation takes place. This mechanism can be responsible for the formation of the polymer alkyl radicals (P·) in the photoinduced initiation step.

In further steps of photooxidation, hydrogen abstraction can be caused by excited ketone and/or aldehyde groups. Hydrogen abstraction by an excited carbonyl group ($C=O^*$) can be considered as a process of energy transfer from an electronically excited n,π^* state of a ketone or an aldehyde group to a stretching vibration involving a nearby hydrogen atom in the methylene group (CH_2). As a result of the transfer of energy, the vibration becomes highly excited. This allows the transfer of the hydrogen to oxygen of the carbonyl group to take place. Whether the reaction is intermolecular or intramolecular, the proximity of the electronically excited center and the hydrogen atom remain essential. Both n,π^* singlet and n,π^* triplets are expected to be reactive by the above mechanism, since either of the two can be an energy donor to a vibration.

The triplet state of carbonyl groups is also associated with formation of the ultimate biradical. The ultimate biradical possesses two independent electrons. The spin components of these two electrons are neither "paired" nor "unpaired"; they are independent. The electrons are spatially separated to such an extent that there is no overlap of their space wavefunctions. The total spin component of the system of two electrons is not defined; only that of each electron is defined. The ultimate biradical formed may abstract a hydrogen atom from the methylene group (CH_2) and produce a polymer alkyl radical (P·). The reactivity of n,π^* carbonyl triplets can also be explained by their biradical-like nature. Hydrogen abstraction reactions by carbonylic compounds have been observed with a large variety of substrates.[28]

The reaction of polymer alkyl radicals (P·) with molecular oxygen (Equation 1) is a very fast reaction. It is generally considered that under normal atmospheric pressure only polymer alkylperoxy radicals exist. They are stable even at room temperature and can be detected by ESR.[12,29] The polymer alkylperoxy radical can abstract hydrogen from the same or from another polymer molecule to form polymer hydroxyperoxides (POOH) (Equation 2).

RESULTS AND DISCUSSION

In order to stabilize a photooxidation process in polydienes it
is very important to scavenge the P·, POO· and PO· radicals. Our
recent studies in our Institute of Polymer Technology, Royal Institute of Technology, Stockholm, were mainly devoted to these problems.
Two groups of chain-breaking compounds, i.e., hindered phenols and
hindered piperidines, were examined.

Hindered phenols stabilize polymer systems by interacting with
free radicals formed in the thermal and/or photooxidation processes
and with singlet oxygen formed in the polymer matrix.[30-32] Hindered
phenols were previously examined as antioxidants in thermal oxidation
of elastomers.[33,34] Hindered phenols react with free radicals forming a non-radical substrate and a phenoxy radical by donation of a
hydrogen atom from the -OH group to the free radical:

In our research hindered phenols with the following groups were examined:

$$R_1 = R_2 = CH_3, \quad C(CH_3)_3 \quad \text{and} \quad R_2 = H, \quad CH_3, \quad C(CH_3)_3$$

If the resulting phenoxy radical is well stabilized, or sterically
prevented from further reactions, it will not participate in the
hydrogen abstraction from methylene groups (CH_2) in polydienes. It
may react with a second POO· or PO· radical in the polymer matrix:

In the case where the R_3 substituent is a hydrogen atom, hindered quinones may also be formed:

$$+ \ HOP \qquad (9)$$

Phenoxyl radicals undergo chemical transformations and may react either as O-radicals (phenoxy radicals) or C-radicals (cyclohexadi-enonyl radicals).[8] The stability and reactivity of phenoxyl radicals are determined by steric effects of substituents in the "ortho" position (R_1 and R_2).[33,35-37] In a series of phenoxyl radicals, lifetimes increase as steric hindrance in the "ortho" and "para" positions increases:

$$(10)$$

Radical life time $\longrightarrow$

$\longleftarrow$ Radical reactivity

The stabilizing effectiveness of the hindered phenols examined in our research program was as follows:

$$(11)$$

    II   I   III   IV

Hindered phenols are also capable of both reacting with and quenching singlet oxygen ($^1O_2$), depending both on their substituents (R_1, R_2 and R_3) and reaction conditions.[38,39] Based on the rate effects and the detection of the phenoxy radicals, the following mechanism of singlet oxygen oxidation of hindered phenols has been proposed:[39-41]

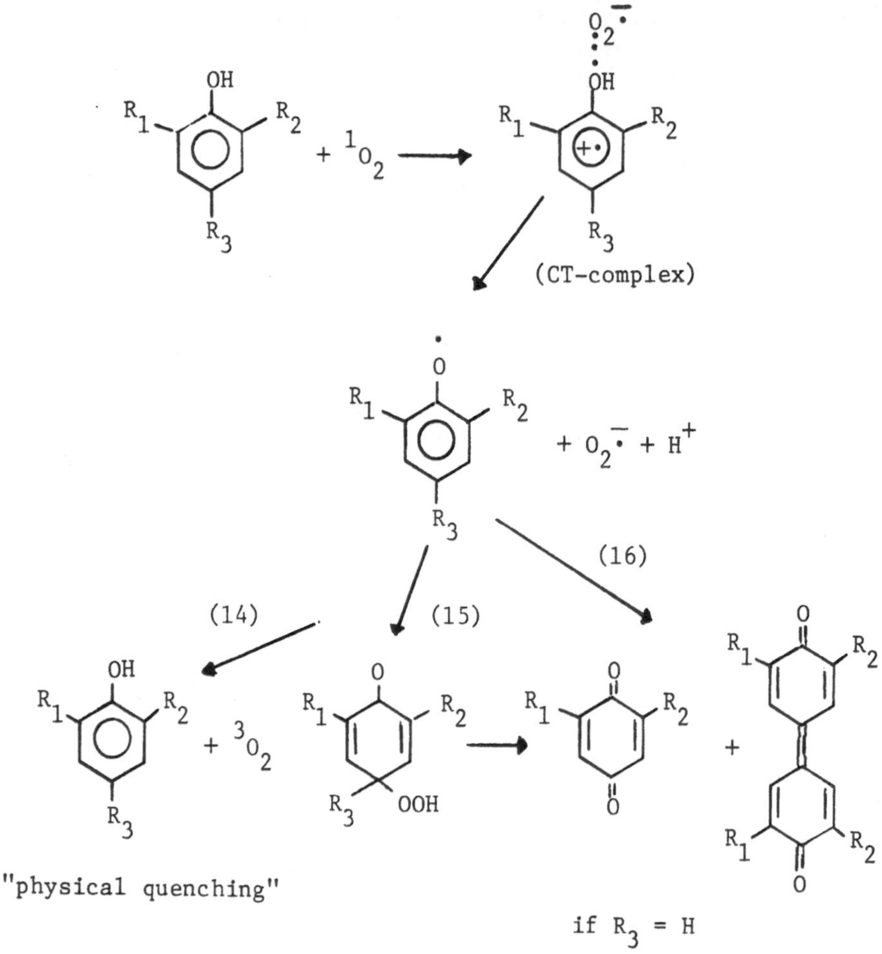

$$\text{(CT-complex)} \tag{12}$$

$$+ O_2^{-\bullet} + H^+ \tag{13}$$

(16)

(14)   (15)

"physical quenching"

if R_3 = H

This complete mechanism considers formation of a charge-transfer
(CT) but not a complete electron-transfer complex (Equation 12).

   Our results show that hindered phenols decrease formation of
hydroperoxy groups (OOH) (Figure 1, UV initiated, and Figure 3,
singlet oxygen initiated) and carbonyl groups (Figure 3, UV initiated). They also decrease chain scission reactions (Figure 4, relative viscosity change). Other results are presented in the original
paper.

   Hindered phenols can be successfully used in stabilizers against
free radical and/or singlet oxygen oxidation of polydienes. The
main disadvantage is that they are not very compatible with non-
polar polydienes and will tend to form crystalline aggregates in the
polymer matrix, leaving large volumes of unstabilized material.
Since the centers of antioxidant activity are, in general, polar

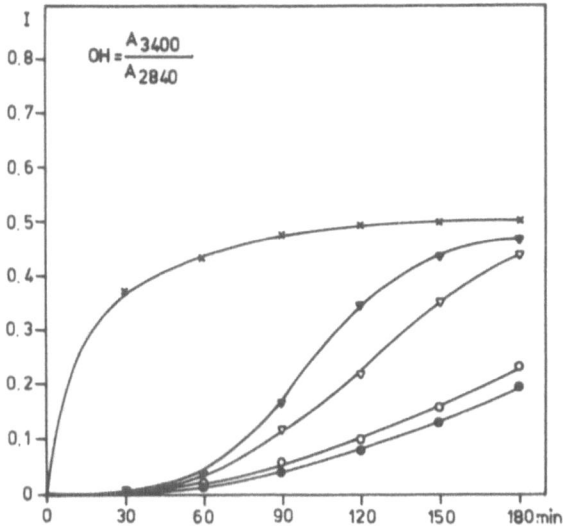

Figure 1. Kinetics of formation of OOH/OH groups in PH after UV
irradiation in the presence of hindered phenols (10^{-3} M):
(O) I; (●) II; (▽) III; (▼) IV; and (X) without.

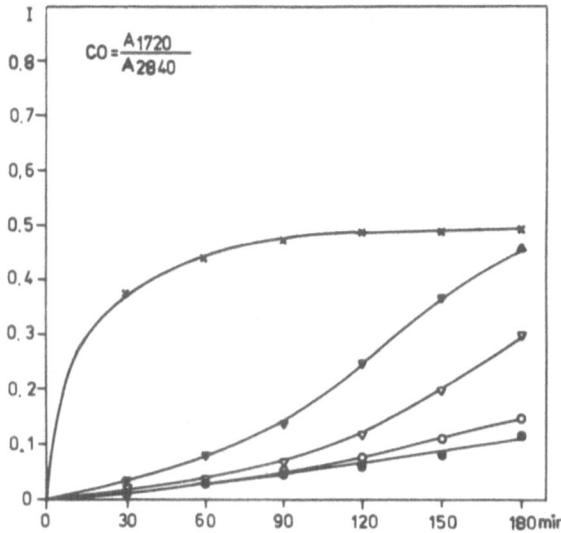

Figure 2. Kinetics of formation of CO groups in PB after UV irradiation in the presence of hindered phenols (10^{-3}):
(O) I; (●) II; (▽) III; (▼) IV; and (X) without.

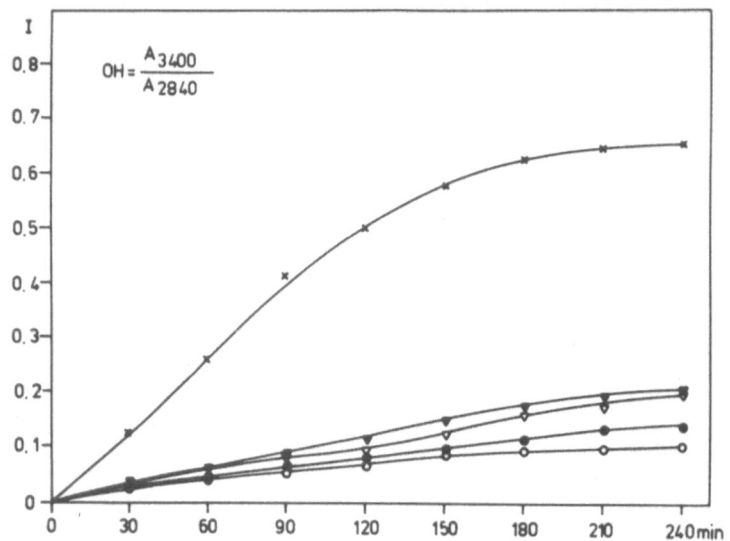

Figure 3.   Kinetics of formation of OOH/OH groups in PB after sing-
            let oxygen oxidation in the presence of hindered phenols
            (10^{-3} M):   (○) I; (●) II; (▽) III; (▼) IV; and (X) without.

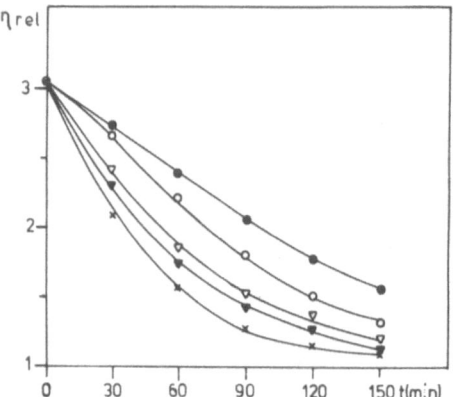

Figure 4.   Change of relative viscosity of PB (1 wt. %) in benzene
            solution in the presence of hindered phenols (10^{-3} M):
            (○) I; (●) II; (▽) III; (▼) IV; and (X) without.

groups (OH) there is a certain incompatibility between polymer and
hindered phenols. In practice this effect can be overcome by incorporating hydrophobic groups into the phenol molecules.

Hindered piperidines stabilize polymeric systems by interacting with free radicals formed in the thermal- and/or photooxidation processes[42-51] and with singlet oxygen formed in the polymer matrix.[52-56] The stabilizing mechanism of hindered piperidines is more complicated than hindered phenols. Hindered piperidines react with free radicals and in the presence of oxygen produce nitroxyl radicals:

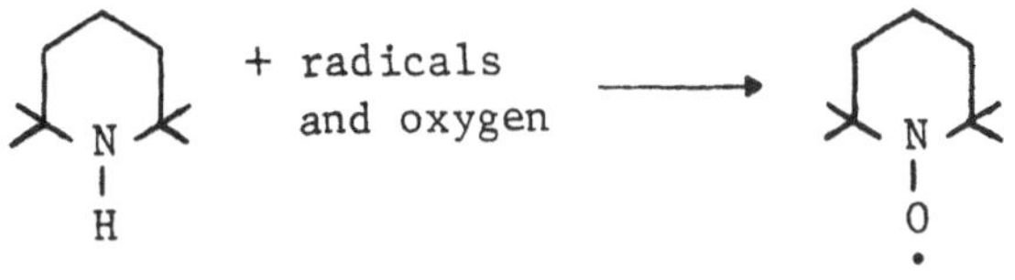

(17)

Nitroxyl radicals may react further with alkyl radicals and even polymer alkyl radicals to which they can be attached:

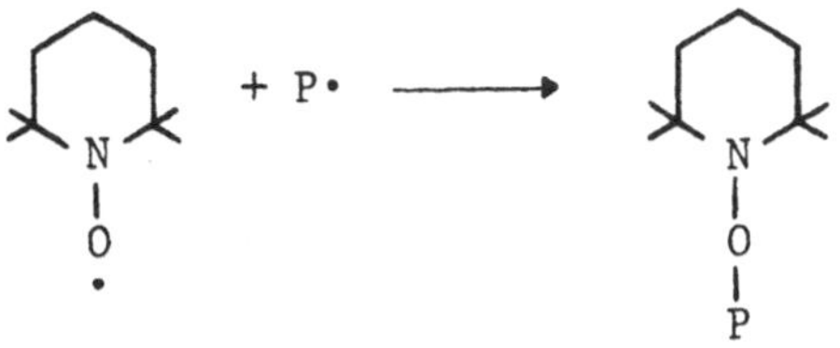

(18)

It is generally accepted that nitroxyl radicals do not react with POO· and PO· radicals, but there is no evident proof for that. It is assumed that POO· and PO· radicals participate only in the regeneration mechanism of nitroxyl radicals according to the following reaction:

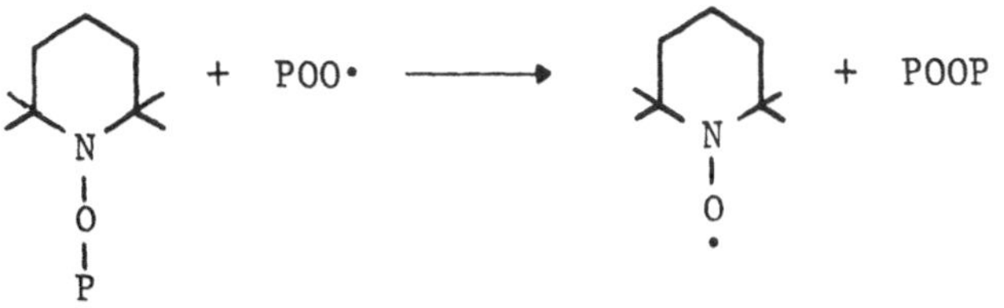

(19)

The formation of nitroxyl radicals (Equation 17), disappearing (Equation 18) and regeneration can be observed by ESR spectroscopy (Figure 5).

It was also suggested that hindered piperidines may decompose hydroperoxide groups formed in polymers during thermal- and/or photo-oxidation, according to the following reactions:

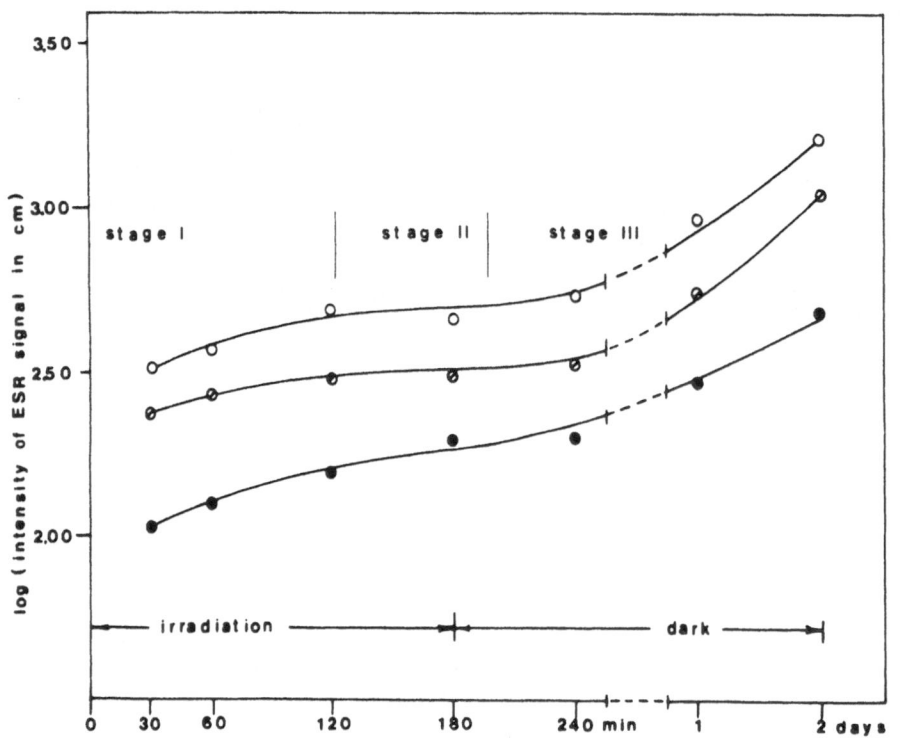

Figure 5.   Kinetics of formation of ESR signals from nitroxyl radicals during UV irradiation: (●) Tinuvin 144; (⊘) Tinuvin
292; (○) Tinuvin 770.

Hindered piperidines are also capable of reacting with singlet oxygen ($^1O_2$);[52-56] during this reaction nitroxyl radicals are formed according to the following recations:

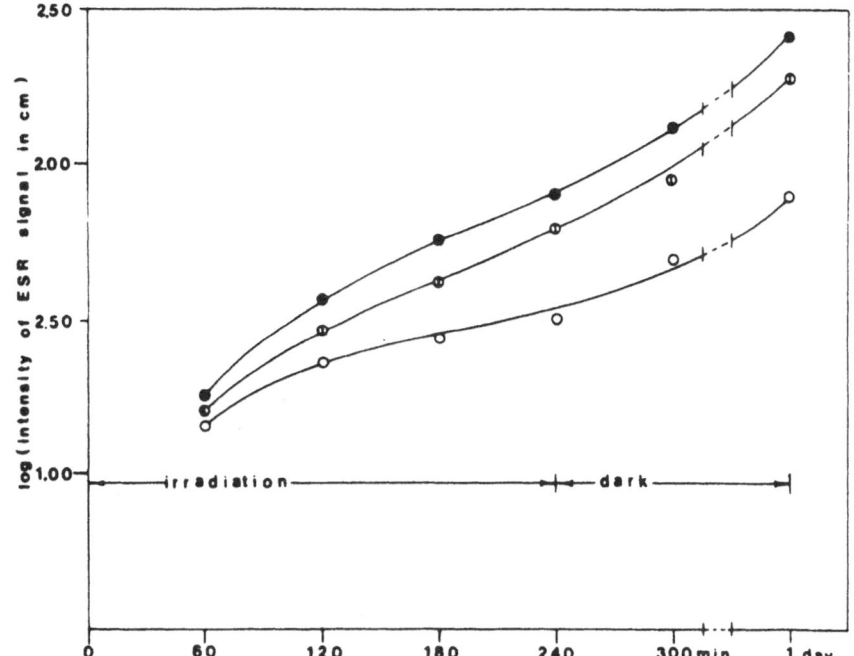

$$(23)$$

The reaction (Equation 23) can be easily observed by ESR spectroscopy (Figure 6) [dye (Rhodamine 6G) sensitized photogeneration of singlet oxygen]. After stopping light irradiation (used for the excitation of dye molecules), the ESR signal still increases and this fact can be interpreted as a result of regeneration of nitroxyl radicals (cf. Reactions 18 and 19).

Figure 6.   Kinetics of formation of ESR signals from nitroxyl radicals during singlet oxygen oxidation: (○) Tinuvin 144; (⊘) Tinuvin 292; (●) Tinuvin 770.

In our research the following commercially available hindered piperidines were examined: Tinuvin 144, Tinuvin 292 and Tinuvin 770. The most effective stabilizer was Tinuvin 144. All of these Tinuvins can be applied as effective stabilizers against free radical- and singlet oxygen photooxidation of polydienes. They decrease formation of hydroperoxy groups (OOH) and carbonyl groups (results not presented here but submitted to the original paper). They also decrease chain scission reaction (Figure 7) and crosslinking reaction (Figure 8). All of the results presented generally here will be covered in more detail in forthcoming publications.[57,58]

SUMMARY

Photochemical reactions involved in the photooxidation of polydienes were discussed in detail on the basis of formation of free radicals, excitation of carbonyl groups, etc. A new mechanism resulting from electronic-vibrational coupling was also presented. For the stabilization of the photooxidation of polydienes, hindered phenols and hindered piperidines were examined.

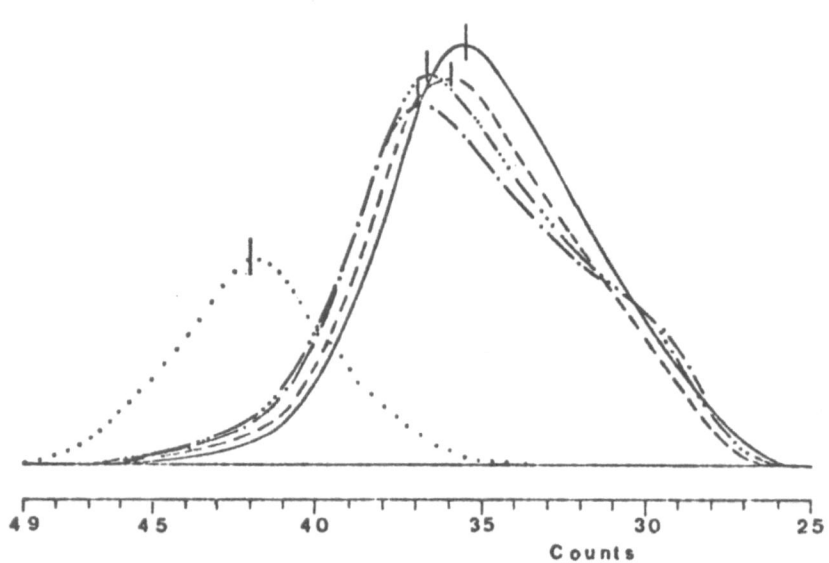

Figure 7.   MWD curves of PB: (——) before UV irradiation and (···) after UV irradiation in the presence of hindered piperidines (5.7 x 10^{-4} M): (---) Tinuvin 144; (——···——) Tinuvin 292; (——·——) Tinuvin 770.

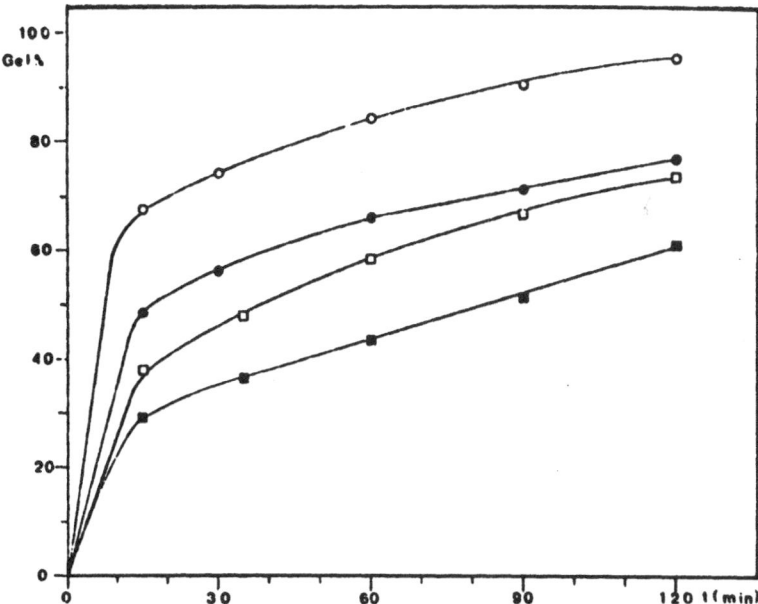

Figure 8.   Change of gel content of PB (%) after UV irradiation in
the presence of hindered piperidines (6.7 x 10^{-4} M):
(■) Tinuvin 144; (□) Tinuvin 292; (●) Tinuvin 770; and
(C) without.

ACKNOWLEDGEMENTS

These investigations are part of a joint research program between the Department of Polymer Technology of the Royal Institute of
Technology, Stockholm, Sweden, and the Institute of Organic and Polymer Technology of the Technical University, Wroclaw, Poland (hindered
phenols) and the Institute of Photographic Chemistry, Academia Sinica,
Beijing, China (hindered piperidines) on "Industrial Use of Photochemical Stabilizers and Stabilizers Against Singlet Oxygen," supported by the Swedish National Board for Technical Development (STU),
help which the author gratefully acknowledges.

References

1. J. F. Rabek in: "Comprehensive Chemical Kinetics," Vol. 14,
       C. H. Bamford and C. F. Tipper, Eds., Elsevier, Amsterdam,
       1975, p. 425.
2. B. Rånby and J. F. Rabek, Photodegradation, Photooxidation and

Photostabilization of Polymers, Wiley, London, 1975.

3. B. Rånby and J. F. Rabek in: "Long Term Properties of Polymers and Polymeric Materials," B. Rånby and J. F. Rabek, Eds., Appl. Polymer Sumposia, No. 35, 1980, p. 243.

4. J. F. McKellar and N. S. Allen, Photochemistry of Manmade Polymers, Applied Science Publ., London, 1979.

5. J. F. Rabek and B. Rånby, Photochem. Photobiol., 28, 557 (1978).

6. J. F. Rabek and B. Rånby, Photochem. Photobiol., 30, 133 (1979).

7. J. F. Rabek and B. Rånby, Rev. Roum. Chim., 25, 1045 (1980).

8. J. F. Rabek and B. Rånby, J. Appl. Polym. Sci., 23, 339 (1979).

9. J. F. Rabek, J. Lucki and B. Rånby, Europ. Polym. J., 15, 1089 (1979).

10. J. F. Rabek and D. Lala, J. Polym. Sci., B18, 427 (1980).

11. D. Lala, J. F. Rabek and B. Rånby, Europ. Polym. J., 16, 735 (1980).

12. D. Lala and J. F. Rabek, Europ. Polym. J., 17 7 (1981).

13. D. Lala and J. F. Rabek, Polym. Degrad. Stabil., 3, 383 (1980/1981).

14. J. F. Rabek, B. Rånby, J. Arct and Z. Golubski, Europ. Polym. J., 18, 87 (1982).

15. M. A. Golub, R. V. Gemmer and M. I. Rosenberger, Adv. Chem. Ser., 169, 11 (1977).

16. M. A. Golub, M. I. Rosenberger and R. V. Gemmer, Rubb. Chem. Technol., 50, 704 (1977).

17. M. A. Golub, NASA Technical Memorandum No. 78604, June, 1979.

18. H. C. Ng and J. E. Guillet, Macromolecules, 11, 929 (1978).

19. H. C. Ng and J. E. Guillet, Photochem. Photobiol., 24, 571 (1978).

20. S. E. Morsi, A. B. Zaki, S. H. Etaiw, W. M. Khalifa and M. Al-Sayed Salem, Polymer, 22, 942 (1981).

21. A. Heller, Mol. Photochem., 1, 257 (1969).

22. S. Califano, Vibrational States, Wiley, N.Y., 1976.

23. H. Ratajczak and W. J. Orville-Thomas, Molecular Interactions, Wiley, N.Y., 1980.

24. B. R. Henry and M. Kasha, Ann. Rev. Phys. Chem., 19, 161 (1968).

25. M. Bixon and J. Jortner, J. Chem. Phys., 48, 715 (1968).

26. D. P. Chock, J. Jortner and S. A. Rice, J. Chem. Phys., 49, 610 (1968).

27. W. Siebrand and D. F. Williams, J. Chem. Phys., 49, 1860 (1968).

28. S. Patai, The Chemistry of the Carbonyl Group, Wiley, N.Y., 1966.

29. B. Rånby and J. F. Rabek, ESR Spectroscopy in Polymer Research, Springer Verlag, Berlin, 1977.

30. T. J. Henman, Dev. Polym. Stabil., 1, 39 (1979).

31. J. Pospišil, Dev. Polym. Stabil., 1, 1 (1979).

32. J. Pospišil, Adv. Polym. Sci., 36, 70 (1980).

33. J. C. Westfahl, C. J. Carman and R. W. Layer, Rubb. Chem. Technol., 45, 402 (1972).

34. J. R. Shelton, Rubber Chem. Technol., 49, 359 (1972).

35. E. R. Altwicker, Chem. Rev., 67, 475 (1967).

36. L. R. Makoney, Angew. Chem., 81, 555 (1969).
37. A. L. Buchachenko, Stable Radicals, Consultant Bureau, N.Y., 1965.
38. L. Taymir and J. Pospišil, Angew. Makromol. Chem., 39, 189 (1974).
39. M. J. Thomas and C. F. Foote, Photochem. Photobiol., 27, 683 (1978).
40. C. S. Foote, M. Thomas and T. Y. Ching, J. Photochem., 5, 172 (1976).
41. B. Stevens, S. R. Perez and R. D. Small, Photochem. Photobiol., 19, 315 (1974).
42. N. S. Allen, Polym. Photochem., 1, 243 (1981).
43. G. Balint, A. Rockebauer, T. Kelen, F. Tüdos and L. Jokay, Polym. Photochem., 1, 139 (1981).
44. P. N. Son, Polym. Degrad. Stab., 2, 295 (1980).
45. N. S. Allen, Polym. Degrad. Stabil., 2, 179 (1980).
46. N. S. Allen and J. F. McKellar, Brit. Polym. J., 9, 302 (1977).
47. N. S. Allen, Polym. Degrad. Stabil., 2, 129 (1980).
48. D. J. Carlsson, D. W. Grattan and D. M. Wiles, Polym. Degrad. Stabil., 1, 69 (1979).
49. N. S. Allen, Polym. Degrad. Stabil., 3, 73 (1980).
50. D. K. C. Hodgeman, J. Polym. Sci., 18, 533 (1980).
51. G. Balint, T. Kelen, F. Tüdos and A. Rehak, Polym. Bull., 1, 647 (1979).
52. V. B. Ivanov, V. Y. Shlapintokh, V. Y. Kvostach, O. M. Shapiro and E. G. Rozantsev, J. Photochem., 4, 313 (1975).
53. D. Belluš, A. Lind and J. E. Watt, J. Chem. Soc. Chem. Comm., 1972, 1190.
54. Y. Lion, M. Delmelle and A. Van de Vorst, Nature, 263, 442 (1976).
55. J. Moan and E. Wold, Nature, 279, 450 (1979).
56. Y. Lion, E. Gandin and A. Van de Vorst, Photochem. Photobiol., 31, 305 (1980).
57. J. F. Rabek, B. Rånby and J. Arct, Polymer Degradation and Stability, (1982).
58. Y. Y. Yang, J. Lucki, J. F. Rabek and B. Rånby, Polymer Photochemistry, (1982).

ADVANCES IN U.V. STABILIZATION OF POLYETHYLENE

F. L. Gugumus

CIBA-GEIGY Limited

CH-4002 Basel, Switzerland

INTRODUCTION

Polyethylene grades account for about 30% of the total plastics production. Low density polyethylene (LDPE) accounts for about two thirds of that amount, high density polyethylene (HDPE) for the remainder. The bulk of HDPE (~ 60 - 80%) is used for blow molding and injection molding, a smaller part (~ 20%) for extrusion. By far the largest amount of LDPE is used for film manufacturing, agricultural applications being a major outlet for LDPE films.

Numerous uses of polyethylene, HDPE as well as LDPE, imply outdoor weathering. The retention of the optical and mechanical properties of polyethylene articles during an extended time period requires adequate stabilization against the effects of u.v. radiation and heat. This stabilization will be a function of the polyethylene type, the specific application and the climatic conditions.

In the following we shall consider successively the advances in u.v. stabilization of HDPE and LDPE. Afterwards, in a more fundamental approach, we shall discuss different aspects of the stabilization mechanisms, comparing literature data with our own results.

EXPERIMENTAL

The polymers used were unstabilized commercial resins.

HDPE: HDPE-1 Ziegler type, MFI (190°C/2.16 kg) ~ 7, d = 0.960
        HDPE-2 Ziegler type, MFI (190°C/2.16 kg) ~ 0,5, d = 0.950
        HDPE-3 Phillips type, MFI (190°C/2.16 kg) ~ 0,2, d = 0.950

LDPE: LDPE-1 high pressure type, MFI (190°C/2.16 kg) ~ 0,2,
                                        d = 0.918
        LDPE-2 high pressure type, MFI (190°C/2.16 kg) ~ 7,
                                        d = 0.917

The additives used were commercial products (Figure 1).   Their
concentrations are expressed in % w/w.

·   Brabender compounding: Unstabilized HDPE powder was plasticized
and homogenized with the additives in an open Brabender plastograph
for 10 minutes at 180°C and 30 rpm.   The homogenized mixture was
taken out of the plastograph and compression molded for 6 minutes at

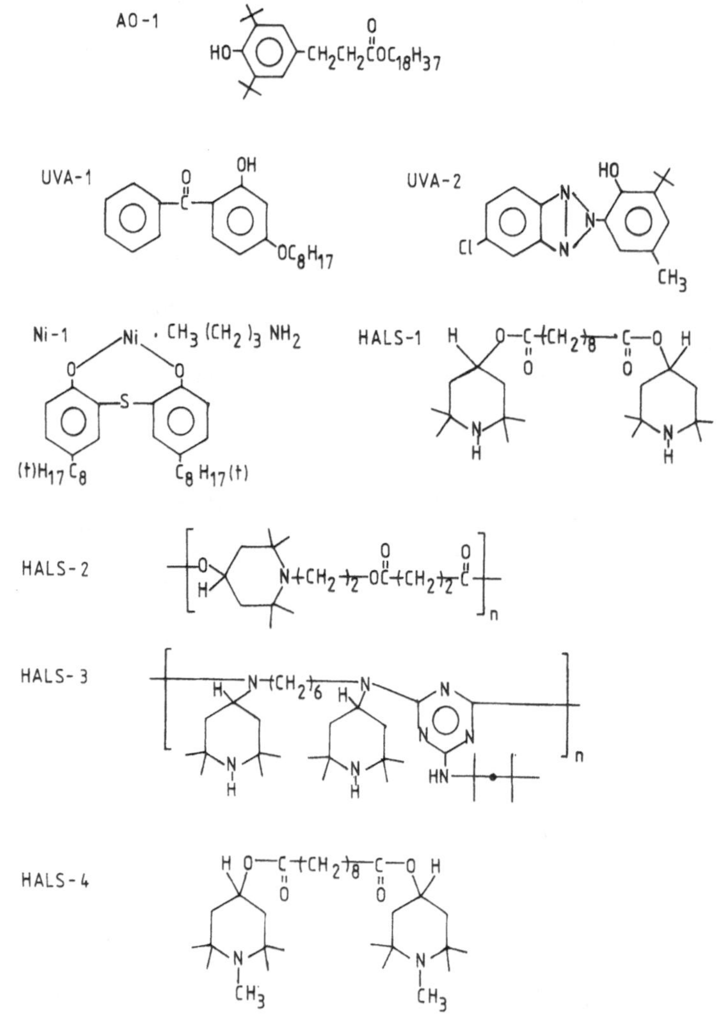

Figure 1.   Antioxidants and light stabilizers used.

210°C into a 0.1 mm thick film which was immediately quenched in cold water. The procedure was the same with LDPE except for the temperature which was 150°C for both mixing and compression molding.

Extrusion: The additives were dry blended into an unstabilized polyethylene powder and extruder compounded at 200°C for LDPE, at 230°C for HDPE. The HDPE tapes were prepared by extruding a film at 260°C and stretching it to a ratio of 1:8.5. The 2 mm HDPE plaques were prepared by injection molding at 250°C (HDPE-1) or 260°C (HDPE-3), the LDPE films by blowing a tubular film (blow head temperature 200°C, blow ratio 1:1.8).

The accelerated weathering devices used include a Xenotest 1200 and a Weather-O-Meter, both equipped with Xenon lamps. When the Weather-O-Meter was used with water spraying the spraying lasted 18 minutes for a total cycle duration of 120 minutes.

The outdoor weathering was performed in Florida, 45° south, direct exposure. Carbonyl development was measured with a Perkin-Elmer 157 IR-spectrophotometer, an unexposed sample being placed in the reference beam.

Mechanical testing, i.e., determination of elongation at break, tensile strength and tensile impact strength was in agreement with the procedures outlined in ASTM D 638, D 882 and D 1822.

DEVELOPMENTS IN THE U.V. STABILIZATION OF HDPE

Despite a fair resistance towards photo-degradation shown by HDPE stabilized against thermal degradation during processing and use, adequate u.v. stabilization is necessary for a prolonged lifetime outdoors.

Initially u.v. absorbers of the benzophenone and benzotriazole types were used to extend the lifetime of HDPE. Their performance in thick sections at concentrations of about 0.2 - 0.3% led to a considerable improvement in u.v. stability. With the subsequent development of Ni-quenchers the stabilization of thin sections was also improved. A further step forward in the u.v. stabilization of HDPE was achieved with the Hindered Amine Light Stabilizers (HALS) developed jointly by CIBA-GEIGY Ltd. and Sankyo Co. Ltd.[1,2]

The improvements in u.v. stability achieved with different stabilizers and measured by retardation of carbonyl development on Xenotest 1200 exposure of compression molded 0.1 mm HDPE films are shown in Figure 2.

The benzophenone type u.v. absorber, UVA-1, and the Ni-quencher, Ni-1, at a concentration of 0.2% lead to a u.v. stability about twice

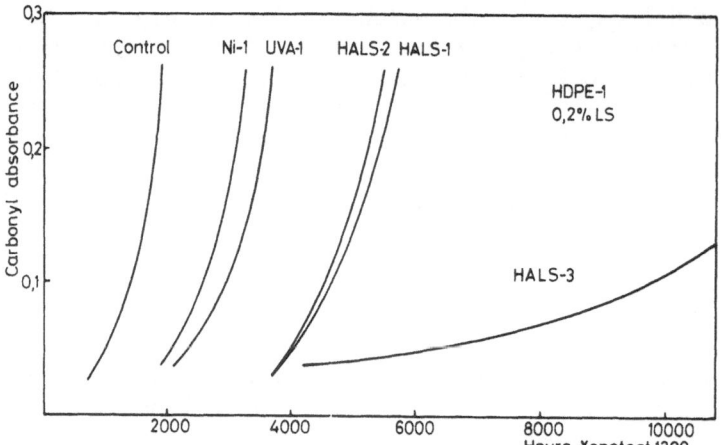

Figure 2. Carbonyl development in 0.1 mm HDPE (Ziegler) films.

that of the control containing only a small amount of a phenolic antioxidant.

The low molecular weight HALS, HALS-1, gives an even better improvement factor of about 3 in this test. In spite of its efficiency,
even more pronounced in polypropylene,[1,3] this compound has nevertheless some shortcomings showing up in special applications where
the relative volatility and limited resistance to extraction and migration caused by the low molecular weight may create problems, e.g.,
in fine fibers. One possibility to overcome these shortcomings is
the use of polymeric HALS such as HALS-2 and HALS-3.

It can be seen in Figure 2 that the polymeric HALS-2 gives about
the same performance in this test as the low molecular weight HALS-1.
Unexpectedly, the other polymeric HALS, HALS-3, in addition to improved secondary properties because of its molecular weight, shows
also a considerably increased effectiveness, about twice that of
HALS-1 and HALS-2. The improvement over the control amounts to
almost an order of magnitude! Finally, with HALS-3, the same improvement in u.v. stability was achieved for HDPE as previously with
HALS-1 for polypropylene.[1,3]

The results obtained on following carbonyl development in 0.1
mm HDPE films are confirmed in even thinner structures, 50 μm thick
HDPE stretched tapes (Table I). The superiority of the HALS over
the u.v. absorber is even more pronounced in tapes, especially if
the much lower concentrations of the HALS are taken into account.

Results obtained on Weather-O-Meter exposure of the unpigmented
tapes show a superiority of about 50% for the low molecular weight
HALS-1 over the polymeric HALS-2. As in HDPE films, HALS-3 shows a

Table I. U.V. Stabilization of HDPE (Ziegler) Thin Sections

HDPE Tales, 50 μm thick
HDPE - 1 + 0.1% Ca-stearate + 0.05 AO - 1
Failure criterion: time or energy to 50% retained tensile
                    strength

	WEATHER-O-METER EXPOSURE		FLORIDA EXPOSURE (NOV, 80)	
u.v. STABILIZATION	T_{50} (HOURS)		E_{50} (KILOLANGLEYS)	
	NATURAL	0.4% TiO_2 (RUTILE)	NATURAL	0.4% TiO_2 (RUTILE)
CONTROL	945	1025	97	94
0.05 % HALS - 1	4280	4560	130	$\geqslant$180 ($\sim$50 %)
0.10 % HALS - 1	6500	9250	$\geqslant$210 ($\sim$50 %)	>210 (60 %)
0.05 % HALS - 2	2920	3770	150	$\geqslant$200 ($\sim$50 %)
0.10 % HALS - 2	4370	6100	$\geqslant$180 ($\sim$50 %)	>210 (55 %)
0.05 % HALS - 3	5850	7200	$\geqslant$210 ($\sim$50 %)	>210 (60 %)
0.10 % HALS - 3	10200	10700	>210 (65 %)	>210 (70 %)
0.30 % UVA - 1	1690	1600	113	100

pronounced superiority over both HALS-1 and HALS-2 with about two
times the effectiveness of the latter. Similar results are obtained
on Weather-O-Meter exposure of white pigmented tapes containing 0.4%
TiO_2 (rutile). In both pigmented and unpigmented tapes the improvement of u.v. stability achieved with 0.5% of the polymeric HALS-3
amounts to an order of magnitude. Almost as big an improvement is
observed in comparison with the tapes stabilized with 3 times as
much UVA-1.

The provisional results of outdoor exposure of the unpigmented
tapes indicate also a considerable improvement of u.v. stability with
the HALS-containing formulations, even if the improvement factors -
ratios of the E_{50} values obtained with u.v. stabilizers to the corresponding E_{50} value of the control - are not as high as on Weather-
O-Meter exposure. Hals-1 and HALS-2 exhibit similar performances at
the test concentrations of 0.05% and 0.1%. Again HALS-3 is much superior, showing the same performance at a concentration of 0.05% as
HALS-1 and HALS-2 at double that concentration. It is too early to
draw definitive conclusions concerning the comparative performances
of the HALS on outdoor exposure of the white pigmented tapes. However, judging from the results available so far, similar conclusions
should hold on a higher level of course (Table I). The u.v. absorber
gives only a negligible improvement over the control, therefore, the
advantage of the HALS over the u.v. absorber used at a much higher
concentration is even more pronounced in white pigmented than in
natural tapes.

As the amount of radiation absorbed by a u.v. absorber is related to its concentration and extinction coefficient as well as to the sample thickness by the Beer-Lambert law, a better performance can be expected from u.v. absorbers in thick sections. This is exactly what is found with 2 mm injection molded plaques in Phillips type HDPE (Table II). The benzotriazole type u.v. absorber UVA-2 leads to a fivefold improvement over the control on Weather-O-Meter exposure. Nevertheless, HALS-1 and HALS-2, at half the concentration of UVA-2, have still a considerable performance advantage.

On outdoor exposure the superiority of the UVA-2 stabilized formulation over the control is much more pronounced and reaches approximately an order of magnitude. Again, the formulations containing HALS-1 and HALS-2 at half the concentration of UVA-2 are superior, but at this time it is neither possible to evaluate the extent of their superiority nor to differentiate between HALS-1 and HALS-2.

In another test series, with titanium dioxide pigmented 2 mm Ziegler type HDPE plaques, differences between the HALS are already showing up (Table III). Because the pigment acts also as a u.v. absorber, it is not astonishing that the contribution of UVA-1 to light stability is rather small. Again, the HALS lead to a considerable improvement of u.v. stability, even at the low concentration of 0.1% used in this study. The provisional results indicate an improvement factor of approximately 5 for the polymeric HALS-2 and of 6 for the low molecular weight HALS-1. The HALS-3 stabilized formulation has not yet failed. Considering the remaining tensile impact strength after 400 kilolangleys, a pronounced superiority over HALS-2 can be expected. Finally, the same ranking between the HALS under consideration is observed in 2 mm plaques as previously in 50 μm thick tapes.

Table II. U.V. Stabilization of HDPE (Phillips) Thick Sections

HDPE - 3 + 0.06% AO - 1, natural
2 mm Injection molded plaques, dumbbells cut out after exposure
Test criterion: time or energy to 50% retained elongation

u.v. STABILIZATION	WEATHER-O-METER (WATER SPRAY)	FLORIDA (AUGUST 1978)
	T_{50} (HOURS)	E_{50} (KILOLANGLEYS)
CONTROL	1200	55
0.30 % UVA - 2	6000	$\geqslant$ 500 (52 %)
0.15 % HALS - 1	16000	$>$ 500 (65 %)
0.15 % HALS - 2	13500	$>$ 500 (65 %)

Table III. U.V. Stabilization of HDPE (Ziegler) Thick Sections

---

HDPE - 1 + 0.1% Ca-stearate + 0.025% AO - 1 + 0.5% TiO$_2$ (rutile)
2 mm Injection molded plaques, dumbbells cut out after exposure
Failure criterion: energy to 50% tensile impact strength

---

u.v. STABILIZATION	FLORIDA EXPOSURE (JULY 1979) E_{50} (KILOLANGLEYS)
CONTROL	60
0.2 % UVA - 1	70
0.1 % HALS - 1	$\sim$ 350
0.1 % HALS - 2	280
0.1 % HALS - 3	> 400 (70 %)

The HALS, especially HALS-2, may not be as efficient in thick
sections of unpigmented Ziegler type HDPE. Then the combination of
a HALS with a u.v. absorber results in a considerable improvement,
so that performances similar to those in pigmented systems can be
achieved.

DEVELOPMENTS IN THE U.V. STABILIZATION OF LDPE

The bulk of LDPE films is not u.v. stabilized. Nevertheless,
for some very important applications of LDPE films, essentially in
agriculture, long lifetimes outdoors are desired. To reach this
goal u.v. stabilization is mandatory.

Initially u.v. absorbers of the benzophenone and benzotriazole
types were used to extend the useful life of LDPE materials. The
following development and use of Ni-quenchers led again to an increased u.v. stability.

A further step forward was expected with the development of
HALS which had permitted considerable improvement of the u.v. stability of PP and HDPE. However, these hopes were dashed with the
low molecular weight HALS available in the early years: all showed
a more or less pronounced blooming after incorporation in LDPE and
a relatively poor performance, especially on outdoor weathering.
With the development of polymeric HALS these difficulties were overcome.

The difference in behavior between a typical low molecular
weight HALS such as HALS-1 and a representative of the polymeric
HALS, HALS-2, is shown in Table IV: the ratio of the times to dis-

Table IV. Compatibility of HALS in LDPE

LDPE - 2 + 0.03% AO - 1
Compression molded 0.1 mm films
Compatibility at room temperature

u.v. STABILIZER	DAYS TO DISTINCT BLOOMING
0.1 % HALS - 1	2 - 3
1.0 % HALS - 1	< 0.1
1.0 % HALS - 2	> 1850 (~5 YEARS)

tinct blooming at the same concentration of 1% amounts to more than
4 orders of magnitude in favor of the polymeric HALS! The performance of the polymeric HALS-2 compared to that of the Ni-quencher
Ni-1 on outdoor exposure of 200 μm blown film is shown in Table V.
It can be seen that in the whole concentration range under consideration HALS-2 shows a considerable superiority over Ni-1. The same
performances as with Ni-1 can be achieved by using half the corresponding amount of HALS-2.

It is clear now that the poor performance of the low molecular
weight HALS in LDPE was not due to a general failure of HALS in LDPE
as a consequence of different photooxidation and/or stabilization
mechanisms but to very unfavorable ancillary properties. This is

Table V. Comparison of a Ni-Complex and HALS

LDPE - 1 + 0.03% AO - 1
200 μm Blown film, plexiglass backing
Test criterion: E_{50} energy to 50% retained elongation

u.v. STABILIZER CONCENTRATION	FLORIDA EXPOSURE (NOVEMBER 1977) E_{50} (KILOLANGLEYS)			
	CONTROL	Ni-1	HALS-2	HALS-1 (INCORPORATED AS A MASTERBATCH)
NO LS	70	-	-	-
0.15 % LS	-	105	175	-
0.30 % LS	-	180	220	-
0.60 % LS	-	230	360	525
1.20 % LS	-	350	500	-

confirmed by the excellent performance of the low molecular weight
HALS-1 when incorporated in LDPE by means of a masterbatch in a polar
substrate (Table V): HALS-1 outperforms not only the Ni-quencher,
but also the polymeric HALS-2! In spite of its better compatibility
by use of this approach, HALS-1 still gives heavy blooming. This
does not impair too much its efficiency but is not acceptable because of the visual aspect which prevented development of this approach for practical stabilization.

The performances of two polymeric HALS are compared to those
of a u.v. absorber and a Ni-quencher in Figure 3. It can be seen
that the effectiveness of both the u.v. absorber, UVA-1, and the Ni-
quencher, Ni-1, increases with their concentration in LDPE film. The
protection provided by the Ni-quencher is somewhat superior to that
of the u.v. absorber. The increase in effectiveness is much higher
if from Ni-1 we change over to the polymeric HALS-2. Within the
whole concentration range under consideration, HALS-2 shows considerable superiority over Ni-1. Nevertheless, HALS-2 does not represent
the optimum for LDPE stabilization. The other polymeric HALS included in the tests, HALS-3, leads once more to a considerable improvement of LDPE's u.v. stability. As a matter of fact, the performance of HALS-2 can be reached by using HALS-3 at half the concentration and even less than that at the higher concentrations.

It can be seen that with HALS-3 it is possible to improve the
u.v. stability of LDPE films by more than an order of magnitude.
Finally, we observe for LDPE the same considerable stability improve-

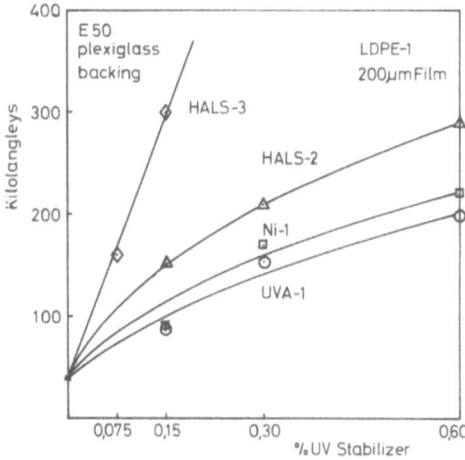

Figure 3. Effect of u.v. stabilizer concentration on LDPE film stability. Outdoor exposure: Florida, 45° south, started
March, 1979. Failure criterion: energy to 50% retained
elongation (E_{50}).

ment already found with HALS-3 in HDPE and previously with HALS-1 in PP. The superiority of HALS-3 over HALS-2 and the other u.v. stabilizers considered in this study is even more pronounced in thin films. This can be seen in Figure 4 where the E_{50} values - energy to 50% retained elongation - obtained for a u.v. stabilizer concentration of 0.15% are plotted as a function of film thickness. The u.v. stability of the control films increases slightly but linearly with film thickness.

With HALS-2 a linear increase in u.v. stability with thickness is also observed. The preliminary results obtained with HALS-3 pointed to a similar behavior. This is not confirmed by the latest results, the stability of the 200 μm thick film is much less than the one expected from a linear extrapolation of the value found with the 50 μm thick film. The behavior of HALS-3 is similar to that of Ni-1 and UVA-1, but on a much higher level.

The differences observed in Figure 4 between the tested u.v. stabilizers concerning the effect of film thickness may not be due only to differences in protection mechanisms. This can be seen best in Figure 5 where the results obtained for 3 concentrations of UVA-1 have been plotted as a function of film thickness: with 0.3 and 0.6% UVA-1 a linear increase of the E_{50} -values with film thickness is observed. It seems reasonable to assume that for the lower concentration of 0.15% UVA-1 a linear relationship holds, too. Therefore, the E_{50} -value found with 0.15% UVA-1 for a film thickness of 200 μm, significantly lower than the one expected from a linear extrapolation, can most probably be attributed to an experimental error.

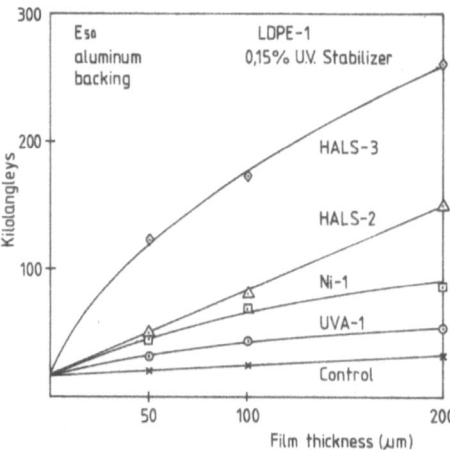

Figure 4. Effect of film thickness on LDPE stability. Outdoor exposure: Florida, 45° south, started May, 1980. Failure criterion: energy to 50% retained elongation (E_{50}).

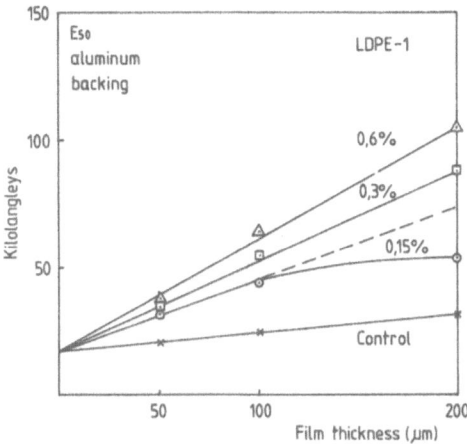

Figure 5.   Effect of film thickness and UVA-1 concentration. Out-
          door exposure: Florida, 45° south, started May, 1980.

     Considering the results observed with UVA-1 it becomes obvious
that no definitive conclusions can be drawn for Ni-1, HALS-2 and
HALS-3 as long as the E_{50} -values for concentrations higher than
0.5% are not available.

     The u.v. stability to be expected at any film thickness below
200μm for 0.15% Ni-1, 0.15% HALS-2 and 0.15% HALS-3 may be determined
graphically from Figure 4. However, the values thus obtained, as
well as those given in the tables, correspond probably to upper performance limits of the u.v. stabilizers in LDPE films because they
were generated under close to ideal test conditions. Of course, this
does not invalidate the comparison between the different formulations. Unfortunately, the values cannot be used as such to predict
the lifetime of LDPE films in actual practice. As a matter of fact,
in greenhouses, e.g., film backing and stress, wind and pollutants
may add their effects to photooxidative degradation and even accelerate it.

DISCUSSION OF STABILIZATION MECHANISMS

     U.v. stabilization may be achieved according to four general
mechanisms. Two of them, specific to photo-stabilization, u.v. absorption and quenching, lead to a reduction of the photo-initiation
rate, the former through a reduction of the u.v. light reaching the
chromophoric groups, the latter by deactivation of the excited states
of these chromophoric groups. The two other stabilization mechanisms
are similar to those encountered in stabilization against thermal
oxidation. Hydroperoxide decomposition involves the decomposition

of the species responsible for chain branching without generating
free radicals. Free radical scavenging finally is based on chemical
trapping of the free radicals responsible for chain propagation.

For details on photo-initiation and photo-stabilization mechanisms the reader is referred to reviews 3 - 7.

In the following we shall consider two specific aspects of stabilization mechanisms, the first related to the behavior of HALS
during processing of LDPE, the second to the u.v. stabilization mechanisms of benzophenone type u.v. absorbers.

BEHAVIOR OF HALS DURING PROCESSING OF LDPE

A prooxidant effect has been reported for HALS-1 on processing
of LDPE in an open mixer.[8] Despite the fact that HALS-1 is not used
in LDPE because of its poor compatibility in this polymer, this behavior seemed astonishing, especially in view of the fact that e.g.
in PP HALS-1 has a small but definitely positive effect during processing.[9] An investigation of the behavior of HALS-1 on processing
of LDPE in a Brabender plastograph was, therefore, undertaken.

The results (Table VI) show no prooxidant effect for HALS-1.
On the contrary, the samples stabilized with HALS-1 show less carbonyl absorbance and, therefore, less oxidative damage than the control samples! It follows that as in polypropylene, HALS-1 acts as
a processing stabilizer in LDPE, too.

In the search for an explanation of the contradictory findings
reported in the literature,[8] we eliminated in succession the influ-

Table VI. Development of Carbonyl with Processing Time

Brabender plastograph (50 EC) at 150°C
HALS - 1:3 • 10^{-3} mol/kg

| L D P E | | CARBONYL ABSORBANCE/100 μM AFTER 10', 20' AND 30' PROCESSING | | | | | |
| | | CONTROL | | | HALS - 1 | | |
		10'	20'	30'	10'	20'	30'
LDPE - 1	1ST TEST	0.000	0.000	0.003	0.000	0.000	0.000
	2ND TEST	0.005	0.007	0.010	0.000	0.000	0.002
LDPE - 2	1ST TEST	0.000	0.008	0.016	0.000	0.000	0.000
	2ND TEST	0.002	0.008	0.016	0.000	0.000	0.002

ence of the polymer batch, of the temperature - no fundamental differences were found at processing temperatures up to 200°C - and also
the influence of residual phenolic antioxidants that may have acted
as processing stabilizers. The much higher values of carbonyl reported in the literature suggested contamination by transition
metals.

To simulate somehow the possible effect of copper that may have
been abraded during mixing, we added copper powder (0.05%) to the
LDPE/stabilizer mixture. This time, everything else being kept constant, the results (Table VII) indicate not only an oxidation rate
superior by about an order of magnitude for the control but also an
oxidation rate in presence of HALS-1 higher than that of the control.
In view of these results it is reasonable to conclude that HALS-1 is
a prooxidant.

To investigate the matter further, all experiments were repeated
with different HALS as well as with a benzophenone type u.v. absorber
(UVA-1) and a phenolic antioxidant (AO-1).

The results of the experiments in absence of copper (Table VIII)
show that of the tested compounds only UVA-1 has almost no effect,
giving about the same oxidation as the control. The tested HALS
and the phenolic antioxidant all act as effective processing stabilizers, the measured carbonyl absorbance being within experimental
error even after a processing time of 30 minutes.

In presence of copper powder (Table VIII) only the phenolic antioxidant AO-1 is superior to the control and prevents effectively
the oxidation of the polymer. Again HALS-1 and in addition HALS-3

Table VII. Development of Carbonyl with Processing Time

Brabender plastograph (50 EC) at 150°C
Presence of copper powder
HALS - $1:3 \cdot 10^{-3}$ mol/kg

LDPE + 0.05% COPPER POWDER		CARBONYL ABSORBANCE/100 μM AFTER 10', 20' AND 30' PROCESSING					
		CONTROL			HALS-1		
		10'	20'	30'	10'	20'	30'
LDPE - 1	1st TEST	0.000	0.002	0.013	0.004	0.012	0.036
	2nd TEST	0.005	0.009	0.013	0.005	0.022	0.044
LDPE - 2	1st TEST	0.003	0.020	0.056	0.012	0.048	0.070
	2nd TEST	0.003	0.023	0.068	0.003	0.047	0.076

Table VIII. Development of Carbonyl with Processing Time

Brabender plastograph at 150°C
Concentration of stabilizing groups: $6 \cdot 10^{-3}$ mol/kg

POLYMER	PROCESSING TIME	CARBONYL ABSORBANCE / 100 µM						
		CONTROL	AO-1	UVA-1	HALS-1	HALS-4	HALS-3	HALS-2
LDPE - 1	10'	0.002	0.000	0.000	0.000	0.000	0.000	0.001
	20'	0.008	0.000	0.008	0.000	0.000	0.000	0.002
	30'	0.016	0.000	0.015	0.002	0.000	0.000	0.003
LDPE - 1 + 0.05% COPPER POWDER	10'	0.003	0.000	0.004	0.003	0.002	0.010	0.003
	20'	0.023	0.000	0.031	0.047	0.020	0.050	0.027
	30'	0.068	0.000	0.083	0.076	0.046	0.084	0.064

favor the oxidation of the polymer in presence of copper. The other
compounds tested, UVA-1, HALS-2 and HALS-4 give rise to effects comparable to that of the control. Thus, the effect observed with some
HALS in presence of copper is not characteristic for HALS but probably associated with the unsubstituted amine groups specific to
HALS-1 and HALS-3. It is well known that amines form complexes with
metallic ions, copper, for example. Hence, the "prooxidant effect"
observed in presence of copper may be caused by some kind of solubilization of copper through complex formation with the amines.

The observed results suggest that if u.v. stability was requested for applications where there are high risks of contamination
by transition metals, e.g., in cables, an adequate solution would
be to use HALS-2 either alone or in combination with AO-1. Usually
a special metal deactivator would be added in such a situation, probably minimizing even the effect of an unsubstituted amine groups containing HALS such as HALS-3.

STABILIZATION MECHANISMS OF BENZOPHENONE TYPE U.V. ABSORBERS

The efficiency of the u.v. absorption mechanism increases exponentially with absorber concentration and distance from surface
exposed to light according to the Beer-Lambert law. We have seen
in Figures 3, 4 and 5 that the efficiency of the benzophenone type
u.v. absorber UVA-1 increases with its concentration and the film
thickness but not exponentially. From these results it can already
be concluded that u.v. absorption cannot account for much of the
protection imparted by UVA-1.

This has been confirmed independently by screening experiments

as shown in Figure 6. From the comparison of carbonyl development
in the control film behind the UVA-1 containing film with carbonyl
development in the unshielded control and UVA-1 containing films it
can be estimated that u.v. absorption accounts probably for less than
10% of the protection provided by UVA-1 to 200 μm thick LDPE films.

From the mechanisms considered for UVA-1 type u.v. absorbers
finally only quenching and free radical trapping can account for
the observed effects. There are good arguments for both of them in
the literature. (For a review see Reference 9). Recently, Winslow[10]
discarded the free radical trapping mechanism for UVA-1 in LDPE on
the grounds that UVA-1 had no significant contribution to ovenaging
of LDPE films at 100°C.

We ran similar experiments and can confirm more or less his
results: no contribution of UVA-1 to ovenaging at 100°C. On the
contrary, even a slight sensitization was observed (Table IX). At
the same time, however, a rapid loss of UVA-1 was observed. The
experiments were repeated at lower temperatures, 90, 80 and 70°C.
The results (Table IX) show that from a slight sensitization at 100°C
and a small positive contribution to ovenaging at 90°C the contribution becomes more and more pronounced with decreasing temperature.

It follows that radical scavenging by UVA-1 seems to become
more and more important with decreasing temperature. In view of the
available results radical scavenging may well account for a major
part of the stabilization effects observed with UVA-1 at the temperatures encountered on outdoor exposure. The benzophenone type

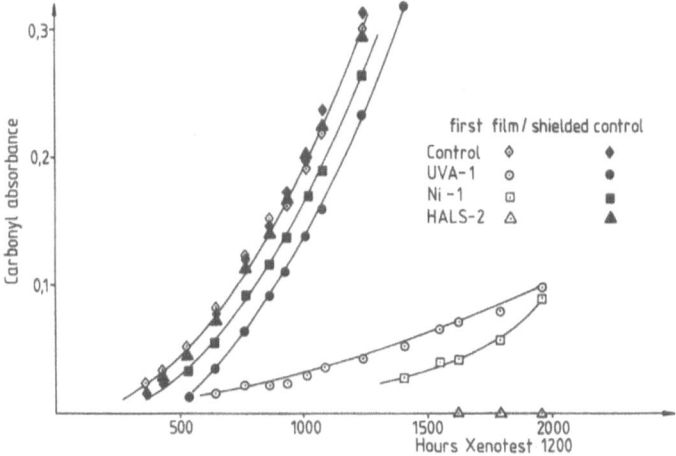

Figure 6.   Comparison of stabilized films with shielded films.
            LDPE-1, blown film (200 μm), 0.03% AO-1 + 0.15% u.v.
            stabilizer.

Table IX.   Effect of UVA-1 on Ovenaging of LDPE Films

LDPE - 2 without AO
200 μm Compression molded films
$T_{0.2}$: time in a draft air oven to 0.2 carbonyl absorbance

---

STABILIZATION	$T_{0.2}$ (HOURS)			
	100°C	90°C	80°C	70°C
CONTROL	77	197	410	810
0.15 % UVA - 1	68	240	510	1000
0.30 % UVA - 1	65	240	600	1200
0.60 % UVA - 1	66	242	780	1600

u.v. absorbers seem to behave in LDPE as a special, light stable type of phenolic antioxidant.

CONCLUSION

The historic development of u.v. stabilization of HDPE and LDPE was shown in some detail. From the first stabilization systems based on u.v. absorbers used alone, different improvements finally led to the optimum stabilization known today, e.g., HALS and polymeric HALS. It was shown that HALS usually do not act as prooxidants but as stabilizers during processing of LDPE and that free radical scavenging may account for most of the stabilization effects observed in LDPE with benzophenone type u.v. stabilizers. However, the present state of our knowledge does not exclude a contribution from quenching of excited states.

ACKNOWLEDGEMENTS

The author appreciates the contribution of experimental data by his colleagues Drs. W. Freitag and E. Pedrazzetti and the discussions with Dr. H. Müller.

REFERENCES

1. F. Gugumus, Paper presented at the 4th European Conference on Plastics and Rubbers, Paris, June, 1974, Caout. Plast. 558, 67 (1976).
2. A. R. Patel and J. J. Usilton, Stabilization and Degradation of Polymers, Adv. Chem. Ser., 169, 116 (1978).

3. F. Gugumus, in: "Developments in Polymer Stabilization-1," G. Scott, Ed., Applied Science Publishers, London 261 (1979).
4. O. Cicchetti, Adv. Polym. Sci., 7, 70 (1970).
5. D. J. Carlsson and D. M. Wiles, J. Macromol. Sci.-Rev. Macromol. Chem., C14, 155 (1976).
6. G. Scott, in: "Ultraviolet Light Induced Reactions in Polymers," S. S. Labana, Ed., ACS Sympos. Ser., 25, 340 (1976).
7. D. M. Wiles and D. J. Carlsson, Polymer Degradation and Stability, 3, 61 (1980).
8. K. B. Chakraborty and G. Scott, Chem. Ind., 237 (1978).
9. F. Gugumus, Paper presented at the 3rd International Conference on Advances in the Stabilization and Controlled Degradation of Polymers, Lucern, Switzerland, June 1-3, 1981.
10. F. H. Winslow, Pure and Appl. Chem., 49, 495 (1977).

HINDERED AMINES AS ANTIOXIDANTS IN UV EXPOSED POLYMERS*

D. J. Carlsson, K. H. Chan, J. P. Tovborg Jensen,
D. M. Wiles and J. Durmis**

Division of Chemistry
National Research Council of Canada
Ottawa, Canada, K1A 0R9

INTRODUCTION

Hindered aliphatic amine light stabilizers have been found to protect polymers including polyolefins, polyurethanes, and styrenic resins against the destructive effects of solar radiation.[1,2] Many of these patented aliphatic amines are based on 2,2,6,6-tetramethyl-piperidine, although the functional group responsible for stabilization would appear to be an aliphatic amine with fully substituted α-methylene groups.[3] Any α-methylenic hydrogens are expected to be vulnerable to rapid free radical attack whereas oxidation at the amine site will produce a nitroxide which can subsequently form a nitrone.[4,5] Both mechanisms can give products capable of ring opening reactions, with destruction of the inhibitor.

The hindered amines are most effective in the stabilization of polymers (PH) which undergo a free-radical chain oxidation after photo-initiation (reactions 1 and 2).

$$P\cdot + O_2 \longrightarrow PO_2\cdot \tag{1}$$

$$PO_2\cdot + PH \longrightarrow POOH + P\cdot \tag{2}$$

Investigations of the mechanisms by which the hindered amine group

---

*Issued as NRCC #00000.
**Research Institute of Organic Technology, CHZJD, 81001
   Bratislava, C.S.S.R.

($>$NH) photo-stabilizes these polymers have implicated reactions 3 to 8.[1,6-8]

$$>NH \xrightarrow[\text{attack}]{\text{radical}} >NO\cdot \tag{3}$$

$$>NO\cdot + P\cdot \longrightarrow >NOP \tag{4}$$

$$>NOP + PO_2\cdot \longrightarrow >NO\cdot + POOP \tag{5}$$

$$>NOP \longrightarrow NO\cdot + P\cdot \longrightarrow >NOH + >C=C< - P' \tag{6}$$

$$>NOH + PPOOH \longrightarrow >NO\cdot + H_2O + PPO\cdot \tag{7}$$

$$>NOH + PPO_2\cdot \longrightarrow >NO\cdot + PPOOH \tag{8}$$

Reactions 4 and 5 represent classical antioxidant steps which will compete with the oxidative propagation steps (reactions 1 and 2) in the polymer. Although the individual reactions 1 and 2 are probably faster than their respective inhibition steps (reactions 4 and 5), we believe that the long kinetic chain lengths involved in the oxidation of polymers such as polypropylene allow effective inhibition to occur by these steps.[6]

Data which support or are consistent with reactions 3 to 8 have largely been obtained from liquid phase work, or work on low molecular weight, volatile, model amines in solid polymers or confined to the semiquantitative identification of only one or two reaction products.[6-8] In this paper we present detailed evidence consistent with the above scheme for a commercial, hindered, secondary amine [bis(2, 2,6,6-tetramethylpiperidyl) decanedioate] (designated HN--NH) in solid isotactic polypropylene (PPH). In addition we present evidence for the involvement of hindered amines in another classical antioxidant reaction - hydroperoxide decomposition - as well as preliminary evidence for the mechanisms by which tertiary amines operate as stabilizers.

EXPERIMENTAL

Bis(2,2,6,6-tetramethyl-4-piperidyl) decanedioate (HN--NH) (Ciba Geigy) was shown to be pure by gas chromatography (GC) and high performance liquid chromatography (HPLC). 4-Hydroxy-1,2,2,6,6-pentamethylpiperidine was esterified with stearic anhydride using p-toluene sulfonic acid as catalyst to give 1,2,2,6,6-pentamethyl-4-piperidyl octadecanoate (R--NCH$_3$). Bis(O-acetyl 2,2,6,6-tetramethyl-

piperidyl) decanedioate was prepared by the reaction of acetyl chloride and the bis-hydroxylamine from HN--NH. Reaction of 4-hydroxy-
2,2,6,6-tetramethylpiperidyl-N-oxyl with stearoyl chloride gave 2,
2,6,6-tetramethyl-4-piperidyl-N-oxyl octadecanoate (R-NO·) which was
also photo-reduced with sodium sulfide under N_2 to 2,2,6,6-tetra-
methyl-4-piperidyl octadecanoate (R-NH).

Hindered amines were diffused into commercial PPH film (25 μm,
Hercules resin, pre-extracted to remove processing antioxidants) by
immersion in iso-octane or hexane solutions of each amine for 15 h,
at 20°C. This method of incorporation of additives allows the use
of large areas of uniform film without the problems of additive modification during melt compounding and extrusion, as well as the use
of pre-photo-oxidized film (thermally unstable) in some cases. After
rinsing in fresh solvent, and vacuum drying, films were exposed to
xenon irradiation (Atlas WeatherOmeter) or oven heat as required and
periodically analyzed for piperidyl species and PPH oxidation products
up to the point of brittle failure. Species that could be removed
by solvent extraction (in an alkane or CH_2Cl_2) were analyzed by HPLC
and/or electron spin resonance (esr). Non-extractables were analyzed
directly or indirectly by e.s.r. Extractable, free-hydroxylamines
were quantified by the $Fe^{3+} \rightarrow Fe^{2+}$ reduction method, followed by formation of the bathophenanthroline complex of Fe^{2+} which was monitored
at 535 nm. Products from model liquid phase reactions were mainly
quantified by HPLC.

PPH oxidation products and piperidyl species in films were estimated from Fourier Transform Infrared (FTIR) Spectroscopy using a
Nicolet 7199 spectrometer, equipped with a mercury cadmium telluride
detector. Usually 200 scans of each sample were averaged.

RESULTS AND DISCUSSION

HN--NH in Photo-oxidizing PPH

To minimize the extremely long lifetimes of PPH films photo-
protected by HN--NH, experiments were performed at a low initial additive concentration (0.03 w%).

Piperidyl compounds from HN--NH which could be unambiguously
identified at these low additive levels are shown in Table I together
with synthetic routes to these compounds. The binitroxide from HN--
NH is designated ·ON--NO· and other compounds are designated by similar abbreviations. Unsubstituted hydroxyl-amines were only characterized indirectly by e.s.r. by way of their well-known rapid quantitative reaction with tert-butyl hydroperoxide (tBOOH). In addition
the doubly grafted species PPON-NOPP was only quantified indirectly
by e.s.r. after treatment with an iso-octane solution of m-chloro-
peroxybenzoic acid (reaction 9). This reaction gave > 90% recovery

Table I.   Preparation of Piperidyl Species and
Their Identification in UV Exposed Film

	Species	Preparation	Identification
I	HN--NH	Commercial product	GC[a]
II	HN--NO•	Oxidation of HN--NH by half equiv. of per acid	HPLC,e.s.r.[a]
III	•ON--NO•	As in II, but excess per acid	HPLC, GC, e.s.r.[a]
IV	•ON--NOH	Reduction of •ON--NO• by half equiv. of 1,2-diphenylhydrazine	HPLC, e.s.r.[a]
V	HON--NOH	As in IV, but excess reductant	Indirect e.s.r., after BOOH addition[a]
VI	PPON--NO•	γ-irradiation of •ON--NO• in O_2-free PPH; exhaustive extraction	Direct e.s.r.[b]
VII	PPON--NOH	————————	Indirect e.s.r., after BOOH addition[b]
VIII	PPON--NOPP	As in VI, but prolonged irradiation	Indirect e.s.r., after per acid treatment[b]

a) On CH_2Cl_2 or iso-octane extracts from PPH films.
b) On exhaustively extracted film.

of •ON--NO• from PPON--NOPP prepared by γ-irradiation of •ON--NO• in
O_2-free PPH film. The PPON--NOPP concentration was derived from the
maximum •ON--NO• level observed over ~ 3h, after allowing for the
(previously estimated) PPON--NO• concentration in the film.

$$PPON\text{--}NOPP \xrightarrow{ClC_6H_4COOOH} \ \bullet ON\text{--}NO\bullet \qquad\qquad (9)$$

The unambiguous identification of nitroxyl products from HN--NH
by e.s.r. proved to be quite difficult (Figure 1). In the solid state
(dissolved or dispersed in PPH film) all nitroxyls appear to give
quite similar, highly anisotropic signals, only the relative proportions of slow and fast modes[9] varying somewhat with the nature of the
N-oxyl. After exhaustive extraction, when a film is immersed in iso-
octane a distorted triplet is observed which is attributed to PPON-
-NO•. Upon penetration into the film the octane allows some motions
of the free >NO• end (to give a triplet-like mono-nitroxyl spectrum) but rotation is restricted because of the additive being
grafted. Very similar spectra have been reported for >NO• spin
labels.[9] The binitroxide •ON--NO• has a complex e.s.r. behavior depending on the solvation of the molecule. In good solvents (e.g.,
CH_2Cl_2) the two >NO• ends are well separated and a triplet is ob-

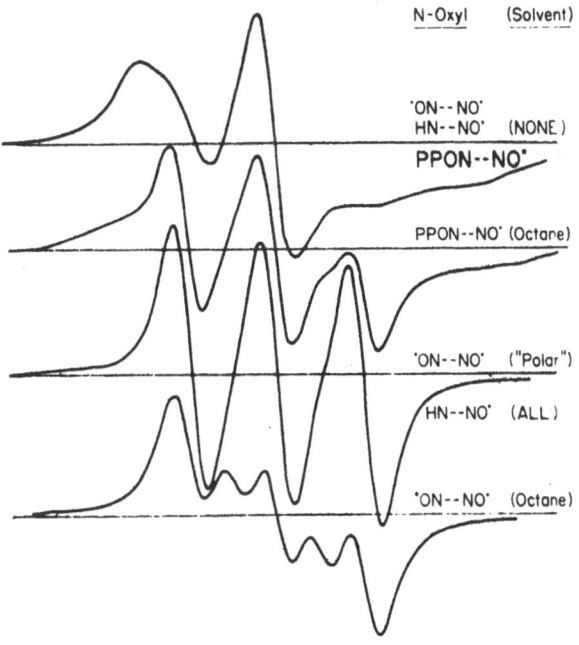

Figure 1. E.S.R. spectra of N-oxyl radicals in various environments.
All samples air saturated, at 22°C. "Solvent free" spectra refer to N-oxyl radicals dispersed in PPH films.

served; in poorer solvents (e.g., iso-octane and other alkanes), intramolecular interactions lead to a 5-line spectrum.[10] However,
triplet spectra were observed for •ON--NO• even in iso-octane if
polar impurities were present. Extractable oxidation products from
oxidized PPH films were sufficient to cause •ON--NO• to show a triplet spectrum in iso-octane. Thus the observation of a triplet e.s.r.
signal cannot be unambiguously attributed to •ON--NH as suggested by
Hodgeman,[8] or •ON--NOH, whereas the observation of a 5-line spectrum
is indicative of •ON--NO•. The e.s.r. technique provides outstanding
sensitivity for nitroxide detection, but only GC and HPLC were found
to reliably identify the differing species from HN--NH.

Analyses of UV exposed films indicated the rapid loss of HN--NH,
and the formation of •ON--NH and PPON--NOPP (Figure 2). •ON--NO•
was detected but only at low levels (<1/10 [•ON--NH]) whereas PPON-
-NO• was always observed at the ~ 0.15 x 10^{-4} mol. kg^{-1} level. Neither •ON--NOH, nor HON--NOH, nor PPON--NOH was indicated when extracts and extracted films were respectively treated with tBOOH. In
addition, UV exposure of HN--NH in air saturated iso-octane gave •ON
--NH rather than •ON--NO• as the dominant nitroxide.

The main conclusions from the analyses of low levels of HN--NH
in UV irradiated PPH are that the parent amine is quickly destroyed

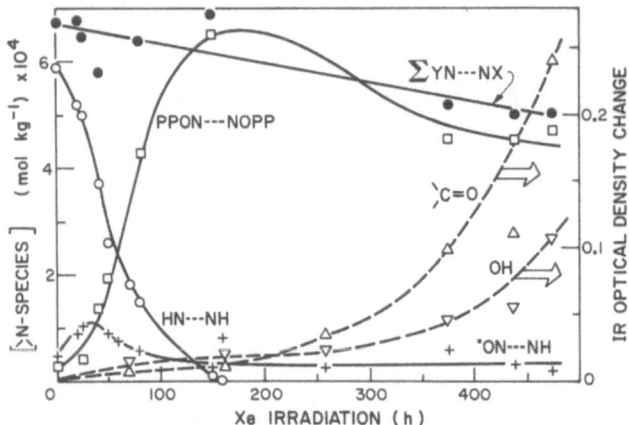

Figure 2. Piperidyl species from HN--NH in photo-oxidizing PPH
           film. Xenon arc irradiation. Initial HN--NH concentra-
           tion 0.03 w%. IR changes measured at 3400 (-OH) and 1715
           cm^{-1} ($>$C=O).

with the initial formation of HN--NO•. The observed products are
consistent with the reaction sequences in scheme 10.

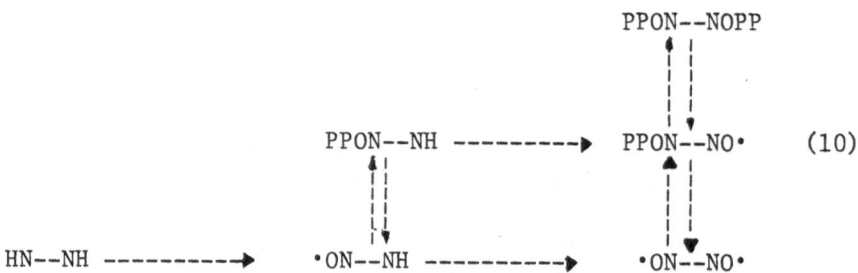

(10)

However, at brittle failure (∿450 h) when PPH oxidation products are
rapidly accumulating, 80% of the initial piperidyl species are still
present in the form of PPON--NOPP (Figure 2), which is suggested to
be an active stabilizer (via reaction 5 and/or reactions 6, 7 and
8). This anomaly could stem either from the low level of additive
used (and perhaps non-uniform distribution throughout the amorphous
regions) or from the lack of selectivity of our method for PPON--
NOPP estimation, that is, peracid can react with several (non-stabilizing) species to generate •ON--NO•. Obviously a direct, unambiguous analysis of the true PPON--NOPP species (and other grafted
species) is required. Only spectroscopic techniques can give direct
information on grafted species in polymer films, and we have attempted Fourier Transform Infrared (FTIR) analysis.

Although FTIR can enhance detection sensitivity beyond that of

conventional IR instruments, species were undetectable at 0.03 w%
in 25 μm PPH films and it was found necessary to go to ~ 0.7 w% HN-
-NH. However, with this level of HALS, film lifetimes in the Weather-
Ometer became impractically long (>5000h). To encourage attack on
the additive in the early stages of irradiation, films were pre-oxidized by xenon irradiation to ~ 0.1 mol. kg^{-1} -OOH groups, prior to
the incorporation of HN--NH. By using the spectral subtraction technique it was possible to identify several piperidyl species in the
films. From comparison with model compounds, the piperidyl $>$ NH
group absorbs at 1236 cm^{-1} whereas the $>$NO-C- group absorbs at ~
1140 cm^{-1}. Nitroxides did not have a convenient unique, IR absorption, and were again estimated by e.s.r. In addition, the sharp
ester absorption at 1748 cm^{-1} could be quantified even in the presence
of the broad carbonyl absorption from PPH oxidation products and used
as an estimation of total stabilizer residues.

    In general our FTIR results at the high initial amine levels
support the conclusions from the less definitive experiments at low
levels in non-oxidized films. The FTIR results indicated a quite
rapid HN--NH loss (based on the 1236 cm^{-1} absorption), an initial
burst of non-grafted $>$NO· and production of $>$NOPP (based on the ~
1140 cm^{-1} absorption) in ~ 80% yield. These changes occurred on a
longer time scale (~ 1000h) than for the very low HN--NH levels used
in our non-oxidized film work; this difference might result from phase
separation at the high HN--NH levels. In addition, ~ 30% of the
$>$ NOPP species were solvent extractable presumably as a result of
the extensive chain scission during pre-oxidation producing short,
PP chains.

    Two other FTIR observations deserve mention. These were forma-
tion of absorptions at 1772 cm^{-1} and 575 cm^{-1} at > 2000h. Although
the 1772 cm^{-1} band could result from γ-lactone formation,[11] this is
unlikely as the levels of other oxidation products ($>$ C=O, -OOH) are
actually reduced during UV exposure in the presence of the piperidyl
species. Another possibility is the formation of $>$NOC(=O)-groups.
Felder et al[12] report an IR absorption at 1748 cm^{-1} for a 1-phenyl-
acetoxy-tetramethyl-piperidine. However, we find that $>$NO-C(=O)CH$_3$
absorbs at ~ 1770 cm^{-1} in iso-octane. The 1772 cm^{-1} absorption suggests that $>$NO-C(=O)PP is formed presumably by scavenging photolysis
products from macro ketones, that it constitutes ~ 30% of the piperidyl species after ~ 400h exposure and that this potentially non-
stabilizing species eventually dominates the products. Felder et al[12]
have reported that $>$NOC(=O)R species are ineffective peroxyl radical
scavengers. The origin of the 575 cm^{-1} absorption is less clear.
This band is indicative of C-ON-O or C-N-O linkages which imply cleavage of the piperidyl ring. However, we detect none of the strong
bands at ~ 1600 cm^{-1} expected from such species in the irradiated
film. Nevertheless, in model liquid phase oxidations of hindered
amines, we have definitely identified nitro products after prolonged
reaction times.

Based on both the high and low concentration studies, HN--NH seems to follow the reaction scheme 10. At prolonged irradiation times inactive (i.e., non-stabilizing) piperidyl products as well as products consistent with the destruction of the piperidyl ring appear to be formed. The progressive loss of the stabilizing piperidyl species together with formation of potential sensitizers such as nitro compounds will cause the eventual failure of the HN--NH stabilized PPH films.

Tertiary Amines in Photo-oxidations

Some N-alkyl substituted hindered amines are reported to be as effective as the unsubstituted amine in photo-stabilization.[1,2] Photo-stabilization by these compounds must be understood before a complete picture of amine protection can be developed. We have investigated the role of 1,2,2,6,6-pentamethyl-4-piperidyl octadecanoate (R-NCH$_3$) in retarding the oxidation of both model liquid systems and PPH or PPOOH films. The long chain substituted piperidine was chosen to minimize loss of the additive by volatilization from films. Typical data for piperidyl species detected during the hydroperoxide photo-initiated oxidation of hexane containing R-NCH$_3$ are shown in Figure 3. The most striking features are the rapid loss of the R-NCH$_3$ and formation of a good yield of the related secondary amine (2,2, 6,6-tetramethyl-4-piperidyl octadecanoate, R-NH). Subsequent products are entirely consistent with the known reactions of secondary amines.

In xenon irradiated PPH or PPOOH films, R-NCH$_3$ showed the same

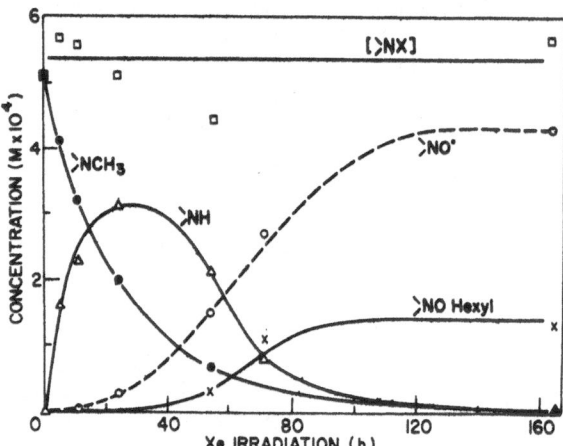

Figure 3.   Piperidyl products from R--NCH$_3$ in hexane. Xenon arc irradiation of 1.0 mm path cells. Photo-initiator tert-butyl hydroperoxide (2.5 x 10^{-3} M). Solutions air saturated. > NX total piperidyl species.

rapid conversion (half life for R-NCH$_3$ destruction < 30h) to give a
60-70% yield of R-NH followed by conversion of the secondary amine
to a low level of nitroxide, but predominantly to the grafted substituted hydroxylamine.

Howard and Yamada[13] have studied various simple tertiary methyl
amines in oxidizing solvents (SH) and suggest amine destruction by
reaction 11. Although we have found formic acid (as the low solubility piperidine formate salts) in our liquid phase systems, we have
not found the expected yield of amine oxide (from reaction 11b).

$$>NCH_3 \xrightarrow[\text{attack}]{\text{radical}} >NCH_2\cdot \xrightarrow{O_2} NCH_2OO\cdot \xrightarrow{SH} >NCH_2OOH \quad (11a)$$

$$>NCH_2OOH \xrightarrow{>NCH_3} >NCH_2OH + >N\!\!\stackrel{O}{\diagdown}_{CH_3} \quad (11b)$$

$$>NH + CH_2O \xrightarrow[O_2]{\text{radicals}} HCOOH \quad (11c)$$

This could imply further rapid reaction of the amine oxide or a more
direct conversion of the $>NCH_2OOH$ species to the piperidine formate,
which does undergo reactions similar to those of the free secondary
amine.

From our liquid and solid state studies, it appears that the
tertiary amines act as precursors to the secondary amines and so feed
into a reaction scheme analogous to reaction scheme 10. The C-H
group α to nitrogen is known to be very reactive towards free radical
abstraction [reaction 11a, (5)], consistent with the rapid loss of
R-NCH$_3$ under our irradiation conditions.

PPOOH Decomposition by Hindered Secondary Amines

As part of the investigation of HN--NH photostabilization mechanisms in the PPH films (HN--NH in Photo-oxidizing PPH, page 37), some
samples were prepared using pre-oxidized PPH film (PPOOH, generation
by UV exposure in the absence of stabilizers). It was quickly realized that these samples were unstable, changes occurring during
ambient storage in the dark. Amines are known to decompose simple
hydroperoxides at elevated temperatures, so that the possibility of
this process occurring for HN--NH at 21°C was examined. From Figure
4, tert. butyl hydroperoxide solutions are completely unreactive towards HN--NH, as shown by the absence of nitroxide even on prolonged
storage. Immersion of pre-oxidized PPH film in HN--NH solution, however, leads to a steady $>NO\cdot$ evolution. The 2,4,6-tri-hydroperoxide
from 2,4,6-trimethylheptane also causes progressive $>NO\cdot$ formation
from HN--NH (Figure 4). The most unexpected result of all was the

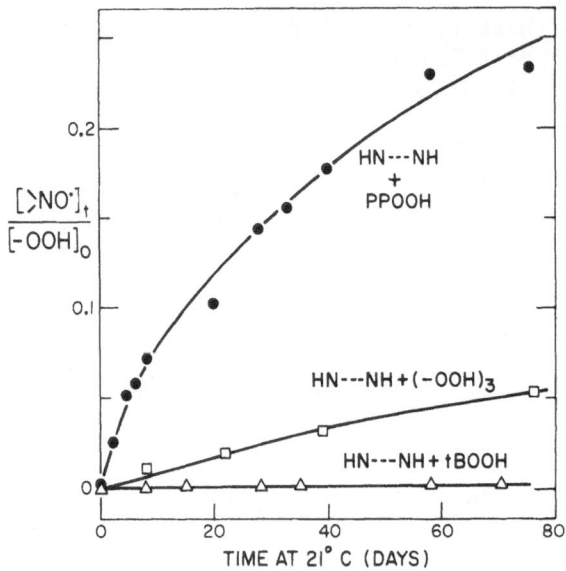

Figure 4.   N-oxyl formation from HN--NH by hydroperoxides. Initial
            amine group concentration 2 x 10^{-2} M in iso-octane.
            Amine ends in ~ 3 fold molar excess over -OOH groups in
            all cases. (-OOH)$_3$ is 2,4,6-trimethylheptan-2,4,6-tri-
            hydroperoxide.

observation that nitroxide evolution from HN--NH in PPOOH in the complete absence of solvent was even more rapid (~ 5 x faster) than in
the presence of iso-octane. Thus for PPOOH (0.03 mol kg^{-1}), ~ 30%
nitroxide yield was observed in ~ 15 days at 21°C from an initial
HN--NH level of 6 x 10^{-4} M in the dry film.

     In order to simplify the investigation of the amine/PPOOH reac-
tion in the solid state, a long chain, involatile, mono-functional
amine (2,2,6,6-tetramethyl-4-piperidyl octadecanoate, R--NH) was
used. This amine was diffused into pre-oxidized (~ 50h Xe exposure)
additive-free PPH film. Films were stored in the dark at room temperature or 64°C and analyses performed for R--NH, R--NO·, R--NOH,
R--NOPP and PPOOH. The macro-hydroperoxide was estimated by iodo-
metry and by IR at 3400 cm^{-1} both before and after SO_2 exposure (to
destroy -OOH and leave only alcohol -OH absorptions[14]). Piperidyl
products were determined by methods similar to those listed in Table
I with the addition that extracted R--NOH was estimated by its ability to quantitatively reduce Fe^{3+} to Fe^{2+}; Fe^{2+} was then estimated
colorimetric as a bathophenanthroline complex.[15]

     At room temperature, R--NH shows a very similar behavior to
HN--NH, and only 64°C data are presented (Figure 5). Similar rates
and reaction products were found for films exposed in air or N_2.

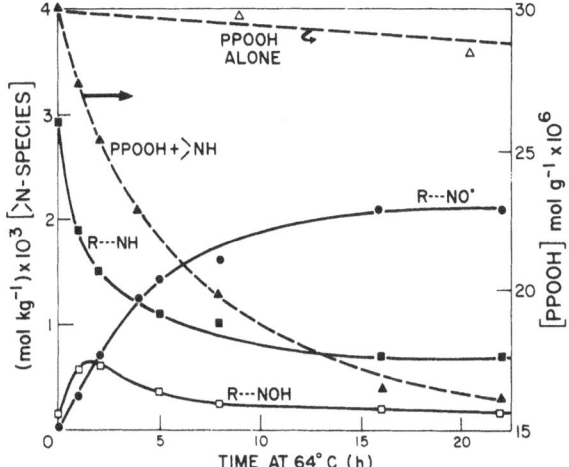

Figure 5.   Reaction of R--NH with PPOOH in the solid state. Initial
PPOOH concentration 0.03 mol. kg^{-1}. Reactions at 64°C
under N$_2$.

At 64°C under N$_2$ PPOOH is slowly lost by direct thermal decomposition
but at a rate ~ 30 x slower than formed in the presence of R--NH
(Figure 5). In the presence of the amine, PPOOH loss was accompanied
neither by -OH nor >C=O group formation.

The rapid, solid state decomposition of PPOOH by R--NH can be
rationalized in terms of a mechanism similar to that proposed by Ball
and Bruice for activated amines (reaction 12).[15]

$$-\text{C-O} \cdots \text{O-H} \cdots \text{O} \cdots \text{C-} \longrightarrow \quad -\text{C-O}\cdot + \cdot\text{O-C-} \quad \xrightarrow{\text{H}_2\text{O}} \quad -\text{COOC-}$$

(12)

This trimolecular process may be favored in the polymer because of
the extensive intra- and intermolecular association of PPOOH groups
and because of R--NH/HOO- association. This latter association is
indicated by the equilibrium solubility of R--NH in PPOOH ([PPOOH]
~ 0.06 mol. kg^{-1}), being ~ 13 times that in PPH (i.e., "non-oxidized"

commercial film [PPOOH] ~ 6×10^{-4} mol. kg^{-1}), using 15 h immersion
in the same R--NH solution in each case to diffuse in the amine.
Hydroperoxide dimerization together with amine-PPOOH association will
make reaction 12 pseudo unimolecular in the solid state; in the liquid
phase, back dissociation will be favored by the diffusion process.
PPOOH is always found to be strongly hydrogen bonded in UV oxidized
PPH, resulting at least in part from the intra-molecular propagation
of the oxidation to give runs of pendant -OOH groups α, γ, ε, etc.,
along the backbone. The relative reactivities of tBOOH (negligible)
and the trihydroperoxide (quite rapid) (Figure 4) imply that the
presence of intra-molecularly hydrogen bonded -OOH groups is essential for reaction scheme 12. The low R--NOH yields observed (Figure
5) will result from hydroxylamine self reactions as well as the rapid
reaction 7 with PPOOH.

PHOTOSTABILIZATION AND ANTIOXIDANT MECHANISMS

Hindered amines appear to act as classical antioxidants in that
they can both radical scavenge and decompose hydroperoxides. Both
the parent amines and their main products differ from antioxidants
such as phenols by being stable to near UV and by not acting as sensitizers. Although the hydroperoxide decomposing ability demonstrated
for the secondary HALS is obviously beneficial for polymer protection,
this process is probably secondary to radical scavenging. For example, macro-hindered amines (molecular weights 2000-2500) are known
to be effective photo-stabilizers although somewhat less efficient
than smaller HALS.[2] The macro-HALS cannot be expected to diffuse to
hydroperoxidized sites in the solid PPH films. During the propagation
cycle (reactions 1 and 2), the peroxyl site "migrates" through the
polymer matrix as a result of the propagation reactions themselves
and segmental motion. Thus the peroxyl sites can be viewed as "mi-
grating" to an (essentially immobile) piperidyl site.[6] Admittedly
-OOH groups will be generated at each step of the peroxyl "migration",
yet the formation of 10-20 -OOH groups may be tolerated if a potential chain of $> 10^3$ steps is prevented.

The role of free hydroxylamines ($>$NOH) from reaction 6 and
reaction scheme 12 is still unclear; although $>$NOH can decompose
hydroperoxide and act as an extremely efficient peroxyl scavenger,
the sensitivity of $>$NOH towards atmospheric oxidation and the above
reactions must prevent its accumulation to a significant level.

REFERENCES

1. F. Gugumus, Ch. 8 in in: "Developments in Polymer Stabilisa-
     tion," Vol. 1, ed. G. Scott, Applied Science, London (1979).
2. F. Gugumus, Research Disclosures, 209, 357 (1981).
3. P. N. Son, Polym. Deg. Stab., 2, 295 (1980).

4. D. F. Bowman, T. Gillan and K. U. Ingold, J. Amer. Chem. Soc.,
    9, 6555 (1971).
5. D. Griller, J. A. Howard, P. R. Marriott and J. C. Scaiano, J.
    Amer. Chem. Soc., 103, 619 (1981).
6. D. J. Carlsson, K. H. Chan, A. Garton and D. M. Wiles, Pure
    Appl. Chem., 52, 389 (1980).
7. J. Sedlar, J. Petruj, J. Pac and A. Zahradnickova, Euro. Polym.
    J., 16, 659, 663 (1980).
8. D. K. Hodgeman, J. Polym. Sci. Polym. Chem. Ed., 18, 533 (1980);
    19, 807 (1981).
9. K. H. Chan, D. J. Carlsson and D. M. Wiles, J. Polym. Sci.,
    Polym. Let. Ed., 18, 607 (1980).
10. A. L. Buchachenko, V. A. Golubev, A. A. Medzhidar and E. G.
    Rozantsev, Teor i Eksp. Khim., 1, 249 (1965).
11. J. H. Adams, J. Polym. Sci., A1, 8, 1279 (1970).
12. B. Felder, R. Schumacher and F. Sitek, Helv. Chim. Acta, 63,
    132 (1980).
13. J. A. Howard, T. Yamada, J. Amer. Chem. Soc., 103, 7102 (1981).
14. D. J. Carlsson and D. M. Wiles, Macromolecules, 2, 597 (1969).
15. S. Ball and T. C. Bruice, J. Amer. Chem. Soc., 102, 6498 (1980).

EFFECT OF VARIOUS ADDITIVES ON THE

PHOTODEGRADATION OF POLYURETHANES

Z. Osawa, E. Tajima, T. Yanagisawa* and K. Suzuki*

Dept. of Polymer Chemistry, Faculty of Engineering
Gunma University
Kiryu, Gunma 376, Japan
*Division of Research and Devenopment, Sennan Factory
Showa Chemical Industry Co., Ltd.
Sennan, Osaka 590-05, Japan

INTRODUCTION

Conventional polyurethanes based on aromatic diisocyanates such as diphenylmethane-p,p'-diisocyanate (MDI) and tolylene diisocyanate (TDI) have been used substantially and world-wide in the manufacture and use of urethane materials because of their excellent physical properties, polymerizability and price. However, they are sensitive to ultraviolet light.[1,2] In order to overcome the shortages of aromatic polyurethanes, some polyurethanes having antiquinoid or non-quinoid diurethane structures, including aliphatic structures, have been developed. The improvement of photo-stability of poly-urethanes was also achieved by the use of various UV stabilizers.[3] We have been studying the photodegradation and stabilization of poly-urethanes,[4-9] and have found that some fluorescent dyes such as stil-bene derivatives and coumarin derivatives were effective retarders for the photodegradation of the polyurethanes. Therefore, various types of related fluorescent dyes were prepared, and the effective-ness of these compounds was determined by the measurement of changes in the residual mechanical strength and the results compared with those of commercial products.

EXPERIMENTAL

Polyurethane

Diphenylmethane-p,p'-diisocyanate (MDI, 178 phr [parts per

hundred by weight]; 0.84 mole) and tolyenediisocyanate (TDI), 22
phr; 0.13 mole) were reacted with a macroglycol, poly(tetramethylene
adipate) glycol, MW. ca. 1000 (500 phr; 0.50 mole) at 80°C for 5 hrs.
under a blanket of nitrogen. After cooling to room temperature, the
contents were dissolved in dry dimethyl formamide (432 parts by
weight) and methylethyl ketone (1296 parts by weight). Then, 1,4-
butanediol (28.5 parts by weight, corresponding to NCO/OH molar
ratio 1.00/0.96) was added and reacted at 70°C for 8 hrs. under a
nitrogen atmosphere. n-Propyl alcohol (7.4 parts by weight) was used
to end the reaction. The resulting polyurethane solution contained
30 wt. % of polyurethane component, and its viscosity was 76,000 cp
at 25°C.

Preparation of Films

A portion of the polyurethane solution (25 g) previously mixed
with a corresponding additive was cast on a silicone-coated paper
and dried in vacuum at room temperature for 48 hrs.

Additives

Additives used in this study are listed in Tables I, II and III,
respectively.

Photo-irradiation

Photo-irradiation of the sample films was carried out with a
Riko Rotary Photochemical Apparatus, RH 400-10 W, and with a Suga
Standard Xenon Fadeometer, FA 25X (discharge voltage 2.1 W). Each
film was placed in quartz glassware 5 cm. away from the light source,
a Riko high-pressure mercury lamp (main wavelength 2537, 2900 and
3650 A) and irradiated without a glass filter. The glasswares, rotating on their axis, were rotated around the mercury lamp and were
immersed in a running water bath.

Methods of Measurement

Stress-strain Property. Stress-strain curves for each sample
(microdumbbell) were recorded on a Tensilon (Toyo Seiki Co., Ltd.)
upright dial gauge-type US-11, and tensile strength at break was
determined (conditions: full scale, 2 kg; crosshead speed, 200
mm/min; chart speed, 200 mm/min). Before testing all samples were
conditioned overnight at 50% relative humidity at 20°C. Five measurements were carried out for each sample and the average values
were calculated. The effect of the additives was estimated by the
residual stress after photo-irradiation.

Ultraviolet Spectra. Ultraviolet spectra were recorded on a
Hitachi double beam type spectrometer, Hitachi Model 124.

Table I. Stilbene Derivatives

No.	Name	Structure	MW	mp (°C)
S-1*	t-Stilbene		180.25	124
S-2	4,4'-Dihydroxystilbene		212.25	278
S-3	4,4'-Diaminostilbene		210.28	231-3
S-4*	4,4'-Diaminostilbene-2,2'-disulfonic acid		342.38	300
S-5	4,4'-Bis(dimetoxytriazinyl) aminostilbene		488.50	248.5
S-6	4,4'-Bis(diaminotriazinyl) aminostilbene		428.46	300
S-7	4,4'-Bis(dimethylaminotriazinyl) aminostilbene		540.67	278
S-8	4,4'-Bis(2-anilino-4-methoxy-1,3,5-triazinyl-6)-diamino-stilbene-2,2'-disulfonic acid triethylamine salt		911.11	300
S-9	2[4-(p-Chlorostyryl)-phenyl]-naphtho[1,2]triazol		381.86	240-2.5
S-10	α-Phenyl-β-(2-methoxy-4-amino-phenyl)-acrylnitrile		250.30	139-40

* :Commercial Products, Others were prepared by
the Showa Chemical Industry Co. Ltd.

Emission Spectra and Fluorescent Lifetime. Emission spectra were recorded using a Hitachi MPF-2A fluorescence spectrophotometer with a xenon source and a photomultiplier.

RESULTS AND DISCUSSION

Effect of Various Additives on the Photodegradation of Urethanes

Breaking strength of urethane samples, before and after ir-radiation of 10 hrs., and residual strength are summarized in Tables IV, V and VI, respectively.

As shown in Table IV, the majority of the stilbene derivatives examined retarded the photodegradation of polyurethane. Especially, S-4 and S-8 were found to be effective and the residual strength was about 80% after 10 hrs. photo-irradiation. Table V also shows that some of the coumarin derivatives are very effective for the

Table II. Coumarin Derivatives

No.	Name	Structure	MW	mp (°C)
C-1*	4-Methyl-7-hydroxycoumarin	4: -Me 7: -OH	176.17	186-8
C-2	4-Methyl-7-diethylaminocoumarin	4: -Me 7: -NEt$_2$	231.98	71-2
C-3	3-Phenyl-7-aminocoumarin	3: -Ph 7: -NH$_2$	237.26	205-6
C-4	3-Phenyl-7-carboethoxyamino-coumarin	3: -Ph 7: -NHCOOEt$_2$	309.32	214-5
C-5	3-Phenyl-coumarinyl-7-urea	3: -Ph 7: -NHCONH$_2$	280.28	330
C-6	N-(γ-Dimethylaminopropyl)-N'-3-phenylcoumarinyl-7)-urea	3: -Ph 7: -NHCONH(CH$_2$)$_3$NMe$_2$	365.43	199-200
C-7	N-(3-Phenylcoumarinyl-7)-maleimide	3: -Ph 7:	317.30	300
C-8	N-(3-Phenylcoumarinyl-7)-succinimide	3: -Ph 7:	319.32	280
C-9	N-(3-Phenylcoumarinyl-7)-glutarimide	3: -Ph 7:	333.34	293-4
C-10	N-(3-Phenylcoumarinyl-7)-phthalimide	3: -Ph 7:	367.36	320
C-11	3-Phenyl-7-cyanocoumarin	3: -Ph 7: -CN	247.25	298
C-12	3-(4'-Aminophenyl)-coumarin	3:	237.25	187.5-9
C-13	3-Phenyl-7-(5-butoxy-6-methyl-benzotriazolyl-2)-coumarin	3: -Ph 7:	429.49	206-8
C-14	3-Phenyl-7-(5-hexyloxy-6-methyl-benzotriazolyl-2)-coumarin	3: -Ph 7:	441.53	197-7
C-15	5,6-Benzocoumarin-3-carbonic acid methyl ester	MeOOC	239.25	187-8.5

* :Commercial Products, Others were prepared by
   the Showa Chemical Industry Co. Ltd.

retardation of the photodegradation of the polyurethane. Among the coumarin derivatives examined, C-5, -6, -7, -8, -9, -10 and -12 were found to be very effective. It was also found that the majority of the evaluated commercial UV stabilizers retarded the photodegradation of the polymer (see Table VI).

Effect of the Concentration of Additives on the Photostabilization of Urethanes

According to the results mentioned above, the representative

Table III. Miscellaneous Compounds

No.	Name	Structure	MW	mp (°C)
M-1*	Melamine		126.12	250
M-2	2,4-Bis(N,N'-diethylamino)-6-chloro-1,3,5-triazine		343.90	117.5
M-3	3-Phenyl-7-aminocarbostyril		236.27	221-2
M-4*	Bis(2,2,6,6-tetramethyl-4-piperidinyl)sebacate (Sunol LS 770)		480.73	81-6
M-5*	2-(2'-Hydroxy-3'-tert-butyl-5'-methylphenyl)-5-chlorobenzo-triazole (Tinuvin 326)		282.36	134-41
M-6*	2,6-di-tert-Butyl-p-cresol (BHT)		220.35	70
M-7*	1,6-Hexamethylene bis(dimethyl-semicarbazide)		288.39	131
M-8*	Zinc Dithiocarbamate		361.90	175

* :Commercial Products, Others were prepared by
the Showa Chemical Industry Co. Ltd.

compounds S-8, C-9 and M-5 were chosen in each series, and the effect of the concentration of the additives was examined in the range of 0.15-1.0 wt. %. As shown in Figure 1, the retardation effect of the representative additives of each series on the photodegradation of the polymer is also apparent in this experiment, and the photostability increases slightly with increasing concentration of additives.

Retardation Mechanism

In order to elucidate the retardation mechanism of the three representative compounds mentioned above, the effect of screening by the additives on the decay of ethylphenylcarbamate (EPC: a model compound of the polymer), emission spectra of the additives, and polymer films with and without the additives were measured.

Decay of the Absorbance Intensity of EPC

The decay of the relative intensity of the absorbance of EPC at ca. 240 nm in each series for the three representative additives, S-8, C-9 and M-5, is shown in Figure 2. In the case of C-9 (left), the decay of the absorbance is remarkably inhibited in the additive system. However, control and screening systems show almost similar

Table IV. Effect of Stilbene Derivatives

No.	Breaking strength(kg/cm^2) Before irrad.	After irrad.	Residual strength(%)	Estimation
Control	446	245	55.0	-
S-1	490	300	61.2	F
S-2	471	323	68.6	F
S-3	447	341	64.3	F
S-4	479	387	80.8	G
S-5	442	331	74.9	F
S-6	346	254	73.4	F
S-7	424	264	62.3	F
S-8	483	391	81.0	G
S-9	412	240	58.3	P
S-10	490	342	69.8	F

Irradiation with a Riko Rotary Photochemical Apparatus for 10 hr

G : Good, F : Fair, P : Poor |Additive| : 0.33 wt%

Table V. Effect of Coumarin Derivatives

No.	Breaking strength(kg/cm^2) Before irrad.	After irrad.	Residual strength(%)	Estimation
Control	446	245	55.0	-
C-1	410	280	68.3	F
C-2	406	278	68.5	F
C-3	398	290	72.1	F
C-4	409	263	64.4	F
C-5	383	329	86.0	G
C-6	343	333	97.1	G
C-7	332	274	82.5	G
C-8	304	255	83.9	G
C-9	317	303	95.6	G
C-10	330	295	89.4	G
C-11	444	213	48.0	P
C-12	487	408	83.8	G
C-13	380	220	57.9	P
C-14	422	222	52.6	P
C-15	434	310	71.4	F

Irradiated with a Riko Rotary Photochemical Apparatus for 10 hr

G : Good, F : Fair, P : Poor |Additive| : 0.33 wt%

decay of the absorbance. The results apparently suggest that there is scarcely any screening effect in C-9. In the case of S-8 (middle), the decay of the absorbance is remarkably retarded in both screening and additive systems. The results imply that there is screening effects in S-8. In the case of M-5 (right), the decay of the absorbance is retarded by the screening, but the retardation effect of the additive system is appreciably higher than that of the screening system. The results suggest that there is a screening effect in M-5 but it does not necessarily represent all of the retardation effects.

Table VI. Effect of Miscellaneous Compounds

No.	Breaking strength (kg/cm^2) Before irrad.	After irrad.	Residual strength(%)	Estimation
Control	446	245	55.0	-
M-1	426	258	51.2	P
M-2	479	307	53.9	P
M-3	432	342	75.0	F
M-4	405	290	71.6	F
M-5	400	364	91.0	G
M-6	455	239	52.0	P
M-7	397	359	90.4	G
M-8	412	337	81.8	G

Irradiated with a Riko Rotary Photochemical Apparatus for 10 hr

G : Good, F : Fair, P : Poor |Additive| : 0.33 wt%

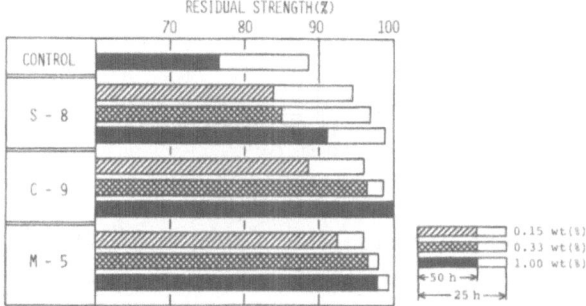

Figure 1. Effect of additive concentration irradiated with xenon fadeometer.

Fluorescence Spectra

     In order to confirm the results mentioned above, fluorescence spectra of the three representative additive systems were measured. As shown in Figure 3, polyurethane films emit fluorescence light at 410 nm with excitation at 366 nm. However, by the addition of C-9 (0.33 Wt. %) the fluorescence emission spectrum from the polyurethane film disappears and at high sensitivity only a weak peak is observed. The results indicate that energy of the excited states of the polymer transfers to the additive, C-9. Accordingly, the results imply that one of the main retardative functions of C-9 in the photodegradation of polyurethanes is a quenching of the excited energy of the polymer.

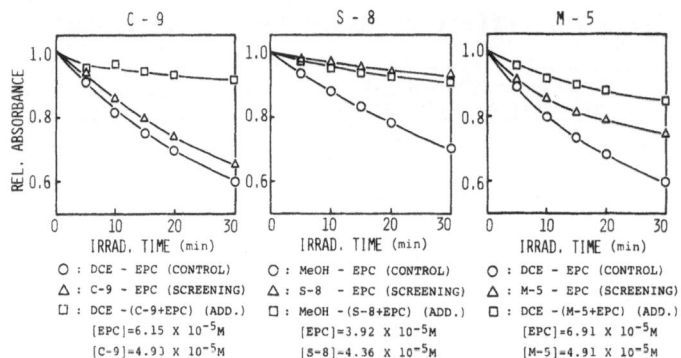

Figure 2.   Decay of the absorption intensity of EPC (at 240 nm).

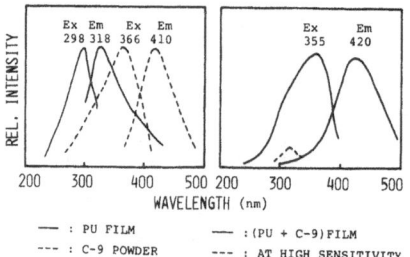

Figure 3.   Fluorescence spectra of PU film, C-9 and PU with C-9.

As shown in Figure 4, S-8 emits fluorescence light at 440 nm with excitation at 400 nm. The fluorescence emission spectrum from the polyurethane film also disappeared by the addition of S-8, and is observed only at high sensitivity. However, from the previous screening experiment, the disappearance of the fluorescence emission spectrum from the polymer is partly ascribed to the screening effect by the additive, S-8.

As shown in Figure 5, the intensity of fluorescence emission spectrum from the polymer is appreciably decreased by the addition of M-5 which did not emit fluorescent light. The results suggest that screening effects of M-5 also plays a very important role in the retardation of the photodegradation of the polymer.

Lifetimes of Excited States

It is well known that with the quenching of the excited states of a substrate, its lifetime is shortened.[10] Therefore, lifetimes of the excited singlet states of the polyurethane and ethylphenyl-carbamate (a model compound of the polymer) systems were measured.

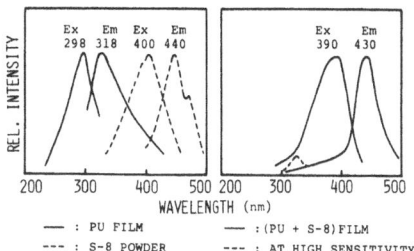

Figure 4. Fluorescence spectra of PU film, S-8 and PU with S-8.

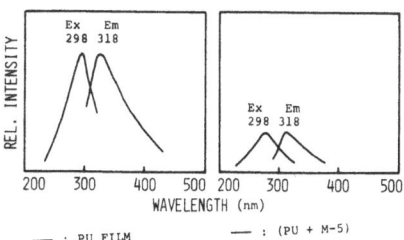

Figure 5. Fluorescence spectra of PU film, M-5 with PU and M-5.

As shown in Table VII, lifetimes of the excited singlet states of the polyurethane and ethylphenylcarbamate are 3.2 and 3.0 n sec, respectively. However, by the addition of C-9, these lifetimes are shortened to 1.8 n sec for the polymer and 1.0 n sec for ethylphenyl- carbamate. The facts also suggest that one of the main retardation functions of C-9 in the photodegradation of the polyurethane is quenching of the excited energy from the polymer.

In a previous paper,[9] we proposed that the photodegradation of the polyurethane initiates from an excited singlet state. There- fore, an energy diagram shown in Figure 6 might illustrate the re- tardation mechanism of the photodegradation of the polyurethane by the additive, C-9. Namely, the excited energy of the polymer is transferred to C-9, thus the photodegradation of the polymer is retarded.

APPENDIX

Preparation of the Representative Additives

   Preparation of 4,4'-bis(2-anilino-4-methoxy-1,3,5-triazinyl-6- diaminostilbene-2,2'-disulfonic Acid Triethylamine Salt (S-8). Into a mixture of cyanuryl chloride (3.7 parts in methanol [50 parts] and water [5 parts]) and 4,4'-diaminostilbene-2,2'-disulfonic acid

Table VII. Singlet Lifetime

	Sample	Lifetime(nsec)
Film	PU	3.2
	PU + C-9	1.8
Soln.	EPC	3.0
(EPA)	EPC + C-9	1.0

PU : Polyurethane
EPC: Ethyl phenylcarbamate

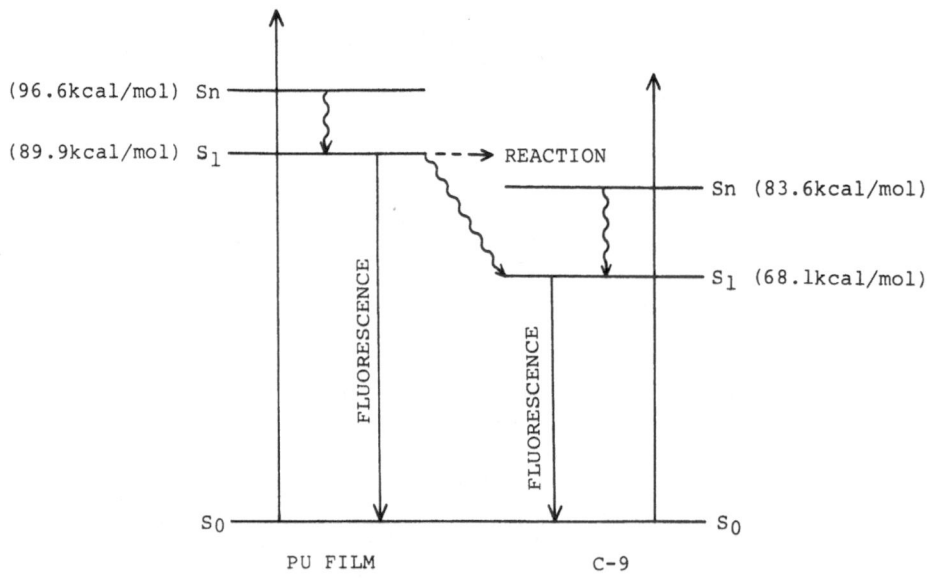

Figure 6. Energy Diagram

(3.7 parts), aqueous sodium carbonate solution (1.06 parts/100 parts
H_2O) was added, and stirred at 25-30°C for 30 min. After the addition of aniline (2.55 parts) temperature of the contents was raised
gradually and methanol was removed. Then the mixture was refluxed
for two and one-half hours and products were salting-out adding
sodium chloride (2.5 parts.)[11]

    4,4'-Bis(2-anilino-4-methoxy-1,3,5-triazinyl-6) diaminostil-
bene-2,2'-disulfonic acid sodium salt (100 parts) thus obtained was
dissolved with water (2000 parts) at 80°C, and then corresponding
sulfonic acid was deposited adding 24 parts of concentrated hydrochloric acid (36%).

    After addition of triethylamine (25 parts) the mixture was
stirred at 65-70°C for ca. 30 min. Then the reaction mixture was

cooled to room temperature and the white crystalline powdery product was separated and washed with water.[12]

Preparation of N-(3-phenylcoumarinyl-7) Glutarimide (C-9).[13]
Into trichlorobenzene (400 parts) 3-phenyl-7-aminocoumarin (23.7 parts) and glutaric acid anhydride (11.4 parts) were added and reacted, refluxing under a nitrogen stream. After the removal of the water produced, the contents were cooled to room temperature and the crystalline products were separated and washed with methanol. The product obtained was recrystallized from o-dichlorobenzene.

SUMMARY

The effects of stilbene derivatives (10 types), coumarin derivatives (15 types) and commercial compounds on the photodegradation of polyurethanes based on polyester diol-diphenylmethane-p,p'-diisocyanate were examined by investigating the changes in residual breaking strengths. Some of the stilbene and coumarin derivatives were found to be effective retarders for the photodegradation of polymers and their effectiveness was comparable to that of commercial stabilizers such as triazole derivatives. The retardation mechanism of the representative compounds was discussed on the basis of the decay of ethylphenylcarbamate, a model compound of the polymer during the photo-irradiation and fluorescence emission spectra of the polymer with and without these additives.

REFERENCES

1. K. C. Frisch, "Advances in Urethane Science and Technology," Vol. I, K. C. Frisch and S. L. Reegen, Eds., Technomic Publishing Co., Westport, Conn., U.S.A., 1973, p. 1.
2. Z. Osawa, Photodegradation and Stabilization of Polyurethanes, in: "Developments in Polymer Photochemistry-3," N. S. Allen, Ed., Applied Science Publ. Co., England, 1982, p. 209.
3. C. S. Chollenberger and F. D. Stewart, Ref. 1, p. 71.
4. E. L. Cheu and Z. Osawa, J. Appl. Polym. Sci., 19, 2947 (1975).
5. Z. Osawa, E. L. Cheu and Y. Ogiwara, J. Polym. Sci., Polym. Letters Ed., 13, 535 (1975).
6. Z. Osawa, E. L. Cheu and K. Nagashima, J. Polym. Sci. Chem. Ed., 15, 445 (1977).
7. Z. Osawa, K. Nagashima, H. Ohshima and E. L. Cheu, J. Polym. Sci., Polym. Letters Ed., 17, 409 (1979).
8. Z. Osawa and K. Nagashima, Kobunshi Ronbunshu (Japan), 36, 109 (1979).
9. Z. Osawa and K. Nagashima, Polymer Degradation and Stability, 1, 311 (1979).
10. T. Matsuura, "Organic Photochemistry," Kagaku-dojin, Kyoto, Japan (1970).

11. Japan Patent Application 3989 (1955).
12. Japan Official Patent Gazette 38097 (1978).
13. Japan Official Patent Gazette 140990 (1981).

PHOTOCHEMICAL DEGRADATION AND BIOLOGICAL DEFACEMENT OF POLYMERS - I

P. D. Gabriele, J. R. Geib, J. S. Puglisi and W. J. Reid

CIBA-GEIGY Corporation
Additives Department
Ardsley, New York, 10502

INTRODUCTION

Photodegradation caused by the ultraviolet portion of terrestrial solar radiation has been studied extensively.[1-6] Polymer photodegradation has been examined mainly through changes in various physical properties such as physical strength, impact resistance, color and gloss retention. Many changes in the physical properties of exposed polymer have been directly related to chemical changes induced by UV radiation.

As polymers continue to be used more extensively in outdoor applications, the necessity for adequate stabilization becomes more critical. This is especially important in applications where the polymer parts must act functionally for long periods of time such as in construction or in automobiles. In this regard, great strides have been made in recent years to upgrade polymer performance outdoors by the resin manufacturers through new polymerization or processing technologies and by the development of new polymers which are inherently more stable. Concurrently, there has been new polymer light stabilizer advances which have dramatically increased the ability of many commodity polymers to function in an outdoor environment for years. For example, the discovery of hindered amine light stabilizers in the early seventies and their utilization in polyolefins has established a new benchmark from which all subsequent systems will be judged.

The purpose of this paper is to discuss the apparent role of ultraviolet light stabilizers in controlling sunlight-induced surface deterioration and the prevention of subsequent microbial infestations such as mildew. Mildew is often mistaken for dirt and to the untrained observer this growth is interpreted as a soiled or weathered

61

surface. This study presents information which indicates that biological defacement of polymers during weathering is a condition prop-
agated through the physical alteration of polymer surfaces. Sunlight
induced erosion and fracturing of polymer surfaces creates a micro-
environment which is conducive to moisture and exogenous carbon source
accumulation. Furthermore, the maintenance of surface integrity
through the use of certain light stabilizers like hindered amines
(HALS) and hydroxy phenylbenzotriazoles has demonstrated dramatic reduction in the accumulation of biological growth.

Biological defacement of plastics and coatings has been well
stated in the literature. This kind of defacement is both aesthetically unpleasant and economically disastrous.[7] The dominant mildew
agent associated with surface defacement is the ubiquitous black yeast
fungus, Aureobasidium pullulans.[8] The characteristic appearance of
this organism is dark green to black, depending upon the nutrient environment from which it feeds. The exact chemical and physical nature
of the environment in which the organism resides are parameters which
affect the characteristics of its growth. These conditions included
substrate chemistry, free moisture, exogenous carbon, and nutrient
sources which accumulate during the weathering process. Isolation of
A. pullulans has been reported[9] from such substrates as painted surfaces, deteriorated greenhouse plastics, tygon tubing, soil, petroleum, fruits and vegetables, and human lymph tissue. The physical environment appears to play a major dynamic role in the organisms'
growth and development. It is now apparent that control of the physical environment is an important factor in the reduction of growth
accumulation on polymer surfaces beyond (or at least as important as)
the classic incorporation of a toxicant to control growth.

EXPERIMENTAL

Preparation and Light Exposure of Polymer Samples

Modified "Weatherable" Styrenic Terpolymer. The granular terpolymer was formulated by having the additives which were dissolved
in minimum amounts of methylene chloride added dropwise to the stirred
polymer. After the solvent had evaporated, the terpolymer was extruded at 430°F to ensure homogeneity of the additives and then was
compression molded at 430°F into 1/8" thick plaques. The plaques were
weathered at 40° facing south in Florida.

Polypropylene. Commercial polypropylene homopolymer pellets
(Profax 6501) were cryogenically granulated. The granulated PP was
formulated by having the additives which were dissolved in minimum
amounts of methylene chloride added dropwise to the stirred PP. After
the solvent had evaporated, the PP was extruded at 450°F to ensure
homogeneity of the additives and then was injection molded at 450°

into 1/8" thick bars. The bars were weathered at 45° facing south 5 in Florida.

Thermoplastic Urethane Films. Commercial thermoplastic polyurethane (Estane 5707 from B. F. Goodrich) was dissolved in a 1:1 (w/w/) mixture of DMF/toluene to form a 20% solution. Light stabilizers were separately dissolved in a minimum of the 1:1 solvent mixture and then were added to the resin solution. Drawdowns of the resin solutions were prepared to obtain a resultant 1.5 mil thick film after drying. The drawdowns were flashed for five minutes at room temperature and then were forced-air dried for 15 minutes at 75°C. Subsequently, the films were exposed at 45° under glass facing south in Florida.

Scanning Electron Microscope (SEM)

Each sample was cemented to a specimen holder and then was coated with approximately 200 Å of aluminum in an EFFA Rotary Vacuum Evaporator. The photomicrographs were obtained from a Cambridge Model S180 Scanning Electron Microscope.

RESULTS

Traditionally, samples returned after aging in Florida have had surface dirt on them which was cleaned off prior to evaluation for impact, color or surface defects. Recently, we decided to take a closer look at these surface defects which then led us to the discovery that some surface discoloration was actually a biological phenomenon. This microscopic examination has identified the presence of various microorganisms. Generally, polymeric systems when unprotected to exterior conditions degrade by forming cracks and fissures on their surfaces. These cracks tend to gather dirt and debris from the environment and so provide a potential organic source for the carbon-scavenging microbes.

Some of the previously observed "dirt" has now been identified microscopically to be these microbes. Additionally, we have found that on the uncleaned, exposed, unstabilized samples that the microbes were present both in the cracks as well as on the surfaces. Optimally stabilized samples did not develop these cracks and fissures and thus the accumulation of the microbes did not occur.

Modified "Weatherable" Styrenic Terpolymer

This polymer system contains styrene, acrylonitrile and a saturated rubber in order to make the polymer more UV light resistant. The material, however, still requires additional UV stabilizer to remain viable for extended periods outdoors. Consequently, ultraviolet stabilizers, e.g., hydroxy phenylbenzotriazoles, benzophenones

and hindered amines, are added to give the necessary additional light
stability.

Unstabilized samples and the various stabilized systems were exposed for four years in Florida (500 kilolangleys). All samples
except the optimally stabilized sample (Figure 1-a) had heavily de-
faced and dirty surfaces within three years of outdoor exposure. The
optimally stabilized sample (0.5% LS 1 + 0.5% LS 2) maintained its
surface integrity with no build-up of dirt or mildew. Closer examination of exposed, unstabilized materials demonstrates widespread contamination by the black yeast Aureobasidium pullulans (Figure 1-b).
The presence of this organism appears to be more highly localized

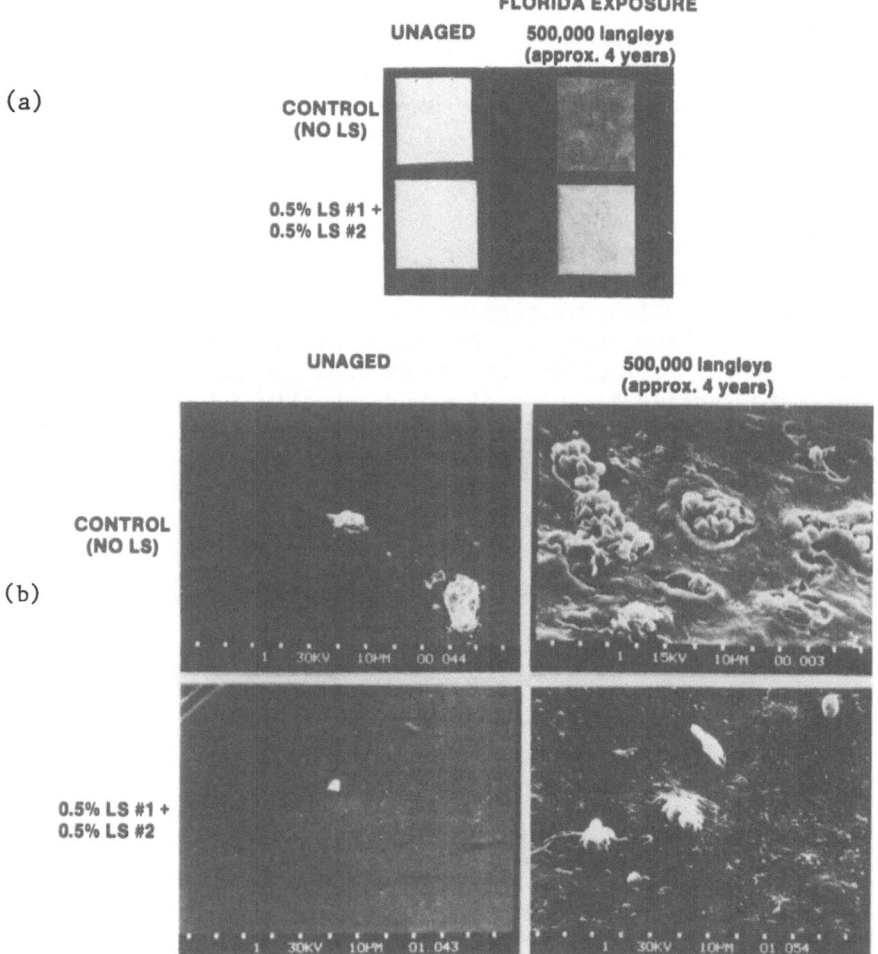

Figure 1. (a) Modified styrenic terpolymer. (b) SEM micrographs
1000 X.

within areas of severe surface erosion. These eroded surfaces increase the surface area by formation of ridges and troughs in which
these organisms thrive because of the accumulation of organic debris.
No such environment exists for the optimally stabilized samples as
their surface has been left intact. Once the organism has become
established, enzymatic degradation of the polymer surface ensues.
This is apparent in Figure 1-b in which these spores appear to have
dissolved the polymer substrate giving the appearance of residing in
a crater. In contrast, the optimally stabilized samples do not exhibit this mode of microbial attack.

The light stabilizers, therefore, maintain the surface integrity
and so prevent biological defacement. However, if the surface is ar-
tifically degraded by processing (e.g., cutting with a saw or blade),
then the light stabilizers (i.e., 0.5% each of LS 1 and LS 2) do not
prevent the mildew build-up. This was established by looking at the
edges of the impact plastics which had been roughened but which had
also been exposed in the same manner as the front face of the tough-
ened plastic (which had maintained its surface integrity for four
years in Florida). These edges had numerous colonies of the black
mildew established within the roughened areas.

Polypropylene

Figure 2-a shows the unstabilized polypropylene samples to be
dark after two and a half years exposures in Florida while stabilized
samples show minimal color change after the same exposure period. On
closer examination, the major surface differences between the two samples becomes apparent. In the unstabilized case, deep troughs have
formed creating cracks for atmospheric pollutants to accumulate. Additionally, moisture can be retained more readily in these structures
thereby creating an ideal environment for microbial habitation. Under
SEM analysis (Figure 2-b), nestled in the trough-like valleys, biological contamination is apparent. Emerging from one of the troughs
is a hyphal structure (part of A. pullulans physiology) as well as
several spore-forms distributed over the surface area of the degraded
polymer. Additional regions examined show pollen spores as well as
hyphae. When stabilized from photodegradation the polymer maintains
its surface integrity eliminating the opportunity for organic accumulation and the subsequent microbial infestation.

Thermoplastic Polyurethane

Figure 3-a shows color changes in unstabilized as well as in stabilized thermoplastic polyurethane samples. Clearly, the change is
more pronounced in the unstabilized case. Additionally, the unstabilized sample shows surface defacement which is exhibited by dark
spotting. In contrast, the stabilized sample shows no surface defacement of this type. On closer examination, it is apparent that
the unstabilized surface is covered by the common mildew organism.

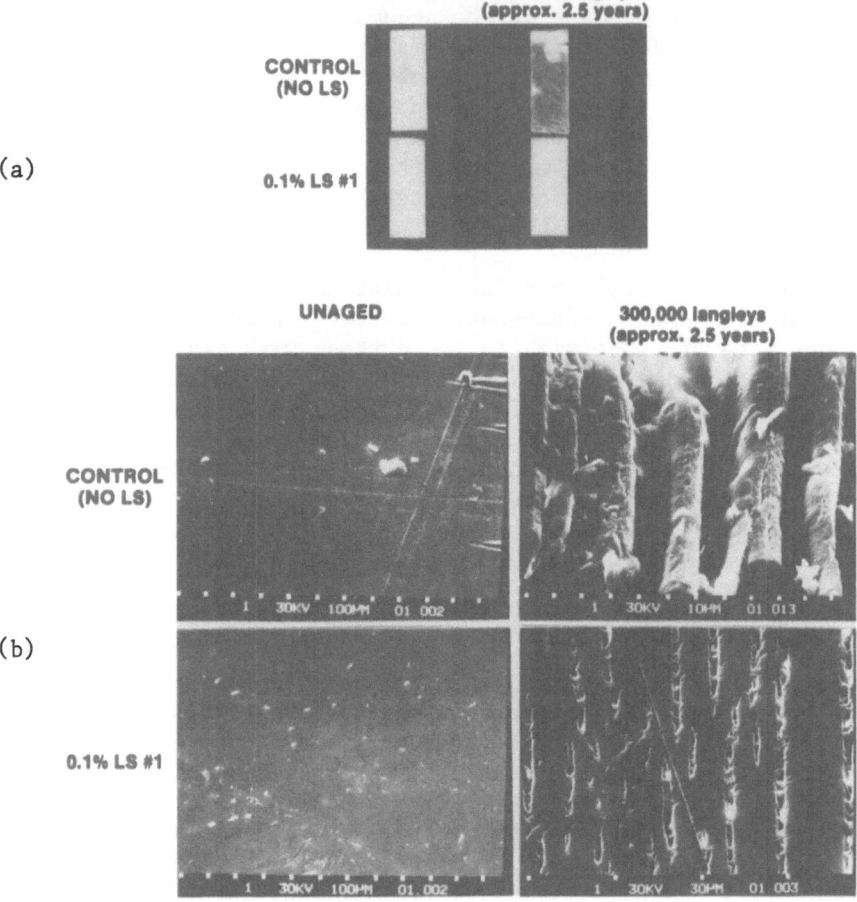

Figure 2.   (a) Polypropylene.   (b) SEM micrographs 100 X.

Again using SEM analysis, differences can be seen in crack propagation between the two samples. In the unstabilized case, severe cracking in the form of fissures is present. The stabilized sample shows
what may be the onset of cracking. Nevertheless, the service life
of a coating and maintenance of the film properties has been extended.
As in the other polymer systems, the greater the surface degradation,
the more attractive an environment for biota. In the case of the unstabilized samples, surface mildew is apparent, with major accumulation surrounding cracks. This is more likely due to the presence of
water within the surface defects. Stabilized samples, in contrast,
exhibit a clean surface with no major accumulation of growth. The
presence of a narrow fissure in the lower right hand corner of Figure
3-b may represent early stages of crack formation. Up to this point,
absence of major surface defects and maintenance of film properties

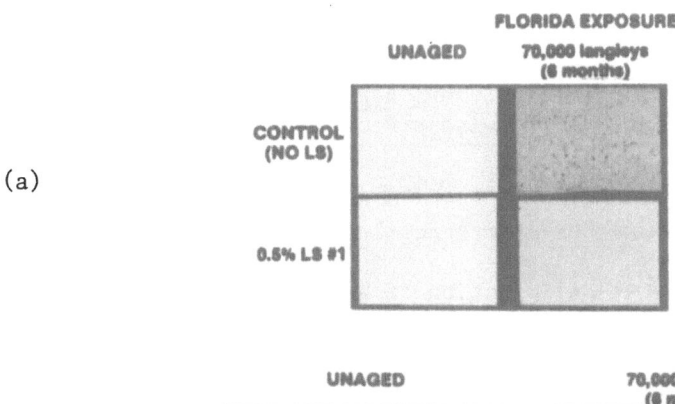

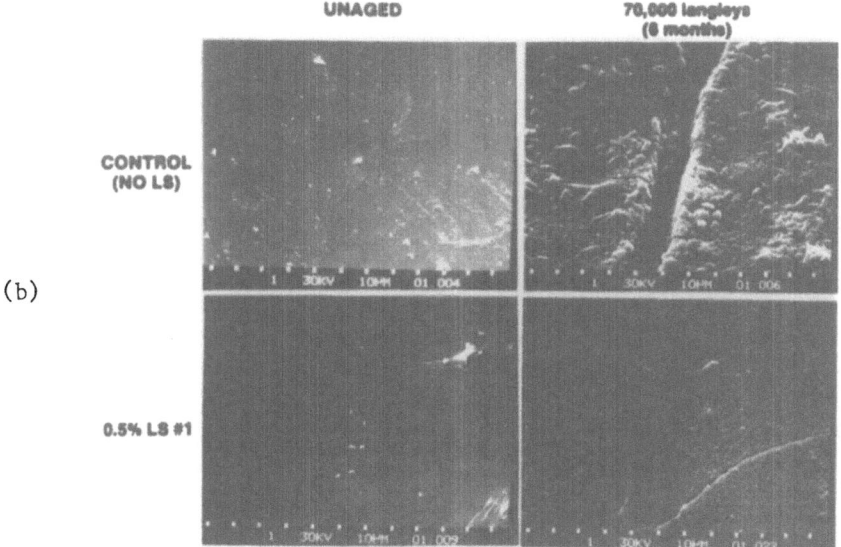

Figure 3.   (a) Thermoplastic polyurethane.   (b) SEM micrographs
1000 X.

prevents the accumulation of biota.

DISCUSSION

Zabel[10] has indicated that in paint systems, clean, dry surfaces
minimize the accumulation of mildew. Moreover, he states that moisture control of surfaces may be the key to microbial control. Film
design may, therefore, be important in reducing surface defects which
should subsequently minimize organic accumulation and free water.
This present study has indicated that photophysical degradation pre-
cedes biological accumulation. We have seen this to be the case with
three different polymer systems. In all three cases, the maintenance
of surface integrity by ultraviolet light stabilizers has dramatically
reduced biological defacement.

Essentially, ultraviolet degradation is a multi-step process leading to free radical initiation and propagation via interaction with oxygen. This free radical propagation will eventually lead to the destruction of many organic polymers both from a physical as well as an aesthetic viewpoint. A stepwise schematic of this process is presented as follows:

$$R \xrightarrow{h\nu} R^* \quad \text{UV absorbance leading to an excited state}$$

$$R^* \longrightarrow R\cdot \quad \text{elimination of absorbed energy through the breaking of bonds and formation of free radicals}$$

$$R\cdot + O_2 \longrightarrow ROO\cdot$$

$$ROO\cdot + RH \longrightarrow ROOH + R\cdot \qquad \begin{matrix} \text{propagation} \\ \text{via reaction} \\ \text{with oxygen} \end{matrix}$$

$$ROOH \xrightarrow{h\nu} RO\cdot + \cdot OH$$

In attempting to control this degradation process, a number of different types of UV light stabilizers have been developed. In our work, we have shown the efficacy of two different types:

bis[2,2,6,6-tetramethyl-4-piperidinyl]sebacate (LS 1)

and

2(2'-hydroxy-5'-methylphenyl)benzotriazole (LS 2).

The question of light stabilizer mechanism remains to be addressed. Ours will be a cursory review in light of the detail already disseminated in other chapters of this book. We have concentrated our study on the HALS and UV absorbers (benzotriazoles and benzophenones).

The hydroxy-phenyl benzotriazoles protect the polymer by preferentially absorbing the harmful UV radiation and converting it into heat energy by means of rapid tautomerism.

The HALS stabilization mechanisms, on the other hand, are complex.[2] Many people believe that HALS work by being oxidized to form nitroxyl radicals (NO·) which in turn react with the polymer radicals to form an amine-ether.

This amine-ether can also terminate peroxy radicals and in the process regenerate the nitroxyl radicals.

Consequently, as long as O_2 is around to form peroxy radicals, the process of stabilization is a perpetual one! Clearly, this free radical scavenger mechanism is too simplistic.

Charge transfer complexes between peroxy radicals and HALS are also proposed to explain stabilization behavior as well as an association through H-bonding of a nitroxyl radical and a hydroperoxide radical.[2]

The type or types of mechanism that are going on may be polymer system specific as well as HALS specific.

There is little dispute, however, that the HALS work very efficiently either alone, for example in the polyolefin case, or in combination with a UV absorber, as in the styrenic example.

CONCLUSION

We have now established that by the addition of certain light stabilizers, polymer surfaces remain intact for many years when exposed outdoors. This surface integrity dramatically reduces the potential for dirt pick-up and the mildew organism's ability to establish a habitat. For example, after 500 kilolangleys (four years) exposure in Miami (Florida), a saturated rubber styrenic terpolymer has retained its color, surface integrity and impact when it was stabilized with 0.5% of bis[2,2,6,6-tetramethyl-4-piperidinyl]sebacate (LS 1) combined with 0.5% of 2(2'-hydroxy-5'-methylphenyl)benzotriazole (LS 2). The unstabilized materials, however, had been biologically attacked since the photochemically degraded surface provided an environment in which the mildew could thrive. Other polymers, such as polypropylene and thermoplastic polyurethane films, are also shown here to be capable of keeping their surfaces intact (and free from mildew attack) by utilizing LS 1.

The unstabilized samples are covered with A. pullulans with their surfaces pitted and distorted.

References

1. B. Ranby and J. F. Rabek, "Photodegradation; Photooxidation and Photostabilization of Polymers," Wiley, New York, 1975.
2. F. Gugumus, "Developments in Polymer Stabilization-2," Gerald Scott, ed., Applied Science Publishers Ltd., Essex, England, Chapter 8, pp. 261-308.
3. D. M. Wiles and D. J. Carlsson, "Polymer Degradation and Stability," 3, 61-72 (1980-81).
4. P. N. Son, "Polymer Degradation and Stability," 2, 295-308 (1980).
5. A. Katbab and G. Scott, "Chemistry and Industry," 573-574, 19 July 1980.
6. R. H. Whitfield, D. I. Davies and M. J. Perkins, "Chemistry and Industry," 418-419, 17 May 1980.
7. B. G. Brand and H. T. Kemp, "Mildew Defacement of Organic Coatings," Paint Research Institute of the Federation of Societies for Coatings Technology, August, 1973.
8. W. B. Cooke, "A Taxonomic Study in the 'Black Yeasts' Mycopathologia," 17 (1), 1-43 (1962).
9. H. D. Hatt, ed., "The American Type Culture Collection Catalogue of Strains," I, 13th ed., 1978, p. 225.
10. R. A. Zabel and F. Terracina, "The Role of Aureobasidium Pullulans in the Disfigurement of Latex Paint Films," Developments in Industrial Microbiology, Vol. 21, 179-190 (1979).

UV STABILIZATION OF

INSTANT COLOR PHOTOGRAPHS

I. O. Salyer and A. M. Usmani

University of Dayton Research Institute

Dayton, OH, 45469

INTRODUCTION

The instant color photography is a useful innovation of modern chemistry. An instant print film is an integral unit with provisions for exposure, open-to-light development, and immediate viewing of the emerging image. The principles of optics and color that apply to traditional color photography apply to instant photography, except processing is automatic and dyes are not produced but required to diffuse to a region designed for viewing them.[1] The viewing area may be in a material containing the light sensitive layers, or it may be integral with them.

In the United States there are two commercial producers of instant color photographic film and cameras, Eastman Kodak and Polaroid. The processes used by these two manufacturers differ significantly, but achieve generally comparable initial color image quality.

The positive color images of instant photography are known to fade away and/or change color substantially when exposed to sunlight or fluorescent light. This paper describes improvement in the long-term performance of instant prints by application of a protective overcoat containing a plasticizing type of UV absorber.

CHEMISTRY OF INSTANT COLOR PHOTOGRAPHY

In Kodak instant photographic material, dyes are attached to carriers that are immobile.[2] Cyan, magenta, and yellow type "dye-releasers" are incorporated in the film associated with red, green, and blue sensitized silver halide layers. Diffusible dyes are re-

71

leased from their carriers in amounts that are inversely proportional
to exposure. After exposure, the highly alkaline activator pouch
is ruptured to spread it between an integral imaging receiver and
the cover sheet. Carbon present in the activator serves as one wall
of the dark room; the other walls are provided by a spacer mask and
black opaque layer, and the white opaque reflective layer. The developing agent in the activator fluid and KOH cause release of dyes
from the "dye-releasers." Thus, the developing agent functions as
an electron-transfer agent and supplies electrons to the developing
silver halide emulsions. It is regenerated when its oxidized form
extracts electrons from "dye-releasers." No exposure causes release
of all three cyan, magenta, and yellow dyes which combine to form
black in the image-receiving layer. Exposure to white results in
no development and consequently, no dye release. Exposure to red
releases yellow and magenta dyes; to green, cyan and yellow dyes;
and blue, cyan and magenta dyes. The dyes are immobilized in the
image-receiving layer by interaction with a polymeric mordant.

Polaroid instant photographic material uses "dye-developers"
that are dyes attached to developing agents.[3,4] Cyan, magenta, and
yellow "dye-developers" are incorporated into the photographic material in close association with its red, green and blue sensitive
silver halide emulsion layers. After exposure, highly alkaline activator fluid spreads between the light-sensitive layers and the
viewing layer. The activator dissolves with "dye-developers" and
carries them along as the fluid diffuses within the light sensitive
layers. In exposed areas, "dye-developers" with the assistance of
other agents also present become oxidized to alkaline-insoluble qui-
nonoid materials and move no further. In less exposed or unexposed
areas, "dye-developers" diffuse to a receiver layer to form positive
color images.

The photographic element is a layered composite of light sensitive layers, dye-developer layers, and spacer layers so that each
"dye-developer" interacts with the layer sensitive to its complement.
Thus, dyes that are complementary to the exposing light are immobilized in the developing areas while the still mobile dyes in the
less exposed areas diffuse to a receiver to form a positive color
image.

A schematic representation of the Kodak instant color photographic process, during and after processing, is shown in Figure 1
and Figure 2.[1] The same type of schematic representation of processed Polaroid SX-70 film is shown in Figure 3.[2]

CHEMISTRY OF UV STABILIZER

Much work has been done to determine chemicals which would
stabilize organic dyes against UV light.[5-8] However, the effective-

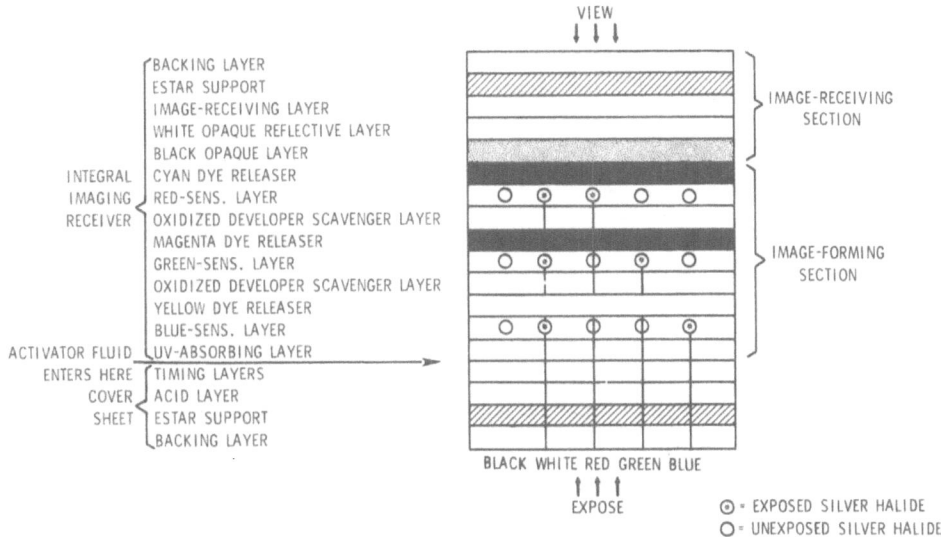

Figure 1. Kodak instant print film schematic section during exposure.

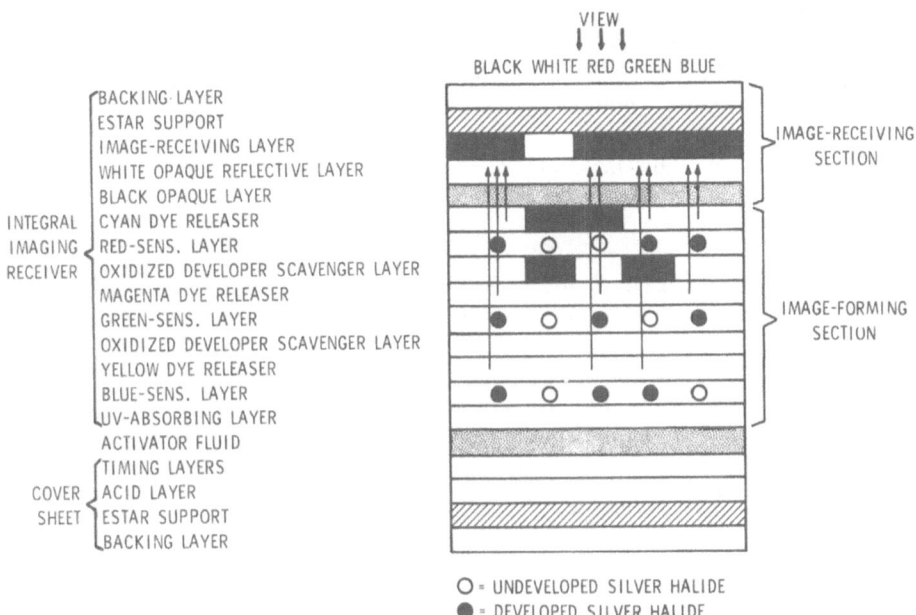

Figure 2. Kodak instant print film schematic section after processing.

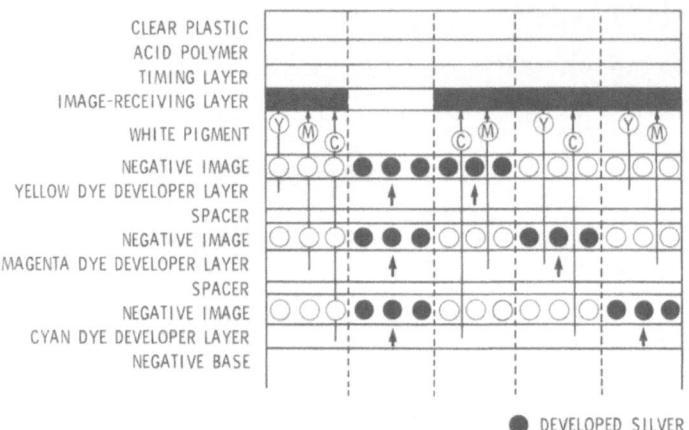

CLEAR PLASTIC	
ACID POLYMER	
TIMING LAYER	
IMAGE-RECEIVING LAYER	
WHITE PIGMENT	
NEGATIVE IMAGE	
YELLOW DYE DEVELOPER LAYER	
SPACER	
NEGATIVE IMAGE	
MAGENTA DYE DEVELOPER LAYER	
SPACER	
NEGATIVE IMAGE	
CYAN DYE DEVELOPER LAYER	
NEGATIVE BASE	

● DEVELOPED SILVER

Figure 3.   Polaroid SX-70 instant print film, schematic section after
            processing.

ness of a given UV stabilizer varies considerably with the dyes to be
stabilized and other characteristics of the positive print viewing
area. It is difficult to predict the degree of effectiveness of a
stabilizer because some dyes, e.g., azo dyes, are more susceptible
than others to UV.

     A number of different types of UV stabilizer compounds have been
previously suggested for stabilization of polymeric materials.[9-14]
However, only four fundamentally different classes have achieved commercial significance.[15] These are the derivatives of 2-hydroxy benzophenone, phenyl esters, substituted cinnamic acid derivatives, and
2-(2H-benzotriazol-2-yl) phenols. The two classes that are pertinent
to this research are the benzophenones and phenyl esters.

Benzophenones

     The chemical structure and absorbance of some common UV screening
agents of the hydroxy benzophenone type are shown in Figure 4.

     The photochemistry of the 2-hydroxybenzophenones has been more
extensively studied than that of the other classes of ultraviolet
absorbers.[5] It is known that (1) is rapidly converted to a "photo-
enol", (2) by absorption of light, and that (2) reverts to (1), with
loss of energy as heat, with almost 100% efficiency.

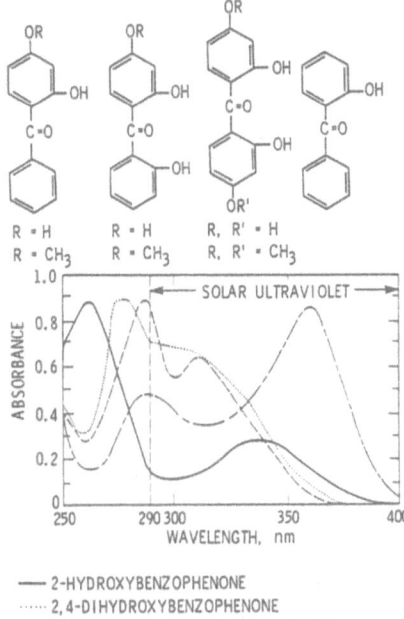

Figure 4.   Derivatives of 2-hydroxybenzophenone: composition and UV spectra.

The existence of the intramolecular hydrogen bond in both (1) and (2) accounts for the rapid and efficient phototautomerism. The only important structural difference between (1) and (2) is the distribution of electrons; hence, interchange of the hydrogen is very rapid.

Phenyl Esters

The most important members of this group are resorcinol mono-benzoate (3) and phenyl salicylate (4a), but substituted aryl salicylates (4b) and diaryl terephthalates (5) or isophthalates (6) are, or have been, sold as ultraviolet absorbers.

These compounds have one feature in comon - very low absorption
in the solar ultraviolet region. However, after exposure to sunlight
for a time, these compounds show an increase in absorption in the
290 to 400-nm region, and, after sufficient exposure, their spectra
resemble those of 2-hydroxybenzophenones. These compounds owe their
effectiveness to a light-catalyzed rearrangement that converts them
to 2-hydroxybenzophenones (4). The products of this photo-Fries
rearrangement are the actual stabilizers; resorcinol monobenzoate,
for example, is converted to 2,4-dihydroxybenzophenone (7), a very
effective ultraviolet absorber.

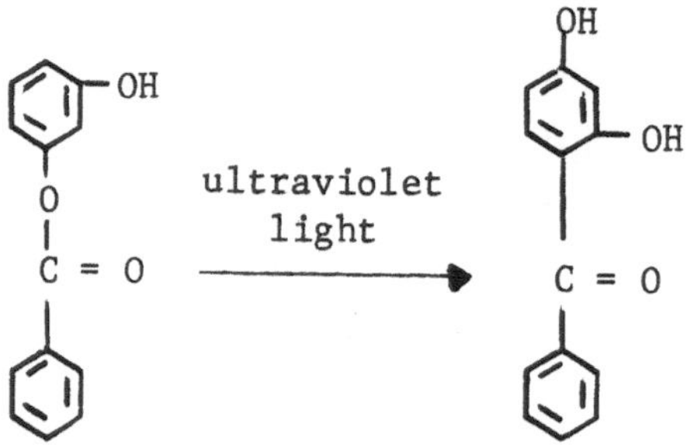

In our present work, di(nonylphenyl) isophthalate was used to
stabilize instant photographic prints. Its chemical formula is as
follows:

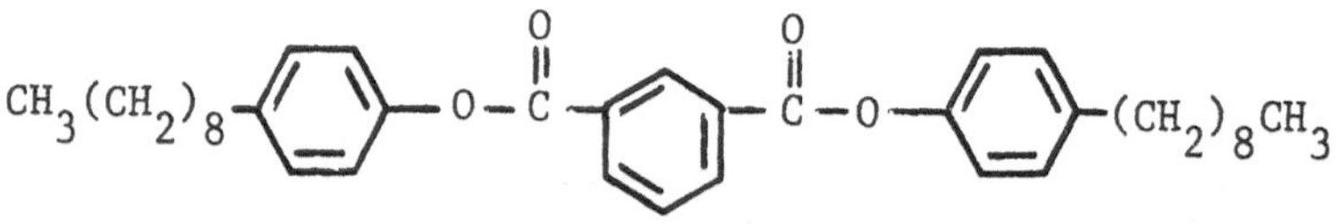

An examination of the formula of di(nonylphenyl) isophthalate
shows that a large number of resonance and hyperconjugated structures
can be drawn. Therefore, di(nonylphenyl) isophthalate can also re-
arrange in situ on exposure to UV to form substituted benzophenones
which are effective UV absorbing compounds. The structure of di-
(nonylphenyl) isophthalate after rearrangement is believed to be as
follows:

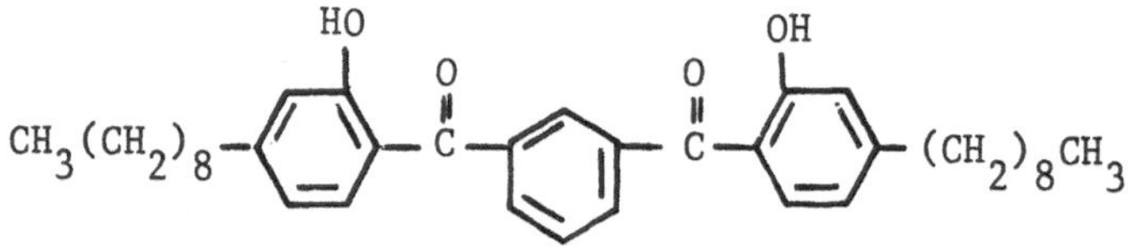

There are several important advantages to "latent" phenyl-ester
type UV screening agents, including:
1. The precursor ester has less initial color (in the visible
    spectrum) than most other types of UV screening agents, and

this makes whiter, brighter products possible.

2. The phenyl ester is more compatible with many polymeric carriers, and can be uniformly incorporated into the vehicle at higher concentrations to provide longer protection from UV.

3. The long-chain aromatic ester structure of the precursor enables it to function as an auxiliary plasticizer in polymeric carriers, such as PVC and cellulosic esters.

4. The ester rearranges in situ upon exposure to UV to form the more active substituted benzophenone UV screening agent on an as needed basis. We have called this property the "reservoir principle."

5. The esters are presumably non-toxic, and (if produced commercially) should be cost competitive with other conventional UV screening agents.

EXPERIMENTAL

Preparation of Instant Color Prints

Approximately, 7.5-cm x 10-cm instant photographic color prints in which the image forming dyes were essentially monoazo derivatives of enolic couplers, e.g., benzylazo-1-naphthols and benzylazo-5-pyrazolones, were developed. All the prints were prepared identically by self-development using Eastman Kodak instant color film and camera. A display board on which were mounted eight, 7.5-cm x 10.3-cm color chips which substantially covered the visible color spectrum was photographed. The procedure was repeated using Polaroid SX-70 color film and camera.

Vehicle and Coating Preparation

The criteria considered in the selection of the vehicles included high optical clarity, water white color, adhesion to the instant print substrate, resistance to moisture, and good solubility. Cellulose acetate (40 percent acetyl content) and coating grade polymethyl methacrylate were found as two suitable vehicles for this application. Two overprint varnished (OPV's) were prepared according to formulations indicated in Table I. The resulting solutions were optically clear and did not require filtration.

Varnishing and Testing

The OPV's were applied onto the prints by knife coating at a thickness between 0.025 and 0.05 mm. The solvent was removed by evaporation. The fading test was an accelerated exposure using a UV lamp (Sunlighter IV Ultraviolet Test Console of Test-Lab Apparatus Co., Amherst, N.H.). The prints were arranged on a flat, 43-cm diameter turntable rotating 14 cm away from a single-bulb source of UV light

Table I. OPV's Formulations

	Formula, %	
	OPV I	OPV II
Partially hydrolyzed cellulose acetate (40% acetyl content)	13.0	--
Tetrahydrofuran (THF)	85.0	75.0
Di-nonylphenyl isophthalate	2.0	2.0
Polymethyl methacrylate (coating grade)	--	23.0
	100.0	100.0

(290-350 nm) of sufficient intensity that about 24 hours of exposure simulates one year of exposure to natural sunlight in Florida. Air temperature inside the cabinet containing the turntable and UV light light source was maintained at 60°C, and the prints were located on the turntable so that each received equal exposure to the UV light. One uncoated (control) and two coated (by OPV-I and OPV-II) prints were tested for 170 hours in the Sunlighter.

RESULTS

Results of continuous exposure of the prints at the end of 70 hours are shown in Table II. These results report usually evaluated colors and "fading coefficients." In Table II, the "fading coefficient" is the percent change from the specified original color in the print to a completely faded color (essentially white) which would be characterized by a 100% "fading coefficient." All colors faded away in the uncoated color prints after 70 hours of UV exposure. OPV-I and OPV-II coated prints remained colorfast, the former being slightly better than the latter. OPV-I and OPV-II protected Kodak prints were stable even after 170 hours exposure to UV in the Sunlighter.

In comparison, it was found that the uncoated Polaroid instant print faded very little after 170 hours in the Sunlighter. However, OPV-I and OPV-II protected Polaroid prints were slightly better than the uncoated print.

A U. S. patent issued December 31, 1981, claims a photographic element, comprising a photographic organic dye image, or precursor thereof, and a stabilizer selected from phenyl and naphthyl diesters

Table 2.   Results of UV Stabilization of Instant Prints

Original Color in Prints	Control (no stabilization)		OPV I		OPV II	
	Final Color	Fading Coeff.(%)	Final Color	Fading Coeff.(%)	Final Color	Fading Coeff.(%)
Dk. beige	Lt. ivory	90	No change	5	No change	5
Maroon	Lt. beige	90	No change	10	Slightly brownish maroon	20
Olive gr.	Lt. ivory	80	No change	5	No change	10
Dk. rose	Lt. ivory	80	Med. rose	15	Orangy rose	20
Yellow	Lt. pale yellow	95	No change	0	No change	0
Dk. red	Lt. ivory	95	Dk.-med. red	10	Med.-dk. red	5
Purple	White	95	No change	10	No change	10
Blue	Lt. brown	80	No change	5	No change	5

of benzene and naphthalene dicarboxylic acids in an amount sufficient
to substantially increase the stability of said dye against ultraviolet radiation.[16]

CONCLUSIONS

Our work illustrates that surface coatings containing di(nonylphenyl) isophthalate, in cellulose ester or polymethyl methacrylate
vehicles, provide an unexpectedly high degree of improvement in stability of the Kodak instant photographs against undesirable UV light
effects such as color fading and color drift. Similar results can
be expected if the stabilizer is incorporated in an external or internal coating on photographic film prior to its exposure in a camera.
Other dye-containing articles that can be stabilized by coatings containing di(nonylphenyl) isophthalate are textile fibers and fabrics,
wall coverings, extruded thermoplastics, and art work, e.g., painting.

Compatibility with many polymeric vehicles, excellent initial
color, and effectiveness as plasticizers (e.g., cellulosics and
vinyls), should enable the "latent" phenyl ester UV screening agents
that convert to even more effective hydroxy substituted "benzophenones" on exposure to ultraviolet light (reservoir principle) to find
broader use than the specific color print protective coating described
herein. As pointed out by Coran and Anagnostopoulous,[9-13] the utility
of this and similar esters has demonstrated to be effective UV stabilizers in a variety of polymers. Specifically, clear polyvinyl-
chloride (PVC) films containing di(nonylphenyl) isophthalate UV
screening agent, in addition to other normal PVC stabilizers, withstood more than ten years of outdoor exposure in Florida without significant discoloration or becoming opaque. This is an impressive
result since, as is well known, it is very difficult to stabilize
initially clear and "colorless" PVC films for anything approaching
a ten-year time period. Collectively, these results provide a strong
argument for commercializing the presently experimental di(nonylphenyl) isophthalate (and similar) UV screening agents.

REFERENCES

1.   J. R. Thirtle, Chem-Tech, 25-35 (Jan., 1979).
2.   W. T. Hanson, Jr., Photogr. Sci. Eng., 20, 155 (1976).
3.   G. Crawley, Br. J. Photogr., 110, 76 (1963).
4.   E. H. Land, Photogr. Sci. Eng., 20, 155 (1976).
5.   W. F. Smith, Jr., and K. L. Eddy, U. S. Pat. 4,042,394 (1977).
6.   W. F. Smith, Jr., and G. A. Reynolds, U. S. Pat. 4,050,938
     (1977).
7.   W. W. Weber, II, and D. W. Heseltine, U. S. Pat. 4,045,229
     (1977).

8. R. F. W. Cieciuch and N. H. N. Schlein, U. S. Pat. 4,025,682 (1977).
9. A. Y. Coran and C. E. Anagnostopoulous, U. S. Pat. 3,024,248 (1966).
10. A. Y. Coran and C. E. Anagnostopoulous, U. S. Pat. 3,255,235 (1966).
11. C. E. Anagnostopoulous and A. Y. Coran, U. S. Pat. 3,256,238 (1966).
12. C. E. Anagnostopoulous and A. Y. Coran, U. S. Pat. 3,284,220 (1966).
13. A. Y. Coran and C. E. Anagnostopoulous, U. S. Pat. 3,284,405 (1966).
14. D. A. Gordon, U. S. Pat. 3,080,339 (1963).
15. Ultraviolet-Radiation Absorbers, in: "Encyclopedia of Polymer Science and Technology," N. Bikales, Ed., Wiley-Interscience, New York, 1971, Vol. 14.
16. I. O. Salyer and A. M. Usmani, U. S. Pat, 4,304,328 (1981).

PHOTOCHEMICAL GRAFTING OF ACRYLATED AZO DYES ONTO POLYMERIC SURFACES
VI. EFFECT OF 1,2-DIPHENYL, 2,2-DIMETHOXY, ETHANONE AS PHOTOINITIA-
TOR ON THE GRAFTING OF SOME ACRYLOXY-SUBSTITUTED AROMATIC DIAZENES
ONTO POLY(PROPYLENE) AND POLY(CAPROLACTAM) FIBERS

Ignazio R. Bellobono, Seba Calgari,* and Elena Selli

Cattedra di Chimica, Facolta di Scienze
University of Milan
20133 Milano, Italy

INTRODUCTION

Photo-induced grafting and graft polymerization onto polymers
promise to become an elegant and powerful method for modifying surface
properties. With the wide variety of vinyl and other monomers available today, photochemical grafting may become an established technical method to impart desired surface characteristics: from antistatic behavior to thermal stability, from adhesion to bacteriocidal
properties, from water repellency to hygroscopicity, from soil resistance to flame retardant properties, from optical imaging to dyeabil-
ity and so on. Photochemistry may consequently play a very important
role in this field, by promoting knowledge, by studying applications,
as well as optimizing conditions of industrial processes, potentially
capable of being developed.

In this presentation we shall focus attention on some photochemical processes for textile treatments, and for textile printing particularly.

A high level of research interest and activity in radiation-
curable treatments for textiles was achieved in the late sixties and
in the seventies.[1,2] These studies were substantially based on photochemically reactive resins and inert (that is, non-photochemically
reactive) pigments. The "non-photochemically reactive" expression
refers to the absence of any desired or programmed photochemical
reaction concerning dyes or pigments. At the same time it is well

---

*Snia Fibre, 20030 Cesano Maderno, Italy.

known that photophysics and photochemistry of pigments and dyes have
an important bearing on the overall properties and behavior of both
UV-curable coatings and textiles. In all these studies, however, no
attention was devoted, as far as we know, to the photochemical reactivity of dyes.

Review of Previous Work

In recent years we have been concerned with studying and developing a process for photochemical grafting of dyes onto polymeric
surfaces (woven fabrics particularly, but not exclusively). During
the course of these studies, we have examined some model systems in
order to understand the fundamental mechanisms which underlay these
processes. Some model azo dyes, (I) - (IV),[3,4] have been function-
alized for photochemical reactivity by introducing an acrylic group
into a suitable part of the molecule, sufficiently isolated from the
electronic system and the chromophores. With one or more of these
dyes, we have studied the influence of some of the most revelant parameters which characterize the process, viz.: (i) the form of contact between the dye and the substrate (crystalline dyes were adsorbed on the surface by evaporation from volatile solutions[3,5,6,7]
or, alternatively, the substrate was irradiated while simply immersed
in dye solutions[4,5]); (ii) monochromatic or polychromatic irradiation;
(iii) kind of polymer [we have been mainly using poly(propylene),
poly(caprolactam), poly(ethyleneterephthalate), and cellulose]; (iv)
dye concentration (when working in solution) or mass of dye deposited
per unit apparent surface (when working with crystalline dyes adsorbed
on films and fibers).

One feature which has been found to be substantially similar in
all investigated cases was the general shape of the graft yield curve
(Figure 1). Two consecutive kinetic processes may be recognized,
which have been shown to be grafting and graft polymerization. The
kinetic law able to describe both processes is very simple (see cap-
tion of Figure 1). When operating in solution,[4,5] the dependency of

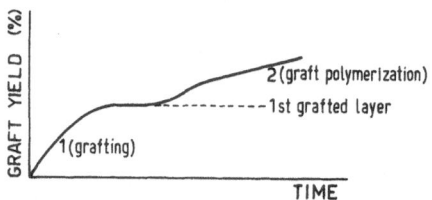

Figure 1.   Typical shape of kinetic curve for grafting (1) and graft
             polymerization (2). Rate equation: $(1/\underline{S})(d\underline{n}/d\underline{t}) = \Phi_{1,2}\underline{I}$
             $\underline{S}$ = apparent surface of polymer; $\underline{n}$ = moles of dye (crys-
             talline or in solution) at time $\underline{t}$; Φ = mean quantum yield;
             $\underline{I}$ = radiation intensity effectively adsorbed.

quantum yields on dye concentration, both for grafting and graft polymerization, closely followed Stern-Volmer plots, by which lifetimes of intermediate excited states for the two kinetic processes (τ_1 and τ_2, respectively) could also be obtained. The marked influence of wavelength on quantum yields and the constancy of the relatively long lifetimes τ_1 and τ_2 at the various wavelengths led to the hypothesis that quantum yields reflect efficiencies of conversion from the states populated by light absorption to reactive upper triplets.[4,5] This observation seems to be quite general for photochemical reactions of compounds which have a carbonyl group and a conjugated double bond capable of twisting in the excited state to a non-planar geometry.[8-10]

The grafting mechanism (first constant rate period of Figure 1), which has been proposed, based on hydrogen abstraction from the reactive sites of the polymer surface, is summarized in the following scheme:

D = dye skeleton-(CH_2CH_2OCO)-

MH = photochemically reactive sites of the polymer surface

$$MH + \underset{D}{CH}=CH_2 \xrightarrow{\Phi_1} M^{\cdot} + \underset{D}{CH}-CH_3 \tag{1}$$

$$M^{\cdot} + \underset{D}{{}^{\cdot}CH}-CH_3 \longrightarrow M-\underset{D}{CH}-CH_3 \tag{2}$$

(grafting of "monomolecular layer")

$$\underset{D}{{}^{\cdot}CH}-CH_3 + \underset{D}{CH}=CH_2 \longrightarrow \underset{D}{{}^{\cdot}CH}-CH_2-(\underset{D}{CH}-CH_2)_x-H \tag{3}$$

$$x=1,2,3,\ldots$$

(growth of oligomeric chains, formation of homopolymer)

$$M^{\cdot} + \underset{D}{CH}=CH_2 \longrightarrow M-\underset{D}{CH}-\overset{\cdot}{CH_2} \xrightarrow{R-CH=CH_2} M-\underset{D}{CH}-CH_2-\underset{D}{CH}-\overset{\cdot}{CH_2} \tag{4a}$$

$$M^{\cdot} + \underset{D}{{}^{\cdot}CH}-CH_2-(\underset{D}{CH}-CH_2)_x-H \longrightarrow M-\underset{D}{CH}-CH_2-(\underset{D}{CH}-CH_2)_x-H \tag{4b}$$

$$x=1,2,3,\ldots$$

(grafting of oligomolecular layer).

Reactions (1) and (2) may occur both in solution and in the solid state, while reaction (4), which describes grafting of an oligomolecular layer on the bare reactive sites of the polymer surface, are practically unfavored when grafting is performed from dye solutions.

As to the second constant rate period of Figure 1, the mechanism is generally outlined in the following scheme:

$$\underset{\underset{D}{|}}{M-CH}-CH_2-\underset{\underset{D}{|}}{(CH}-CH_2)_x^{-}H + \underset{\underset{D}{|}}{CH}=CH_2 \xrightarrow{\Phi_2} \underset{\underset{D}{|}}{M'-\overset{\displaystyle\cdot}{C}}-CH_3 + \underset{\underset{D}{|}}{\overset{\displaystyle\cdot}{C}H}-CH_3 \qquad (5)$$

$$x = 1,2,3,\ldots$$

$$(M' = \underset{\underset{D}{|}}{M-CH}-CH_2-\underset{\underset{D}{|}}{(CH}-CH_2)_{x-1}^{-})$$

$$\underset{\underset{D}{|}}{M'-\overset{\displaystyle\cdot}{C}}-CH_3 + \underset{\underset{D}{|}}{\overset{\displaystyle\cdot}{C}H}-CH_3 \longrightarrow \underset{\underset{D}{|}}{M'-\overset{\overset{\displaystyle CH_3}{|}}{C}}\!-\!-\!\underset{\underset{D}{|}}{\overset{\overset{\displaystyle CH_3}{|}}{C}}-H \qquad (6)$$

(termination)

$$\underset{\underset{D}{|}}{M'-\overset{\displaystyle\cdot}{C}}-CH_3 + \underset{\underset{D}{|}}{CH}=CH_2 \longrightarrow \text{graft polymerization} \qquad (7)$$

It simply consists on activation of the first grafted layer and grafting, as well as graft polymerization, on the latter.

Another important feature of the process is represented by the surface density of grafted molecules at the end of the first constant rate period of the graft yield kinetic curve in Figure 1. The influence of dye concentration is determinant on this value. Of course, when operating in solution this parameter is the concentration of the dye properly. When operating with adsorbed dyes (dyes deposited on the polymer surface), on the contrary, the parameter which may be varied is the mass of dye deposited per unit apparent surface. In order to compare the two cases, consequently, concentration has to be expressed, for both, as n/S, that is mass (in moles n) per unit apparent surface S. Schematically, as it is represented in Figure 2, two types of behavior can be distinguished.

When operating with crystalline dyes adsorbed on the polymer surface, the surface density of grafted molecules is dependent upon the kind of polymer as well as on n/S. This latter dependency has been found fairly linear, but the slope is independent of the kind of polymer. This clearly expresses the contribution of grafting of an oligomolecular layer in the mechanism given for the first constant rate period of the graft yield curve.

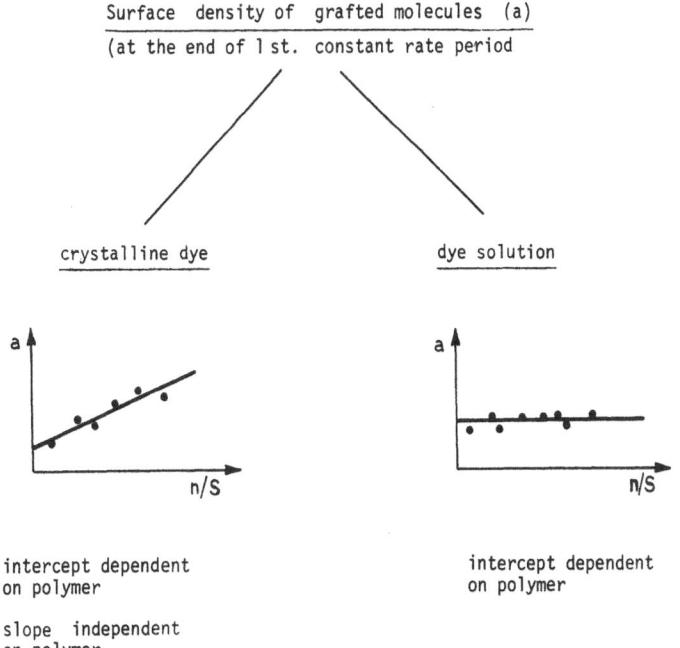

Figure 2. Typical shapes of the plot of surface density of grafted molecules (at the end of the first constant rate period) as a function of n/S.

When operating in solution, on the contrary, (schematic plot to the right of Figure 2) the concentration of dye has apparently no effect on the surface density of the grafted molecules in the first grafted layer and this supports the idea of a nearly monomolecular layer for this stage, when grafting is carried out from dye solutions.

The influence of the chemical nature of the polymer has been confirmed in the order poly(propylene) < poly(caprolactam) < poly(ethyleneterephthalate) for quantum yields of grafting (Φ_1). Quantum yields of graft polymerization (Φ_2), on the contrary, have been found to be independent on the nature of the polymer.

EXPERIMENTAL

Poly(propylene) and poly(caprolactam) fibers (Meraklon 160/22, Montedison and Lilion 60/18, Snia Viscosa, respectively) were used in the form of a ribbon obtained from a woven fabric. They were purified through a mold cleaning treatment with sulfonated lauryl alcohol (5g/1) for 30 min. at 60°C, thoroughly rinsed with water at room tem-

perature, and finally Soxhlet extracted with 50% v/v cyclohexane-
ethanol.

    Dyes (I) - (IV) were prepared and purified as previously re-
ported.[3,4]

    Grafting and graft polymerization were carried out as described
in a preceding paper,[3] the only difference being represented by the
further addition of 1,2-diphenyl - 2,2-dimethoxy ethanone (V) as
photoinitiator to the toluene solutions of the dye, in the desired
ratio $\underline{R}$ with this latter. Photoinitiator (V) was a commercial product (Ciba Geigy) used without further purification. 2 ml of the toluene solutions containing photoinitiator (V) and 0.002-0.10 M acry-
lated dye in the ratio $\underline{R}$ of their concentrations (0.05 < $\underline{R}$ < 10) were
uniformly deposited on 5.0 x 10.0 cm^2 of ribbon, and the solvent evaporated at room temperature. Irradiations were then carried out with
a 500 W high-pressure mercury arc lamp (Italquartz), in such a way
as to have a constant impinging photoenergy on the sample of 90±5
mW cm^{-2}, with no exclusion of air during experiments. The temperature
of the system increased to 45±2°C during irradiation. No dark reaction was, in any way, observed at this temperature for exposure times
used in the kinetic curves. These latter (see Figure 1) were obtained
by irradiating the sample at various times. After the desired irradiation time, the sample was Soxhlet extracted with acetone, and the
acetone solution of the dye and photoinitiator analyzed spectrophoto-
metrically for the dye. The graft or graft polymerization yield was
calculated as the percent ratio between unextracted dye and that initially deposited on the polymer ribbon. Reflectance spectroscopy of
the photochemically dyed surface showed that grafted dyes were substantially in their trans-azo form, the contribution of the cis-form
being negligible (< 4%).

RESULTS AND DISCUSSION

    In the present work, the effect of 1,2-diphenyl - 2,2-dimethoxy-
ethanone (V), as photoinitiator, on grafting and graft polymerization
of dyes (I) - (IV) onto poly(propylene) and poly(caprolactam) fibers
has been systematically examined, by employing crystalline dyes and
photoinitiator (V) uniformly deposited on the polymer fibers. Irradiation was carried on polychromatically. Mean quantum efficiencies
for grafting (Φ_1) and graft polymerization (Φ_2) have been evaluated
from graft yield curves, such as that schematically represented in
Figure 1, by following the same procedure already described.[3]

    An example of this study, relative to acrylated dye (IV), is
reported in Figure 3. In this figure mean quantum yields for grafting Φ_1 (ordinate to the left) and for graft polymerization Φ_2 ordinate to the right) onto poly(propylene) and poly(caprolactam) ribbons

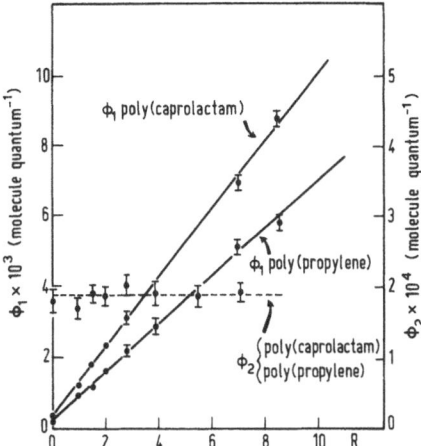

Figure 3.   Quantum yields for grafting (Φ_1) and graft polymerization
(Φ_2) of acrylated dye (IV), as a function of the R ratio
between n/S (μmoles/cm^2 of apparent surface) of photoinitiator ($\overline{V}$) and the corresponding values of dye (IV), onto
poly(propylene) and poly(caprolactam) ribbons (energy
input 90±5 mW/cm^2)

are reported as a function of ratio R between n/S (μmoles cm^{-2} of
apparent surface) of photoinitiator (V) and the corresponding value
of dye (IV). A similar behavior was observed for the other dyes.
It may be readily noticed that Φ_2 values are practically unaffected
by the presence of the photoinitiator. Φ_1 values, on the contrary,
are markedly influenced: the acceleration factor brought about by
the photoinitiator is around two powers of ten, when R is of the
order of ten. The effect, anyway, is not correlated with the absolute values of n/S (μmoles cm^{-2} of apparent surface) of either photoinitiator or dye. It rather depends, as may be seen in Figure 3, on
the ratio R between n/S values of photoinitiator (IV) and n/S of
acrylated azo dyes. This is quite reasonable, since this ratio is
a convenient measure of the photoinitiator "concentration" in the
crystalline mixture adsorbed on the polymer surface.

Most interestingly, if, by linear regression analysis of experimental plots of Figure 3 (Φ_1 vs. R), following equation:

$$\Phi_1 = \Phi_1^\circ + \underline{m} \; \underline{R} \qquad\qquad (8)$$

where Φ_1° denotes the quantum yield in absence of photoinitiator, the
ratio between the slope $\underline{m}$ and Φ_1° is evaluated (see Table I), this
latter ratio results independent, within the limits of experimental
uncertainty, both on the kind of polymer and on the molecular structure of acrylated dyes (I) - (IV).

Table I. Parameters of Equation (8) for Photochemical Grafting of
          Acrylated Azo Dyes (I) - (IV) onto Poly(propylene) (PP)
          and Poly(caprolactam) (PC) Ribbon at 45°C (Impinging
          Photoenergy 90 ± 5 mW/cm^2).

DYE	$\Phi_1^{\circ} \times 10^4$ (molecule quantum^{-1})		m / Φ_1°	
	PP	PC	PP	PC
I	1.7 ± 0.4	2.1 ± 0.4	2.9 ± 0.4	3.0 ± 0.4
II	2.0 ± 0.2	2.3 ± 0.3	3.2 ± 0.4	2.8 ± 0.3
III	1.9 ± 0.2	2.8 ± 0.2	3.2 ± 0.2	3.2 ± 0.3
IV	2.6 ± 0.2	3.4 ± 0.2	2.9 ± 0.2	3.0 ± 0.2

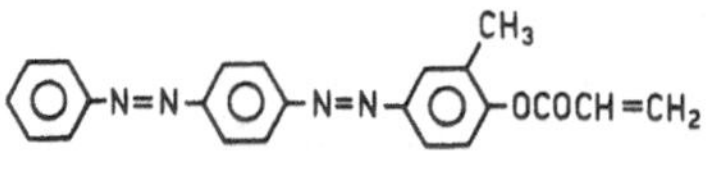

⟨C₆H₅⟩-N=N-⟨C₆H₄⟩-N=N-⟨C₆H₃(CH₃)⟩-OCOCH=CH₂

I

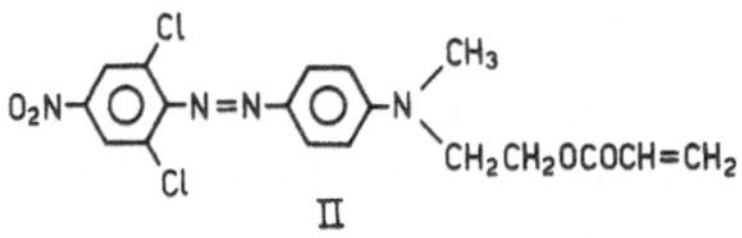

O₂N-⟨C₆H₂(Cl)(Cl)⟩-N=N-⟨C₆H₄⟩-N(CH₃)(CH₂CH₂OCOCH=CH₂)

II

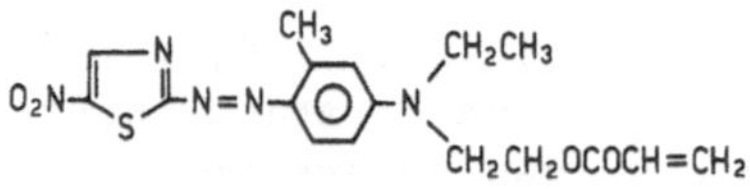

O₂N-⟨thiazole(N,S)⟩-N=N-⟨C₆H₃(CH₃)⟩-N(CH₂CH₃)(CH₂CH₂OCOCH=CH₂)

III

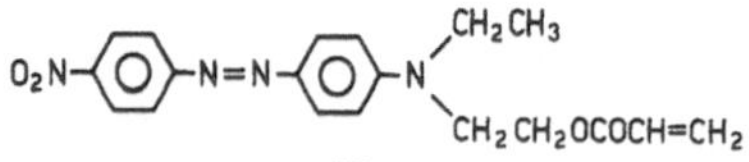

O₂N-⟨C₆H₄⟩-N=N-⟨C₆H₄⟩-N(CH₂CH₃)(CH₂CH₂OCOCH=CH₂)

IV

It has been previously underlined that the surface density of
grafted molecules at the end of the first constant rate period of the
graft yield curve of Figure 1 is an important aspect of the process.
From its knowledge valuable information about the oligomeric fraction
of the first grafted layer can be obtained.[3,7] In the presence of
the photoinitiator this oligomeric fraction strongly increased, as
much as the R ratio increased, as may be seen in some examples reported in Figure 4. In this figure the variation of the number of
grafted molecules per unit apparent area at this stage of the process
(end of the first constant rate period, corresponding to the plateau

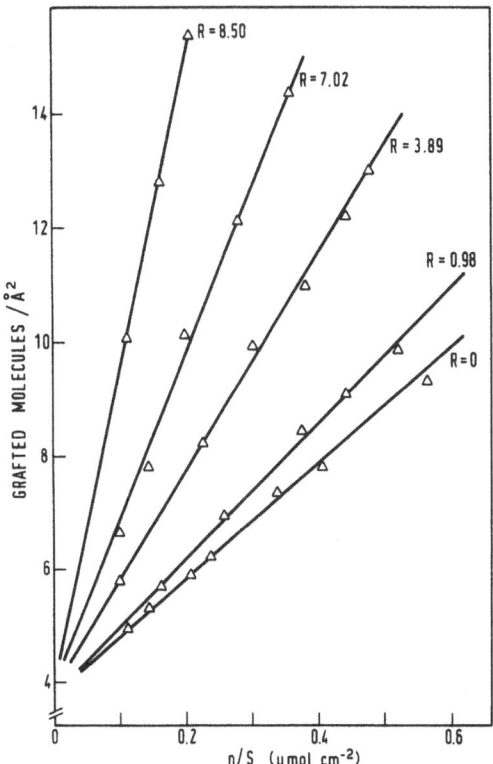

Figure 4. Grafted molecules of dye (IV) per apparent unit area (Å²)
at the end of the first constant rate period (plateau of
the graft yield curve) as a function of n/S (μmol/cm²) of
deposited dye for poly(propylene) ribbon (energy input 90
mW/cm²) at various values of ratio R between photoinitiator and dye.

of the kinetic curve of graft yield vs. time, schematically represented in Figure 1) is reported as a function of n/S at various R
values, relative to grafting of dye (IV) onto poly(propylene) ribbon.
Extrapolation of the lines of Figure 4 at n/S→0 gives a reasonably
constant, common value of about 3.8 molecules/Å², which may represent
the surface density of a "monomolecular" layer on the poly(propylene)
ribbon. Values of surface density higher than 3.8 molecules/Å², for
the experimental conditions tested, thus reflect the fraction of olig-
omers, which clearly grows with n/S, and, at constant mass of acrylated dye deposited per unit apparent surface, also with R. It must,
however, be noted that with R > 0.1 a progressive build-up of radical
species generated by photolysis of photoinitiator was clearly observ-
able, e.g., from variation of the reflectance spectrum of the grafted
dye on the surface.

The overall behavior of photoinitiated grafting is thus consistent, on one side, with the important role which is played in these processes by radical cage reactions and with the Norrish-type I cleavage shown by the initiator (V),[11] on the other with the proposed mechanism for grafting and graft polymerization. Reactions within a radical pair, photoinitiated by (V), may actively contribute to growth of oligomeric chains and consequently to efficient grafting of an oligomolecular layer. In the second constant rate period, on the contrary, which is rather a step-by-step polymerization on the first grafted layer (see equations 5,7), the photoinitiator is not efficient at all: its radicals, which do not react in the cage, may form escape products by hydrogen atom abstraction or other reaction paths.

The know-how of these investigations allows to proceed in gaining some technologically important achievements. On one side, quantum yield measurements afford a detailed information about rates, which may prove extremely useful for the mechanistic interpretation as well as for studying the medium effect. A systematic investigation of the effect of the presence of other monomers or radiation-curable binders in the system, as well as of the influence of the prepolymer structure may be carried out by these methods. On the other side, quantitative information about the surface coverage provides an idea on reactive sites available and on the molecular characteristics of the grafted layer.

The main advantages of these photochemical processes are not only in requiring significantly less energy than the conventional dyeing and printing on textiles, and polymers in general, which is notoriously a typical advantage of ultraviolet curing. Some new advantages may be added by the described processes:

  i) the possibility of grafting the print coating to the substrate, that is to product chemical rather than physical bonding;

 ii) the possibility of obtaining curing times low enough (of the order of the tenth of a second) to realize a mobile irradiating head capable of writing or drawing on the polymer surface (non impact printing);

iii) the possibility of treating by the same process different kinds of polymers, both natural and synthetic, and/or their mixtures;

 iv) the possibility of employing highly automatized and computerized systems for printing.

REFERENCES

1. W. K. Walsh, A. Makati and E. Bittencourt, Text. Chem. Color., (1978), 10, 220.
2. K. Park, R. L. Frame and G. M. Bryant, Text. Chem. Color., (1979), 11, 107
3. I. R. Bellobono, F. Tolusso, E. Selli, S. Calgari and A. Berlin, J. Appl. Polym. Sci., (1981), 26, 619.

4. E. Selli, I. R. Bellobono, S. Calgari and A. Berlin, J. Soc. Dyers Colour., (1981), 97, 438.
5. E. Selli, I. R. Bellobono, F. Tolusso and S. Calgari, Ann. Chim. (Rome), (1981), 71, 147.
6. I. R. Bellobono, S. Calgari, M. C. Leonardi, E. Selli and E. Dubini Paglia, Angew. Makromol. Chem., (1981), 100, 135.
7. S. Calgari, E. Selli and I. R. Bellobono, J. Appl. Polym. Sci., (1982), 27, ...
8. E. F. Ullman and N. Baumann, J. Am. Chem. Soc., (1970), 92, 5892.
9. I. R. Bellobono, L. Zanderighi, S. Omarini, B. Marcandalli and C. Parini, J. Chem. Soc., Perkin Trans. 2, (1975), 1529.
10. I. R. Bellobono, E. Dubini Paglia, B. Marcandalli and M. T. Cataldi, Gazz. Chim. Ital., (1979), 109, 697.
11. A. Borer, R. Kirchmayr and G. Rist, Helv. Chim. Acta, (1978), 61, 305.

CRITICAL PHENOMENA IN THE INHIBITED OXIDATION OF POLYMERS

Yu. A. Shlyapnikov

Institute of Chemical Physics, USSR
Academy of Sciences
117334, Moscow, USSR

The free radical generation in polymer oxidation is a result of chain initiation in the interaction of non-oxidized monomeric units of polymer or admixtures to it with oxygen, and of chain branching in the secondary reactions of oxidation intermediates.[1] In the developed reaction the rate of chain branching is proportional to the concentration of free radicals $RO_2^{\cdot}$ (x), the chain termination proceeds in interaction of free radicals or in their reactions with inhibitors (IH)

$$RO_2^{\cdot} + IH \longrightarrow ROOH + I^{\cdot} \qquad (1)$$

where $I^{\cdot}$ is a free radical of low reactivity. The equation for active radical balance in the branched chain reaction is

$$\frac{dx}{dt} = w_o + f x - k i x - k_r x^2 \qquad (2)$$

where w_o is the rate of chain initiation, i is the concentration of IH, and k is the rate constant of reaction (1). The rate of free radical interaction $w_r = k_r x^2$ is rapidly decreasing when the radical concentration decreases and in the presence of the inhibitor it may be neglected.

If the inhibitor effectively retards the oxidation, that is $dx/dt = 0$, the free radical concentration according to (2) will be

$$x = \frac{w_o}{k - f} \qquad (3)$$

The latter expression is of a definite meaning only when its numerator

95

is higher than zero, i.e., if the inhibitor concentration exceeds a certain value, called critical:[1,2]

$$i_{cr} = \frac{f}{k} \qquad\qquad (4)$$

If $i < i_{cr}$, the free radical concentration x will increase until limited by the radical interaction neglected in deriving (3).

As long as below the critical inhibitor concentration the oxidation is fast and self-accelerated, whereas above this concentration it is slow, the curves of induction period vs. inhibitor concentration display a pronounced bend corresponding to i_{cr}, which bend enables ready appreciation of the critical concentration (Fig. 1). The induction period, i.e., the time needed to reach a certain value of the reaction rate or of the free radical concentration is often used as a measure of the oxidation rate.

In some cases the rates of secondary reactions resulting in chain branching increase with inhibitor concentration. A result of it may be the absence of critical concentration (such inhibitors are weak antioxidants) or the appearance of an upper critical concentration, over which the inhibitor does not suppress polymer oxidation. Beginning from this concentration the induction period either is constant or even decreases (Fig. 2). The theory of this phenomenon

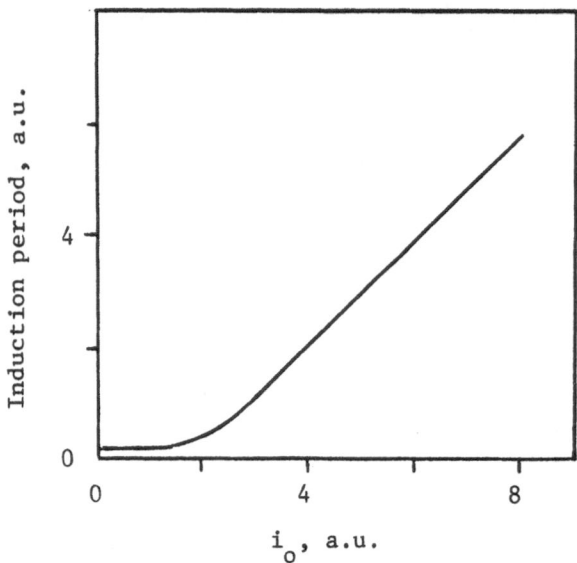

Figure 1.   Induction period of the branched chain reaction as a function of initial inhibitor concentration (calculated from equation (2) and equation of inhibitor consumption $- di/dt = kix$), a.u. - arbitrary units.

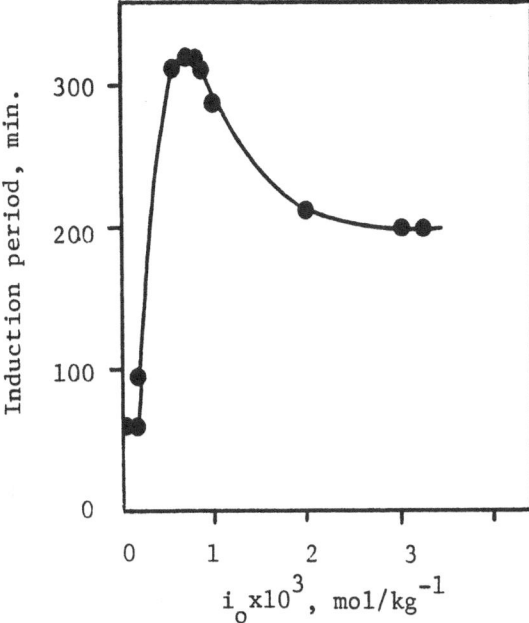

Figure 2. Induction period of oxidation of inden-coumarone resin as
a function of initial inhibitor concentration. 150°C,
oxygen, 300 mm Hg, inhibitor 2,2'-methylene bis (4-methyl-
6-tert.-butylphenol).

is discussed in Reference 2.

The increase in the free radical concentration in chain branching
usually is the result of generation and subsequent decomposition of
hydroperoxide groups ROOH:

$$RO_2^{\bullet} + RH \longrightarrow ROOH + R^{\bullet} \qquad (5)$$

$$ROOH + RH \longrightarrow (R^{\bullet} + RO^{\bullet} + H_2O) \xrightarrow{(RH)}$$

$$\longrightarrow \delta R^{\bullet} + \text{inactive products} \qquad (6)$$

The compounds which decompose hydroperoxide groups without free
radical generation can suppress chain branching. These compounds
involve organic sulfides and phosphites. In the presence of hydroperoxide decomposers the weak antioxidants display a critical concentration[2] and the critical concentration of other inhibitors becomes smaller (Fig. 3). The hydroperoxide decomposers display no
critical concentration and they are only seldom used as individual
antioxidants.

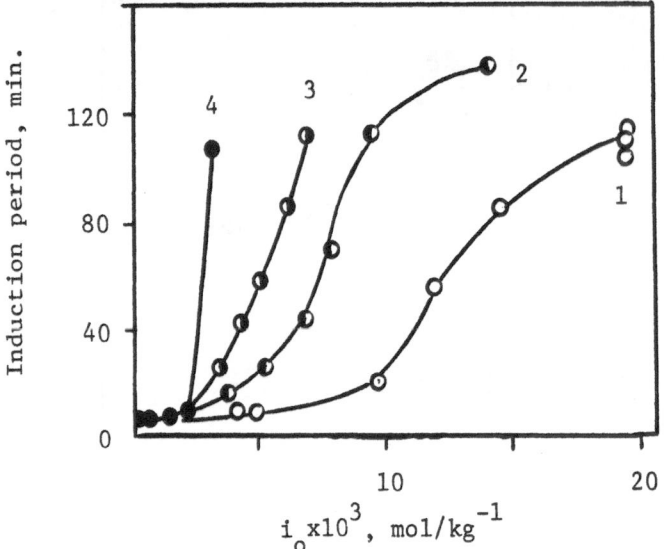

Figure 3. Induction period of oxidation of γ-irradiated polyethylene
as a function of inhibitor concentration. 200°C, oxygen
pressure 300 mm Hg, radiation dose 100 Mrad, inhibitor
2,2-methylene-bis(4-methyl-6-tert-butylphenol). Hydroperoxide decomposer dilaurylthiodipropionate, concentrations: 0 (1), 0.025 (2), 0.05 (3), and 0.10 mol/kg (4).

The value of critical inhibitor concentration may be used as a
quantitative measure of the antioxidant effectivity in the given
polymer. The less is critical concentration, the more effective is
inhibitor and the slower is the oxidation process in the presence
of it.

On the other hand, the critical inhibitor concentrations may
be used as tools for investigation of the polymer oxidation mechanism
because these concentrations are sensitive to any change in the chain
branching rate or in inhibitor reactivity. For example, it has been
found that a ferrous salt admixture increases the critical concentration of the effective inhibitor in polypropylene.[3] This showed that
ferrous and ferric ions increase the free radical yield in hydro-
peroxide group decomposition. The supposed mechanism is: $ROOH + Fe^{2+} \longrightarrow RO^{\cdot} + Fe^{3+} + OH^{-}$; $ROOH + Fe^{3+} \longrightarrow RO_2^{\cdot} + Fe^{2+} + H^{+}$.

The free radicals $R^{\cdot}$ and $RO_2^{\cdot}$ formed in polymer oxidation are
mostly of low mobility. To terminate such a radical the inhibitor
must move in the polymer bulk. The polar groups in the polymer which
can form complexes with inhibitor molecules decrease the number of
mobile molecules and, consequently, the ability of inhibitor to terminate chains. This results in a marked increase of the critical

concentrations of phenolic inhibitors in polyamide compared to polyethylene.[4]   Another result of the complex formation is an increase
of inhibitor solubility in polyamide-12, compared to polyethylene.
The solubility of non-polar diphenylmethane in both polymers is virtually the same.

The critical concentration of the same inhibitor in solid polymers depends on the method of sample preparation (Fig. 4).[5] This
dependence cannot be unambiguously interpreted because the method of
sample preparation affects both the inhibitor mobility and the polymer reactivity.

Most inhibitors are readily oxidizable compounds and free radicals are formed in their oxidation. In the presence of such inhibitors the rates of chain initiation and of inhibitor consumption at
$i \gg i_{cr}$ are directly proportional to the inhibitor current concentration

$$- \frac{d\,i}{d\,t} = k_{eff}i \tag{7}$$

whereas according to 3 when w_0 is proportional to i the free radical
concentration is independent of it.[6] Assuming that during the period
$\tau - \tau_{cr}$ the inhibitor concentration change is first order within the

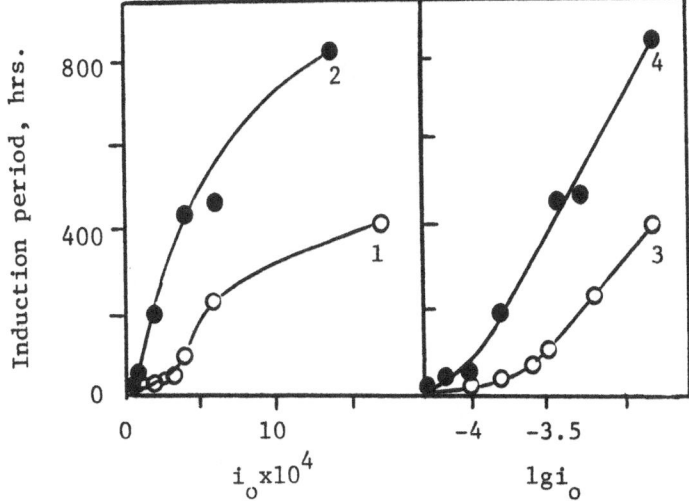

Figure 4.   Induction period of solid polypropylene oxidation as a
            function of inhibitor concentration. 100°, oxygen, 300
            mm Hg, inhibitor - methyl ester of (3,5-di-tert. butyl-
            4-hydroxy phenyl) propionic acid. 1, 3 - rapid cooling
            of melt, 2, 4 - slow cooling. R.H. - the same in semi-
            logarithmic coordinates.

range i_0 to i_{cr}, that is

$$i = i_0 \exp(-k_{eff}t) \qquad (8)$$

we get

$$\tau = \tau_{cr} + \frac{1}{k_{eff}} \log \frac{i_0}{i_{cr}} \qquad (9)$$

Here τ_{cr} is the value of induction period τ for $i_0 = i_{cr}$, which usually coincides with that in the absence of the inhibitor. According to (9), in many cases the experimental dependences of τ on i_0 when plotted in semilogarithmic coordinates "τ vs log i_0" become a straight line (Fig. 4, rhs).

According to (4), the critical inhibitor concentration depends only on chain branching factor f and on the rate constant of chain termination k, and does not depend on the rate of chain initiation w_0, and on the rate of inhibitor oxidation. For this reason carbon black, the catalyst of inhibitor oxidation, when added to polyethylene does not change the critical (i.e., maximum ineffective) inhibitor concentration, but causes a marked decrease in induction period above i_{cr} (Fig. 5).[7]

It follows from (9) that a decrease in i_{cr} will result in prolongation of induction period, i.e., of the antioxidant action time. The critical concentration can be decreased by addition of hydroperoxide decomposers (usually of organic sulfides R_2S) so that a mixture of the inhibitor of a radical scavenger type and a peroxide decomposer often retards the oxidation more effectively and for a longer period than does each of the components. In these cases distinct maxima are observed in the curves "induction period vs antioxidant mixture composition (Fig. 6).[2,8] Such an increase in the effectivity (synergism) is often used in the polymer stabilization. As mentioned above, in the absence of a radical scavenger the hydroperoxide decomposer shows no critical concentration.

Some other critical phenomena are encountered in polymer inhibited oxidation. There are: the critical composition values in binary mixtures limiting the range of high antioxidant effectivity, the limiting size of the oxidizing sample part which may be deprived of the inhibitor without making it a center of fast reaction[2] and some others.

The theory of critical phenomena is the essential part of the theory of polymer stabilization.

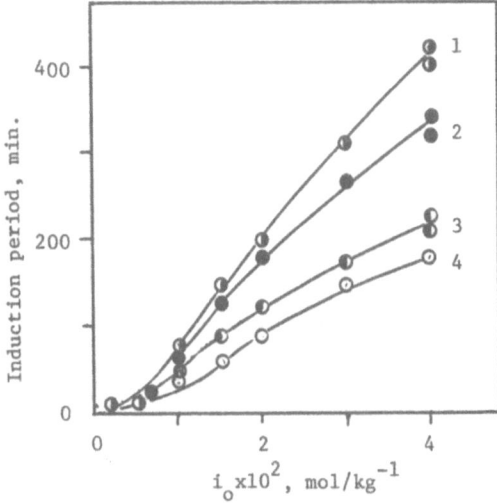

Figure 5. Induction period of low-density polyethylene oxidation
           as a function of inhibitor concentration in the absence
           (1) and in the presence of carbon black: 1% (2), 3% (3),
           and 5% (4). 200°C, inhibitor 2,2-methylene-bis(4-methyl-
           6-tert.butylphenol).

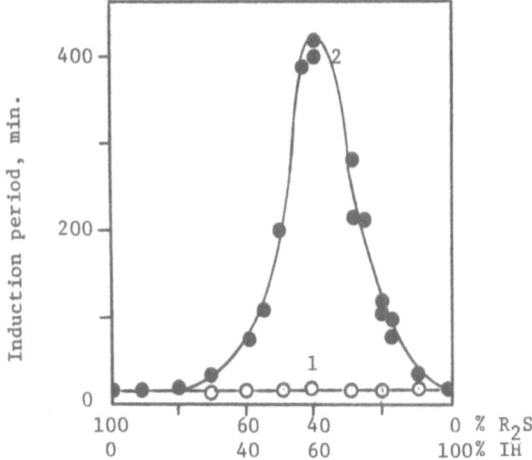

Figure 6. Induction period of polypropylene oxidation as a function
           of antioxidant mixture composition. 200°C, oxygen pres-
           sure 300 mm Hg, IH = phenyl-β-naphtylamine; R_2S = dilauryl-
           thiodipropionate, total antioxidant concentrations are
           0.005 mol/kg (1), and 0.01 mol/kg (2).

REFERENCES

1. N. N. Semenov, Chemical Kinetics and Chain Reactions, Oxford,
    Clarendon Press, 1935.
2. Yu. A. Shlyapnikov, Pure Appl. Chem., 57, 337, 1980.
3. L. L. Yasina, V. B. Miller, Yu. A. Shlyapnikov, Vysokomolek.
    Soedin., B11, 467, 1966.
4. A. P. Maryin, I. V. Yatsenko, S. R. Avetisyan, Yu. A. Shlyap-
    nikov, M. S. Akutin, Doklade Acad. Sci. USSR, 253, 159, 1980.
5. T. V. Monakhova, T. A. Bogaevskaya, Yu. A. Shlyapnikov, Vysokomo-
    lek, Soedin., B16, 840, 1974.
6. B. A. Gromov, Yu. A. Shlyapnikov, Vysokomolek, Soedin., A12,
    2637, 1967.
7. Ya. P. Kapachauskene, Yu. A. Shlyapnikov, Plasticheskiye Massy,
    1964, No. 12, p. 3.
8. I. A. Shlvapnikova, V. B. Miller, Yu. A. Shlyapnikov, Vysokomo-
    lek, Soedin., B14, 526, 1972.

CRITICAL TEMPERATURE FOR SOLUBILITY

OF A PHENOLIC ANTIOXIDANT

Valerie Kuck

Bell Laboratories
600 Mountain Ave.
Murray Hill, NJ, 07974

INTRODUCTION

Antioxidants are commonly added to polyethylene to retard oxidative degradation. The effectiveness of these antioxidants is dependent on both their chemical and physical properties. Thus an antioxidant must possess the chemical capability to retard the oxidation of the polymer, and in addition must have the physical properties which allow it to remain in the polymer bulk at a sufficient concentration to insure long term stabilization. The amount of antioxidant remaining in the polymer bulk will be dependent on: the equilibrium solubility of the antioxidant in the polymer, the rate of diffusion of the antioxidant and the rate of volatilization of the antioxidant from the polymer surface.[1] For long term stability, an antioxidant should have a high equilibrium solubility, a low rate of diffusion and a low volatility. Knowledge of these three factors would be helpful in screening chemically similar antioxidants. However, obtaining these data, especially near room temperature, is quite difficult because antioxidants usually have low solubilities (< 0.1 weight %) in hydrocarbon polymers and low vapor pressures (< 10^{-3} mm). In the current work we have attempted to develop a simple method for the measurement of solubility.

EXPERIMENTAL

Union Carbide's low density polyethylene (density = 0.92) and tetrakis[methylene-3(3',5'-di-tert.-butyl-4'-hydroxyphenyl)propionate] methane, THPM, from Ciba Geigy were used as received.

A laboratory mill was used to prepare a master batch containing

103

1% by weight of THPM and an aliquot was then diluted with unstabilized polyethylene to obtain the desired 0.1% weight concentration
of antioxidant in polyethylene. The polyethylene formulations were
milled at 120°C for 3 minutes. Films, 13 mils thick, were made by
compression molding between aluminum sheets for 1 minute at 150-
160°C and a pressure of 70 lbs/in^2. The molten films, held between
the two aluminum sheets, were rapidly cooled to room temperature by
immersion in a 10°C water bath for approximately 5 minutes. This
quick quench was chosen as similar to that used in wire production.
After removal from the aluminum sheets, the films were cut into 1"
x 2-1/2" samples and placed between two glass plates. The ends of
the glass plates were then secured with tape. The two large surfaces
of the sample were sealed in this way to reduce the volatilization
of antioxidant. The sandwiched samples were placed in jars under
nitrogen and placed in forced air ovens maintained at 40°, 60° and
80°C. Several samples were also aged at room temperature.

    After a lapse of time a portion (~ 5 mg) of the film was removed
and washed with isopropanol (~ 2 ml) to remove any antioxidant on
the polymer surface. The antioxidant concentration in the washed
films was determined by differential thermal analysis using a DuPont
990 Thermal Analyzer equipped with a DSC cell base. The analyses
were performed using aluminum dishes, a flow rate of 100 cc/min for
the purging gas and a heating rate of 20°C/min. In this procedure
the sample is heated at the programmed rate to 200°C under a nitrogen
atmosphere, then isothermally exposed to oxygen and the time to oxidation onset noted. The concentration of antioxidant remaining in
the samples was indirectly determined by comparison with a correlation curve of antioxidant concentration vs. induction time (Figure
1) obtained with freshly milled samples of low density polyethylene
containing THPM. The aged films were judged to have reached the

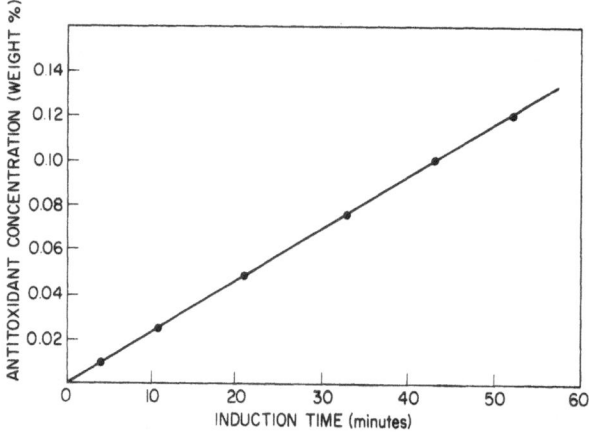

Figure 1. The relationship of Irganox 1010 concentration on polymer
           stability as measured by induction time.

equilibrium when there was little or no further change in antioxidant concentration over a period of time. It should be noted that initial results using high pressure liquid chromatography appear to agree well with results obtained by differential thermal analysis.[2]

In order to observe the migration of excess antioxidant to the polymer surface, a Reichert light microscope operated in the incident light position was used during the aging of the samples.

RESULTS

The films aged at 40°, 60° and 80°C initially showed a rapid decrease in antioxidant concentration; however, after 1-3 weeks the antioxidant concentration remained constant for the duration of the test (Figure 2).

The equilibrium solubilities of THPM did not increase linearly with increasing temperature (Figure 3). A discontinuity was observed between 40° and 60°C as the solubility fell from the 0.075 weight percent value obtained at 40°C to 0.055 weight percent at 60°C.

The film stored at room temperature was occasionally monitored, and after 4 months the antioxidant concentration in the polymer bulk had decreased to 0.07 weight percent and was unchanged after a further 6 months.

It should be noted that no correction of the observed equilibrium solubilities has been made for the increase in crystallinity of the polyethylene due to annealing at elevated temperatures.

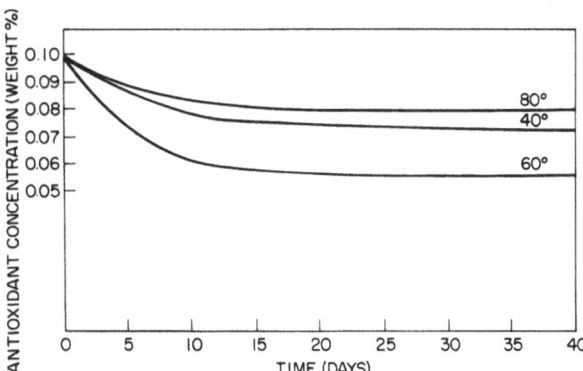

Figure 2. The decrease in antioxidant concentration in the polymer bulk of films aged at 40°, 60° and 80°C with time.

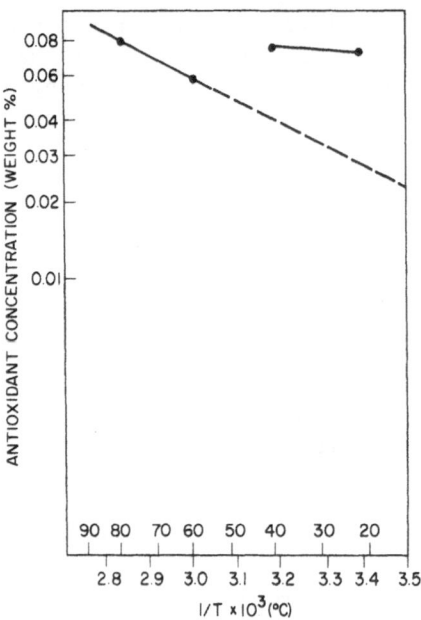

Figure 3.   The relationship of equilibrium solubility with temper-
            ature.

DISCUSSION

    A time-lag technique has been used previously to determine the
equilibrium solubility of an antioxidant in low density polyethylene.[3,4,5]   In this method antioxidant from a super-saturated sheet
of polyethylene is allowed to migrate into a stack of unstabilized
sheets of polyethylene and the equilibrium solubility determined by
extrapolation.   The present method requires the molding of only one
single sheet rather than a number of flat, uniformly thin films and,
therefore, is much simpler than the time-lag procedure.   The values
for the equilibrium solubility of THPM in low density polyethylene,
using the new procedure, were indirectly determined and found to be
0.075, 0.055 and 0.08 weight percent at 40°, 60° and 80°C, respectively.   These values are significantly higher than the 0.0029, 0.0091
and 0.0111 values observed by Roe et al[3] at 56°, 66° and 76°C, respectively, using the time-lag technique and thermal gravimetric analysis
for the determination of the extent of antioxidant migration.   The
approximate equilibrium value of 0.07 weight percent determined by
the new procedure is orders of magnitude larger than the 0.0001 weight
percent extrapolated by Roe et al.   However, the values obtained
herein are in fairly good agreement with the 0.003, 0.013, 0.056 and
0.20 values measured at 23°, 40°, 60° and 80°C, respectively, by
Moisan using the time-lag technique and ultraviolet spectroscopy for
the determination of the extent of antioxidant migration.[4,5]   The
greatest disagreement with this data occurs at the lower temperatures

and reflects the discontinuity in equilibrium solubility at ~ 50°C
observed using the new procedure.

Moisan[4,5] observed that the diffusion and solubility coefficients for several antioxidants did not increase linearly when plotted against reciprocal temperature as would be expected from the Arrhenius relationship:

$$k = k_o e^{-\Delta H/RT}$$

where k = the solubility constant at T for a given polymer, k_o = the
solubility constant in the completely amorphous polymer and ΔH = the
heat of solution of the substance.[6] Moisan found that antioxidants,
which had melting points in the temperature range at which the diffusion and solubility coefficients were measured, evidenced anomalous
behavior. It should be noted that THPM was one of the antioxidants
studied; however, it was not cited as evidencing a discontinuity in
its diffusion and solubility coefficients but having "enhanced" solubility due to its molecular structure. This relationship between
the melting point and anomalous behavior cannot be extended to explain the discontinuity observed herein at 50°C since the melting
range for THPM is 107-115°C.

A review of the melting behavior of pure THPM showed that after
melting, recrystallization does not occur upon cooling to room temperature (Figure 4) but a glass is formed. Allowing the glass to
stand for one week at room temperature does not result in crystallization.

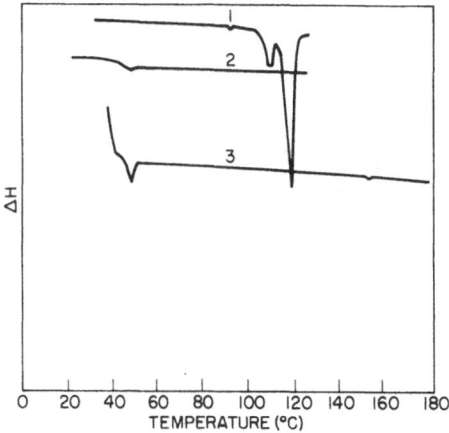

Figure 4. The DSC thermogram of THPM. (1) Initial heating; (2)
          second heating after quick quench and; (3) third heating
          after slow cool down (amplification 2 1/2x).

Examination of low density polyethylene films stabilized with 0.1 weight percent of THPM which had been aged for 3 months at 40°C, show the presence of small globules on the polymer surface (Figure 5). However, similar films aged for 1 week at 50°C had both globules and small groupings of needle-like crystals on the polymer surface (Figure 6). Based on this observation, a sample of neat THPM was melted, stored at 60°C for 1 week and then reheated in the DSC cell.

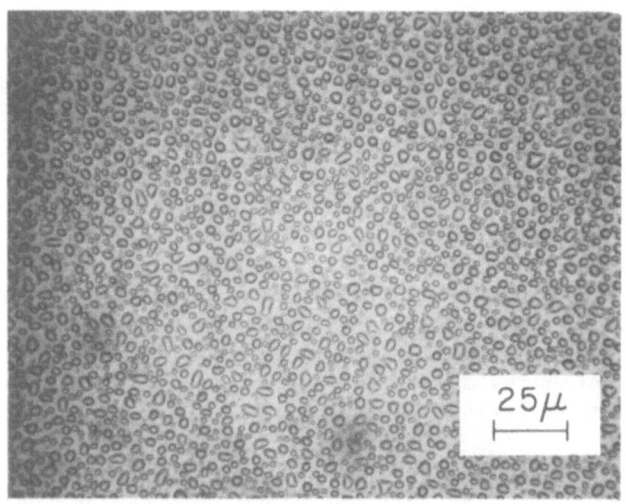

Figure 5.   Formation of globules on film aged at 40°C.

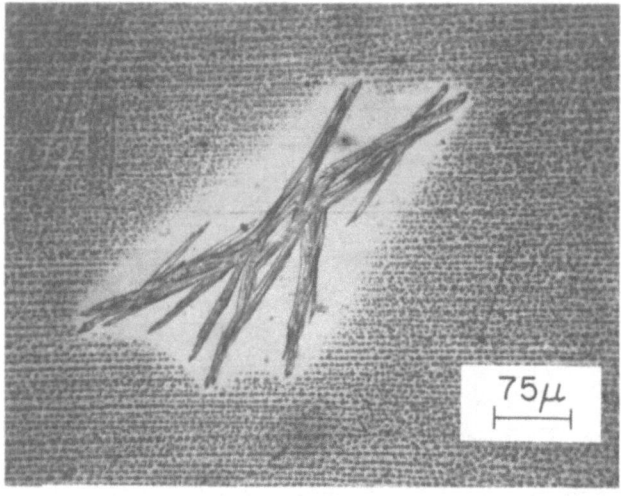

Figure 6.   Formation of both needle-like crystals and globules on a film aged for 1 week at 50°C.

The sample was found to have recrystallized (Figure 7). This recrystallization of THPM in the 50-60°C temperature range coincides
with the discontinuity in the equilibrium solubilities observed
herein. Discontinuous solubility curves have been previously found
for some aqueous salt solutions and the discontinuity has been attributed by Mee[7] to a change in the character of the solute as it
changes from one state to another. It appears that the discontinuity
in solubility which THPM exhibits at 50°C is reflecting its change
from a glass to a crystalline material.

Billingham has stated that the solubility of an antioxidant in
a polymer is dependent on the change in free energy associated with
the dissolution of the antioxidant from its equilibrium state as a
pure material.[8] The crystalline form would thus be expected to be
less soluble than the amorphous form.

The enhanced solubility of the amorphous form can be used to
explain the results observed by Howard[9] during the oven aging of
insulated 19 AWG aluminum wire samples. The low density polyethylene
insulation was stabilized with 0.1 weight percent THPM and N,N'-di-
benzal oxalyldihydrazide and aged at 25°, 50°, 70° and 100°C. The
samples aged at 25°, 50° and 100° were found to have greater stability than the samples aged at 70°C.

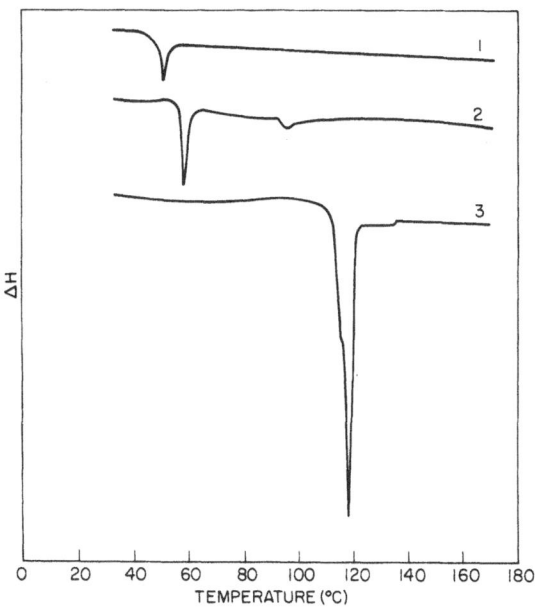

Figure 7. The DSC thermogram of previously melted THPM. (1) second
           heating; (2) heating after standing 5 months at ambient
           temperature and; (3) heating after standing at 60°C for
           1 week.

The fact that at lower temperatures (< 50°C) THPM exists in a more soluble form leads to an increased equilibrium solubility at room temperature. After 10 months of aging at room temperature, the concentration of antioxidant remaining in the polymer bulk was 0.07 weight percent and this has important implications for polymer stability.

CONCLUSION

The equilibrium solubility of tetrakis [methylene-3(3',5',-di-tert.-butyl-4'-hydroxyphenyl) propionate] methane in low density polyethylene was measured between room temperature and 80°C. At 40°, 60°, 80°C and ambient temperature the observed equilibrium solubilities were 0.075, 0.055, 0.08 and 0.07 weight percent, respectively. The discontinuity occurring between 40° and 60°C was attributed to a change in physical state of the antioxidant from a more soluble amorphous form to a crystalline solid. The equilibrium solubility values especially at the lower temperatures were significantly higher than previously reported and underscores the effect the physical state of the antioxidant has on solubility.

REFERENCES

1. P. D. Calvert and N. C. Billingham, J. Applied Polymer Science, 24, 357 (1979).
2. V. J. Kuck and I. P. Heyward, to be published.
3. R. J. Roe, H. E. Bair and C. Gieniewski, J. Applied Polymer Science, 18, 843 (1974).
4. J. Y. Moisan, European Polymer Journal, 16, 979 (1980).
5. J. Y. Moisan, Ann. Telecommunic., 34, 53 (1979).
6. A. S. Michaels, H. J. Bixler and H. L. Fein, J. Applied Physics, 35, 3165 (1964).
7. A. J. Mee, Physical Chemistry, 422, Aldine Publishing Company, Chicago, 1964.
8. N. C. Billingham, P. D. Calvert and A. S. Manke, J. Applied Polymer Science, 26, 3543 (1981).
9. J. B. Howard, Polymer Engineering and Science, 13, 429 (1973).

AN AUTOMATED, MULTISTATION, OXYGEN UPTAKE SYSTEM

J. C. Wozny

Borg-Warner Chemicals, Inc.
Technical Centre
Washington, WV 26181

INTRODUCTION

The consequences of the reactions of polymers, such as poly(buta-
diene), ABS, and polypropylene, with atmospheric oxygen range from
changes in color, flow, impact strength, etc., to explosion and fire,
i.e.,:

substrate + oxygen ⟶ oxidative degradation.

Since oxidation is a common and generally undesirable event, particularly at elevated temperatures, a variety of methods have been developed to determine the relative oxidative stability of polymers and
the effect of additives on the thermal oxidative stability of a substrate.[1] Oxygen uptake measurement is one of those methods. The
intent in using that method is to directly measure the consumption
of one of the reactants, oxygen, as a function of time under a set
of experimental conditions.[2] Many oxygen uptake techniques have been
devised, some of which include data recording with a strip chart recorder.[3,4,5] The oxygen uptake system described in this paper is
designed to provide a relative measure of oxidative stability which
being operationally and mechanically simple and relatively inexpen-
sive, so as to allow the manageable, simultaneous operation of many
stations and facilitate data handling and analysis.

SYSTEM DESCRIPTION

The design of the system is schematically presented in Figure
1. In this variable pressure, constant volume system, each station
is composed of a thin walled, disposable reactor fitted with a pres-

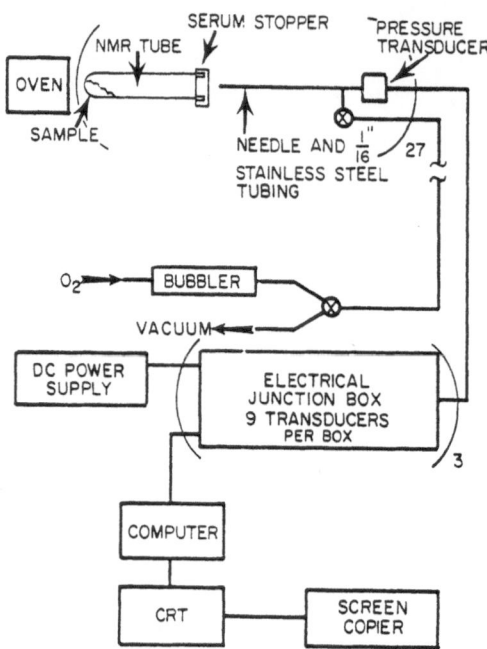

Figure 1. Diagram of the oxygen uptake system.

sure transducer. Pressure readings as a function of time are automatically collected and stored using the in-house computer system.
Each station is independently controlled, including frequency of data
point collection which can be varied during the course of the experiment. Data from any station can be displayed (digitally or graphically) at any time during, as well as after, an experiment via a
graphic cathode ray tube (CRT) computer terminal. Analysis and manipulation of the data, such as determination of stabilization time,
multiple overlay comparison, etc., as well as filing, is conducted
with the assistance of RSI software.

The computer control organization, as displayed in Figure 2, is
composed of three major sections. The first is a series of interactive Fortran programs, directly accessed from the Menu, that control data acquisition, maintain files, and display the status of each
transducer in the system. Access to the second and third group of
programs is described in the Menu.

The second section is composed of two interactive Fortran programs that utilize PLOT 10 subroutines to generate a station vacuum
integrity check plot and a snap-shot plot of the data collected on
a station. The third section is the RS/1 software package into which
the data is copied. RS/1 is a very easy to operate and flexible program that is employed to assist the experimenter in all facets of
data analysis, manipulation, filing, etc.

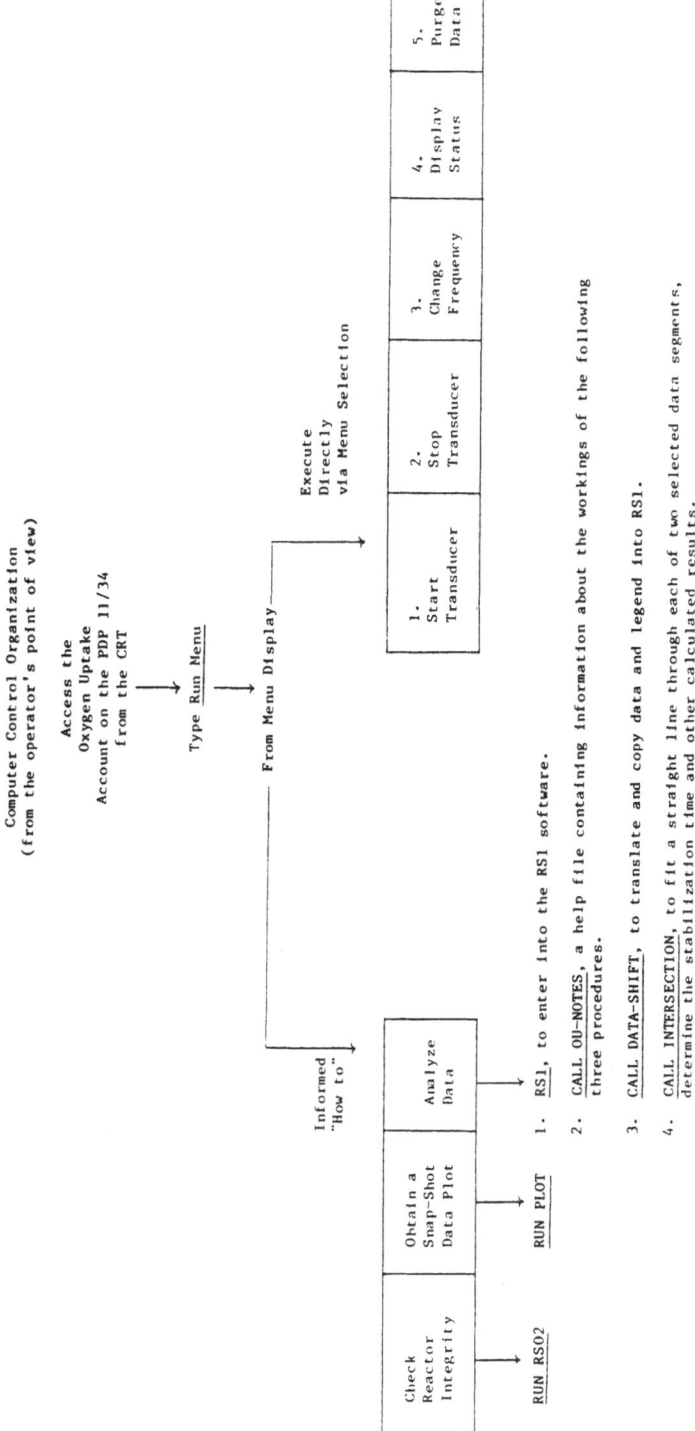

Figure 2. Computer control organization.

MAJOR EQUIPMENT

1. Pressure transducers, Trans-Metrics Inc., Solon, OH, Model
    P21AA, absolute pressure transducer, 0-800 torr range.
2. DC power supply, Elpac Electronics Inc., Santa Ana, CA,
    Elpac Power System, Type OLV-120-24 with overvoltage pro-
    tection, type OVP-3.
3. Computer, Digital Equipment Corporation, Marlborough, MA,
    PDP 11/34 with appropriate analog to digital signal con-
    version.
4. Graphic CRT, Tektronix Inc., Beaverton, OR, Model 4006.
5. Computer Terminal Screen Copier, Tektronix Inc., Model 4611.
6. Software: (a) PLOT 10, Tektronix, Inc.; (b) RS/1, Bolt
    Beranek and Newman Inc., Cambridge, MA.

SAMPLE EXPERIMENT

The following is a step-by-step description of an oxygen uptake
experiment including the associated computer displays. (In order to
clarify figures that are screen copies of CRT displays, the operator
entries are underlined - they are not underlined on the screen.)

1. BHT at a loading of 0.2 pph was incorporated into polypropylene resin, Profax 6501. The dry resin was placed in a clean, pre-
weighed, disposable, 5 mm diameter NMR tube such that the tube contained approximately 100 mg of resin (weight determined to five dec-
imal places in grams). The tube was sealed with a serum stopper,
labeled, and connected to pressure transducer 1B via a stainless
steel syringe needle and stainless steel tubing.

2. After logging into the oxygen uptake account on the computer
with the appropriate passwords, the Menu program was called. Menu
is the master data acquisition and handling program. Upon calling
the Menu program, the main menu was displayed as shown in Figure 3.
From this one display an operator either is informed how to or can
directly perform all the tasks associated with the system.

3. In order to check whether or not data was being collected
from transducer 1B and if data associated with 1B was in a Fortran
file, the statut of the transducers was displayed by hitting 4 (followed by return) on the terminal in accordance with the menu instructions. The resulting display, shown in Figure 4, indicated that
transducer 1B was ready to be used. (For simplicity the status of
only 10 stations is displayed in Figure 4). Note that the display
showed the current status of the transducers and that 1A, 1D-G, were
accumulating data, whereas 1H and 2A were inactive but did have data
in a Fortran file. In order to reduce the load on the computer, the
system is designed to restrict the amount of data stored. For example, a transducer such as 1H that has data in a Fortran file cannot

```
1    INITIATE SAMPLING OF A TRANSDUCER
2    END SAMPLING OF A TRANSDUCER
3    CHANGE SAMPLING FREQUENCY
4    DISPLAY TRANSDUCER STATUS TABLE
5    PURGE TRANSDUCER DATA
*    INTEGRITY CHECK: EXIT, THEN RUN RS82
*    TO PLOT RESULTS: EXIT, THEN RUN PLOT
*    FOR ANALYSIS: EXIT, ENTER RSI, AND CALL OU_NOTES
9    EXIT THIS PROGRAM.

ENTER # BESIDE THE DESIRED FUNCTION SHOWN ABOVE:
```

Figure 3.   The O_2-uptake Menu.

TO	STATUS	FREQ. (Mins)	DATA FILE STATUS
1A	ACTIVE	1.00	DATA LOADED. DO NOT USE
1B	INACT	0.00	NO DATA LOADED
1C	INACT	0.00	NO DATA LOADED
1D	ACTIVE	0.25	DATA LOADED. DO NOT USE
1E	ACTIVE	5.00	DATA LOADED. DO NOT USE
1F	ACTIVE	5.00	DATA LOADED. DO NOT USE
1G	ACTIVE	5.00	DATA LOADED. DO NOT USE
1H	INACT	0.00	DATA LOADED. DO NOT USE
1I	INACT	0.00	NO DATA LOADED
2A	INACT	0.00	DATA LOADED. DO NOT USE

HIT RETURN KEY TO GO BACK TO MENU. (99 WILL EXIT PGM):

Figure 4.   The transducer status table.

be restarted until that data is erased (purged) from the system.   In
general, only one batch of data can be associated with a pressure
transducer at a time.   Note, too, that the frequency of data point
collection can be different from each transducer.   If the operator
desires, the frequency of data point collection can be changed during
the course of an experiment, as indicated in the menu.   Changing the
frequency of data point collection is a useful technique to both
obtain higher resolution in break areas, as well as to reduce the
amount of stored data from relatively uneventful periods in an experiment.

    4.   In order to check that the station associated with trans-
ducer 1B was free of leaks, the integrity of the station was checked.
To do that, the Menu program was exited and RS02 was called; for
dialog see Figure 5.   After entering the number of the pressure transducer, a pressure vs. time grid was displayed.   One reading per second
was placed on the grid while vacuum was applied to station 1B.   The
unit was then sealed and the display was watched for a few seconds.
The lack of a pressure increase indicated that the station was free
from serious leaks.

```
I    INITIATE SAMPLING OF A TRANSDUCER
2    END SAMPLING OF A TRANSDUCER
3    CHANGE SAMPLING FREQUENCY
4    DISPLAY TRANSDUCER STATUS TABLE
5    PURGE TRANSDUCER DATA
*    INTEGRITY CHECK: EXIT, THEN RUN RSO2
*    TO PLOT RESULTS: EXIT, THEN RUN PLOT
*    FOR ANALYSIS: EXIT, ENTER RSI, AND CALL OU_NOTES.
9    EXIT THIS PROGRAM

ENTER # BESIDE THE DESIRED FUNCTION SHOWN ABOVE: 9
> RUN RSO2

THIS PROGRAM RUNS THE INTEGRITY CHECK ON A TRANSDUCER
READINGS ARE TAKEN ONCE PER SECOND

TO STOP THE PROGRAM, HOLD DOWN THE CTRL KEY AND HIT C
WHEN THE MCR> APPEARS, TYPE ABORT

ENTER NAME OF TRANSDUCER TO CHECK (99 WILL EXIT): IB
```

Figure 5.   The Menu together with the dialog for checking the
integrity of a station.

5.   Oxygen was then admitted into the station. The process of
pulling vacuum on the system followed by filling with oxygen was repeated several times in order to insure complete replacement of air.
The integrity program was exited as per the instructions displayed
on the screen. The screen display associated with the end of the
integrity check is shown in Figure 6.

6.   The Menu program was recalled and the dialog for starting
data acquisition was entered by hitting 1 (followed by return). The
first part of the dialog involved entering a description of the sample
and experimental details, as shown in Figure 7. The final response
was to hit the return key which automatically marked time zero and
started data collection. Before the final response, the reactor was
inserted into a hot block (190°C) such that the sample was under a
blanket of oxygen. The pressure increase that occurred due to the
expansion of gas caused by the temperature increase was vented out
the bubbler portion of the Firestone valve. After allowing one minute
for thermal equilibrium to be reached, the station was sealed and
data collection begun. (Alternate starting conditions are discussed
later in this report.) After making the final response in the start-
up dialog, the computer automatically returned to and displayed the
menu. However, since no other stations were to be used, the menu
program was exited.

7.   In order to determine the progress of the experiment, the
Plot program was run. (Access instructions for this program are
listed in the menu display.) The Plot program began with a short
dialog in which the transducer number was requested. Also requested

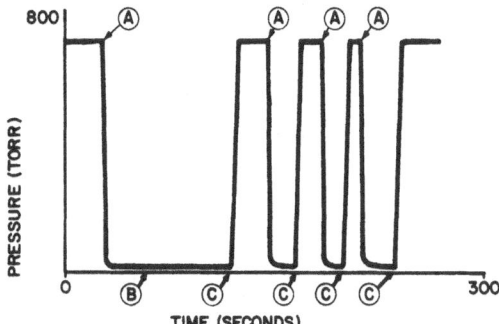

(Data Points Appear on the Screen in X-Y Recorder Fashion)

INTEGRITY CHECK

To stop hold down CTRL KEY AND HIT C.
When MCR PROMPT APPEARS, TYPE ABORT

Ⓐ   Vacuum applied to the station.

Ⓑ   Station isolated (sealed) to check for leaks.

Ⓒ   Oxygen admitted to the station.

Figure 6.   The integrity check display (data points appear on the screen in X-Y recorder fashion).

```
ENTER NAME OF TRANSDUCER TO BE START: IB
ENTER SUBSTRATE (MAX 12 CHARS): PP 6501
ENTER ADDITIVE 1 (MAX 12 CHARS): BHT
ENTER CONCENTRATION 1 (MAX 12 CHARS): 0.2 PPH
ENTER ADDITIVE 2 (MAX 12 CHARS):
ENTER CONCENTRATION 2 (MAX 12 CHARS):
ENTER ADDITIVE 3 (MAX 12 CHARS):
ENTER CONCENTRATION 3 (MAX 12 CHARS):
ENTER ADDITIVE 4 (MAX 12 CHARS):
ENTER CONCENTRATION 4 (MAX 12 CHARS):
ENTER ADDITIVE 5 (MAX 12 CHARS):
ENTER CONCENTRATION 5 (MAX 12 CHARS):
ENTER SAMPLE NO. (MAX 12 CHARS): 24-3
ENTER TRIAL NO. (MAX 12 CHARS): 1-3
ENTER OPERATOR INITIALS (MAX 4 CHARS): JCW
ENTER SAMPLE WEIGHT IN MG. (MAX 12 CHARS): 107.00
ENTER TEMPERATURE (MAX 8 CHARS): 190
ENTER SOLVENT (MAX 16 CHARS)
ENTER ATMOSPHERE (MAX 12 CHARS): O2
ENTER ADDITIONAL SPECIFICATIONS (MAX 1 LINE):
1 MIN. O2 THEN SEAL AND START
ENTER DESIRED SAMPLING FREQUENCY (IN MINS)
MUST BE A MULTIPLE OF 0.25 : .5
HIT RETURN KEY TO MARK ZERO TIME AND START SAMPLING:
```

Figure 7.   Dialog for starting a transducer (the legend of the experiment).

was how much of the data was to be displayed, i.e., all of it, every
tenth point, etc. A plot resulting from this dialog is shown in
Figure 8. Note that these plots are self-scaled by the computer; the
operator has no control of the format - except via the system pro-
grammer if some permanent change is to be made.

8. When the experiment was judged to be over, a hard copy of
the snap-shot, Fortran plot was made for archival storage of all the
data by simply hitting the copy button on the CRT. (A paper copy of
the screen display emerges from the screen copier that is adjacent
to the CRT.) The Menu program was called. Data acquisition was
stopped (2, return).

9. To analyze the data and store the results, the RS/1 software
was used. First, as indicated in the menu, the Menu program was
exited. The RS/1 program was called. (Once inside the RS/1 software, OU-Notes could be called. This program is a help file containing information about the programs that we have written in RS/1
especially to assist analysis of oxygen uptake data and store the
experimental results and history.)

10. The following tasks were performed from within the RS/1
software:

(a) The data and legend of the experiment were translated
into RS/1 by calling the procedure DATA-Shift. After calling the

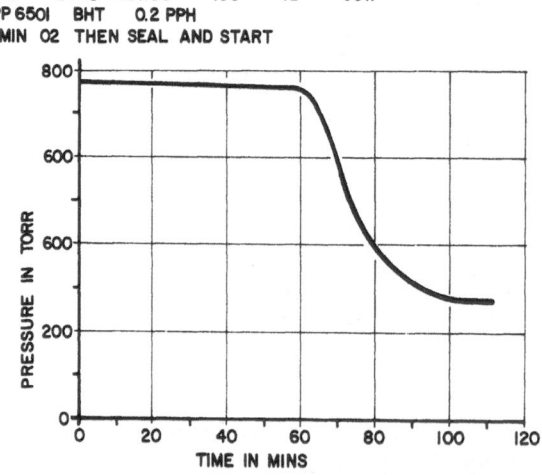

Figure 8. A "snap-shot" plot.

program, a short dialog was entered to identify the batch of data and legend to be copied into RS/1. After several screen displays that describe the status of the translation task, the final output of the program was a graphic display of the data (graph name, TD1BG) as shown in Figure 9. Data-Shift also created in RS/1 a table of the data (table name, TD1B) and a table containing the legend (table name, TD1BL). The general name format is TDXX for data, TDXXG for the graph of the data, and TDXXL for the legend of an experiment where XX stands for the transducer number.

(b)   The data can now be manipulated by the operator as desired using simple RS/1 commands. Manipulation usually includes abridging the data so as to display only the portion of interest. Not, for example, that the display in Figure 9 was slightly altered by the operator in comparison to Figure 8, i.e., the x-axis maximum was reduced to 110 min. from 120, and the y-axis spans 200 to 800 torr instead of 0 to 800 torr.

(c)   The stabilization time in an oxygen uptake experiment is the period from time zero up to the dramatic, catalytic consumption of oxygen indicated by the pressure drop. In order to determine that "break point," two straight lines are drawn through selected segments of the data. Those two lines represent the initial relatively slow rate of oxidation and the rapid, catalytic oxidation, respectively. The time corresponding to the point of intersection of those two line segments is the stabilization time. To assist in the fitting of least squares straight lines to segments of the data, the program, Intersection, was called. The program sequentially asked for the intervals of data through which to fit a straight line. Once the in-

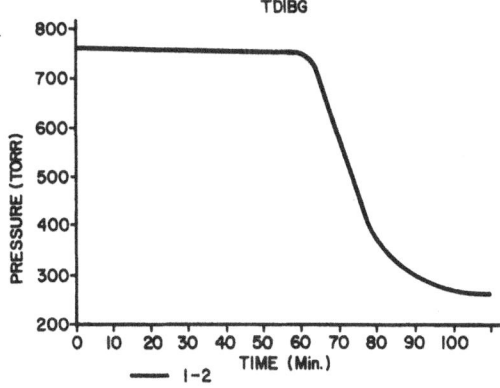

Figure 9.   The final output of the Data-Shift program.

terval was supplied, the program redisplayed the data together with
the line fit to the selected data, as shown in Figures 10 and 11.
If in the judgement of the operator the fit (actually the chosen interval) is a good representation of that period of the experiment,
then the equation of that fit line is saved and the analysis continues, otherwise the fit is erased and a new interval is chosen. Both
straight lines were fit to the data, the stabilization time was determined and displayed on the screen together with the graph containing the fitted data, as shown in Figure 11. A hard copy of this
display was made for archival reference. The program also automatically prepares a table of the results containing the equations of
the lines fit to the data, the stabilization time and other quantities
derived from the data. The results table is shown in Figure 12.

        (d) The program, Summarize, was next called. This program
combined and filed for later recall the legend with the results of
the experiment into the last row of the OU-Summary table. The row
of the OU-Summary table for this experiment together with a previous
(and duplicate) experiment is shown in Figures 13 and 14. For sake
of orderliness and again to reduce the load on the computer, the Summarize program also deleted the data, graphs, and other tables associated with the experiment that were in RS/1.

    11. The experiment would be considered essentially complete at
this point. A few housekeeping details remained to be cleaned up.

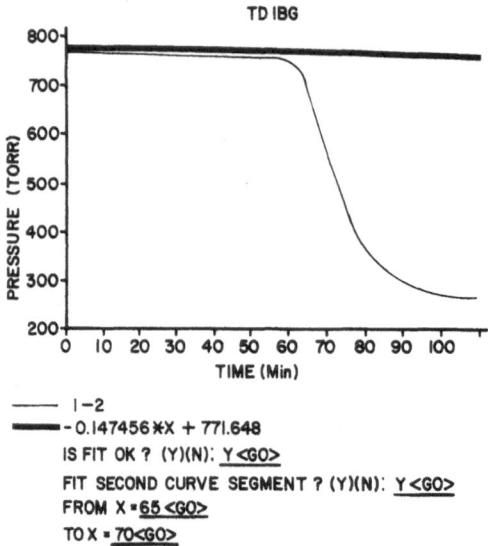

Figure 10.   First line fitted to the data using the program
             intersection.

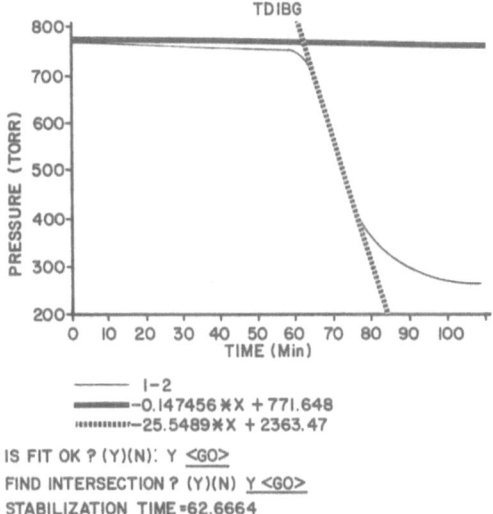

IS FIT OK ? (Y)(N): Y <GO>

FIND INTERSECTION ? (Y)(N) Y <GO>

STABILIZATION TIME =62.6664

SEE AND SAVE THE RESULTS ?(Y)(N):

Figure 11. Second line fitted to the data (stabilization time calculated from the equations of the straight lines).

TDIB_RESULTS 7C X |R   09/23/81

		1 MI (TORR/MIN)	2 M2 (TORR/MIN)	3 STAB_TIME (MIN)	4 P AT BREAK (TORR)
I.	I-2	-0.147456	-25.5489	62.6664	762.407

		5 NET P AT BREAK	6 BI (TORR)	7 B2 (TORR)
I.	I-2	9.24054	771.648	2363.47

TO ENTER THE RESULTS AND LEGEND INTO THE OU_SUMMARY TABLE
CALL SUMMARIZE.

(Each block is a separate screen display. Currently the
Transpose of this table is displayed, i.e. one column of
seven rows.)

Figure 12. The results table.

Namely, the NMR-tube reactor was withdrawn from the hot block. After
cooling, the reactor was discarded. The needle portion of the transducer was rinsed with solvent (acetone) and gently blown dry. Finally, the menu was recalled and the data associated with 1B was purged
from the Fortran file. Note that after translating the data into
RS/1 with Data-Shift, two copies of the data existed; one in Fortran
and the other in RS/1. This was a safety device. If the operator
damaged or "lost" the data in RS/1, then the data (and legend) re-

OU_ SUMMARY 28 C X 2R   09/23/81

		I STAB TIME (MIN)	2 MI (TORR/MIN.)	3 SUBSTRATE	4 ADD I
I.	I-I	63.118	-0.101769	PP 6501	BHT
2.	I-2	62.6664	-0.147456	PP 6501	BHT

		5 CONC I	6 ADD 2	7 CONC. 2	8 SAMP NO
I.	I-I	0.2 PPH			24-3
2.	I-2	0.2 PPH			24-3

		9 SAMP. WT (MG)	IO MI/SAMP WT (TORR/MIN/MG)	II TEST TEMP(C)
I.	I-I	107. 37	-0.000947835	190
2.	I-2	107. 00	-0.0013781	190

		12 M2 (TORR/MIN)	13 P AT BREAK (TORR)	14 NET P AT BREAK	I5 SOLVENT
I.	I-2	-22.5495	762.548	6.42346	
2.	I-2	-25.5483	762.407	9.24054	

(Each block is a separate screen display)

Figure 13.   Part of the OU-Summary table.

maining in RS/1 could be deleted, and the experiment retranslated
into RS/1 from the Fortran file using Data-Shift.

At the conclusion of data usage, the operator must purge the
data from the Fortran file because the same stipulation applies to
the Fortran file of a transducer as in RS/1. Namely, only one batch
of data per transducer can exist at any one time in the Fortran lan-
guage. Therefore, in order to restart 1B, the old data must be purged
via the Meny program. Flowing from this stipulation is the requirement for an experimenter to deal with and delete data; failure to do
so on a timely basis essentially removes the station from the system.

CURRENT OBSERVATIONS AND TECHNIQUES

In oxygen uptake experiments of polypropylene, the expected increase in oxidative stability was measured when the loading of antioxidant (BHT) was increased, as shown in Figure 15. However, BHT is
known to be a relatively volatile antioxidant, and therefore of limited benefit to a substrate exposed to a flow of air at high temperature, e.g., in high temperature oven aging experiments or in testing
via the differential scanning calorimeter (DSC). In fact, one of
the criticisms of sealed tube measurements of oxidative stability is

OU_ SUMMARY 28C X 2R   09/23/81

		16 ATM	17 OPER	18 TRANS. NO	19 TEST DATE
I.	I-I	O2	JCW	IA	23-SEP-81
2.	I-2	O2	JCW	IB	23-SEP-81

		20 ADD 3	21 CONC 3	22 ADD 4	23 CONC 4
I.	I-I				
2.	I-2				

		24 ADD 5	25 CONC 5	26 BI (TORR)	27 B2 (TORR)
I.	I-I			768.972	2185.83
2.	I-2			771.648	2363.47

		28 ADDL SPECS
I.	I-I	I MIN. O2 THEN SEAL AND
2.	I-2	I MIN. O2 THEN SEAL AND

(Each block is a separate screen display.)

Figure 14.   Part of the OU-Summary table.

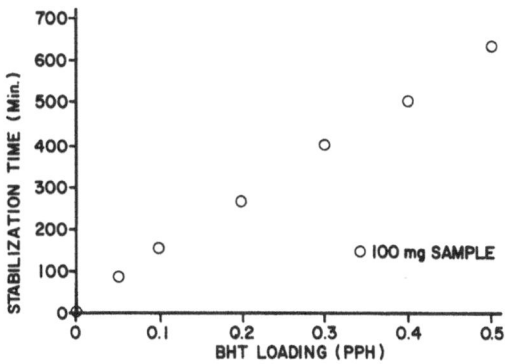

Figure 15.   Plot of the effect of BHT loading on the oxidative
stability of polypropylene ($1O_2$, 175°C, 100% O_2).

that substrates stabilized with BHT look relatively good in contrast
to their relative instability in high temperature, circulating air
over aging experiments. We have found that sealed tube oxygen uptake

experiments on the automated system can be used to provide information
concerning both the activity of the antioxidant if constrained to stay
in the system (the normal sealed tube condition) and the volatility
of the antioxidant. The volatility of an antioxidant (note the dis-
tinction here from oxidative products of the original additive or of
the substrate) is observed by use of different starting technique.
Namely, vacuum is applied to the sample at room temperature. The
reactor is inserted into the hot block with the vacuum maintained
for a measured but variable period of time. Oxygen is then admitted;
the pressure equilibrated; the system sealed; and the pressure change
measured as usual. The results of such an experiment, displayed in
Figure 16, show that in spite of prolonged vacuum at high temperature,
the sample containing Irganox 1010 (a high molecular weight, relatively non-volatile antioxidant) retained the oxidative stability
that was expressed without vacuum heating. In contrast, the oxi-·
dative stability of the sample containing BHT, although much more
stable than the sample containing an equal weight of 1010 when simply
sealed under oxygen, diminished to that of unstabilized polypropylene
after only a few minutes of vacuum at high temperature prior to seal-
ing under oxygen.

    As mentioned previously, multiple overlay plots can easily be
generated using the RS/1 software. Such graphs can be used to clearly display the effects of additives in a polymer. An example is provided in Figure 17. It should be noted that the RS/1 software was
used to normalize the data from each experiment prior to creation of
the multiple overlay graph.

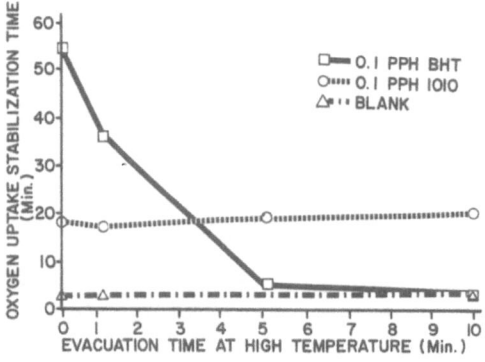

Figure 16.   Effect of high temperature evacuation on the oxidative
             stability of polypropylene stabilized by antioxidants
             of differing volatility (190°C, 100% O_2).

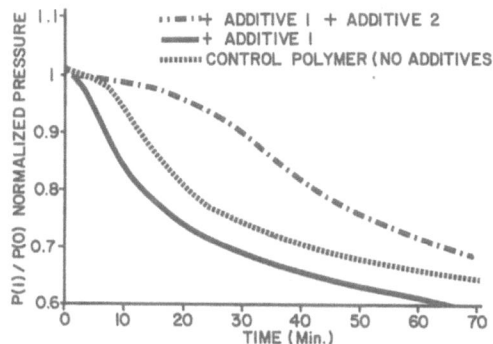

Figure 17. Effect of additives on polymer thermal oxidative
            stability.

As a further test of the oxygen-uptake system, the oxidation of
initiated cumene solutions with and without primary antioxidant (BHT)
was determined. A 3.5 ml vial equipped with a "rice grain" magnetic
stirring bar and a teflon septum functioned as the reactor. After
addition of 0.5 ml of the test solution, the reactor was stirred and
purged with oxygen, venting being achieved via an additional needle
attached to a bubbler. The increase in oxidative stability of the
solution with increasing concentration of BHT is shown in Figure 18.

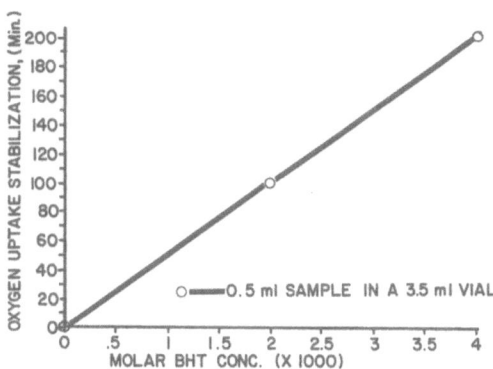

Figure 18. Oxidative stability of an initiated liquid sample (2.5M
            cumene, 6 x 10^{-2}M AIBN in chlorobenzene at 60°C).

CONCLUSION

This oxygen uptake system makes possible a more complete and
rapid examination of the relative oxidative stability of solid and
liquid substrates and of the phenomena involved by combining the
features of numerous stations, automated data collection at maximum
sensitivity, and computer assisted data analysis and manipulation.

ACKNOWLEDGEMENTS

The encouragement and support of Borg-Warner Chemicals, Inc.,
is gratefully appreciated. The thought and efforts of numerous BWC
personnel contributed to the development of this system. Among them
are: A. Allman, J. Biber, M. Brown, R. Coffin, E. Chambers, E. Dye,
B. Ferrell, C. Gray, J. Jones, D. Kulich, D. Paul, L. Paul, G. Pooler,
W. Ray, Y. Shears, D. Wood, R. Woodbury, J.Wooddell. In addition the
advice of Drs. D. J. Carlsson and J. C. W. Chien was most helpful.

REFERENCES

1. W. Lincoln Hawkins in "Polymer Stabilization," W. Lincoln Hawkins,
      Ed., Wiley-Interscience, NY, 1972, Chapter 10.
2. J. R. Shelton, Rubber Chemistry and Technology, 30, 1251 (1957).
3. R. A. Krueger, J. Appl. Polym. Sci., 17, 2305 (1973); and refer-
      ences cited therein.
4. D. W. Grattan, D. J. Carlsson and D. M. Wiles, Chem. Ind.
      (London), 228 (1978).
5. D. D. Davis and K. L. Stevenson, J. Chem. Ed., 54, 394 (1977).

RECENT DEVELOPMENTS IN PHOSPHORUS STABILIZERS

Elyse Lewis

Borg-Warner Chemicals, Inc.

Washington, West Virginia 26181

INTRODUCTION

One class of compounds that has been employed as stabilizers
in many polyolefin applications is phosphites. In particular tris-
nonylphenyl phosphite (P-1) and distearyl pentaerythritol diphosphite (P-2) have found widespread usage in these polymers. These
two phosphites (see structures in Figure 1) are used in polyolefins
to protect them against melt flow and color change during processing;
to retard discoloration in end use applications; and to reduce degradation and discoloration due to U.V. exposure. The face that they
are FDA sanctioned for use in olefin polymers has also contributed
to their success.

The improvement of stability that can be obtained by using just
a small amount of phosphite is significant. For example, the melt
index of polypropylene increased substantially when it was ground
and reprocessed. Such an increase is represented by the "No Phosphite" curve in Figure 2. The addition of 0.10 parts of tris-nonylphenyl phosphite (P-1) or distearyl pentaerythritol diphosphite
(P-2) resulted in almost no melt flow change.

Phosphite	P-1	P-2
Structure	$(H_{19}C_9-\langle O \rangle-O)_3 P$	$H_{37}C_{18}-O-P\begin{smallmatrix}OCH_2\\OCH_2\end{smallmatrix}C\begin{smallmatrix}CH_2O\\CH_2O\end{smallmatrix}P-O-C_{18}H_{37}$
Melt Range	Viscous Liquid	40-70°C

Figure 1. Commercial phosphite stabilizers

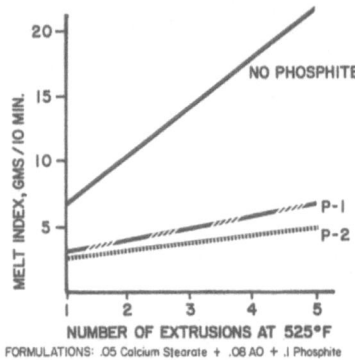

Figure 2. Polypropylene melt flow stability.

Both of these phosphites are also used extensively to reduce
discoloration due to heat in processing and during end use. Polypropylene samples (Figure 3) were extruded five times and then suspended
in an oven for 400 hours at 150°C. The color of the samples is indicated by Hunter L-b values. The lower the number the more severe
the discoloration. P-1 and P-2 reduced the amount of discoloration
encountered during processing. [Compare 74 (P-1) and 76 (P-2) to
73 (control); also 70 (P-1) and 74 (P-2) to 69 (control).] They
also provided significant color stability during oven aging. [Com-
pare color values 63 (P-2) and 59 (P-1) to 56.]

Even though these two stabilizers have enjoyed widespread usage
over the years, their physical properties have imposed some limitations on their use. Tris-nonylphenyl phosphite is a liquid. It
can easily be used in plants which are designed to handle liquid
additives. Distearyl pentaerythritol diphosphite (P-2) has been the
phosphite of choice when solid additives are used. Like slip agents,
DSTDP and other low melting solids, distearyl pentaerythritol diphosphite is used primarily in feeding systems which subject the
additives to minimal heat and shear.

Therefore, despite the excellent improvements in polymer sta-

POLYPROPYLENE COLOR STABILITY

| | HUNTER L-b VALUES | | |
	1ST EXTR.	5TH EXTR.	5TH EXTR. + 400 HRS. AT 150°C
CONTROL	73	69	56
0.10 P-1	74	70	59
0.10 P-2	76	74	63

FORMULATIONS: .05 CALCIUM STEARATE + .08 AO

Figure 3. Polypropylene Color Stability.

bility provided by these two phosphites, several suppliers are offering new compounds which possess good stabilizing activity but are
easier to handle. The new phosphorus stabilizers which have appeared
on the market are free flowing solids with high melting points.
Their properties will be discussed in this paper.

RESULTS AND DISCUSSION

     The structures of these new phosphorus stabilizers are shown
in Figure 4 and their physical characteristics are compared in
Figure 5. All three of the new compounds have high melting points
which improve their handleability. One of these new compounds is
a phosphonite. P-3 is tetrakis [2,4-di-t-butylphenyl] 4,4'-biphenyl-
ylenediphosphonite. The new phosphites are more sterically hindered
around the phosphorus atom than those previously available. P-4 is
tris-2,4-di-t-butylphenyl phosphite and P-5 is bis-(2,4-di-t-butyl-
phenyl) pentaerythritol diphosphite.

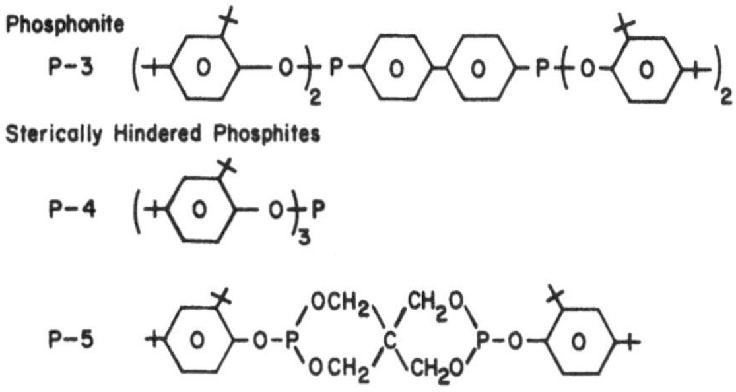

Figure 4. New phosphorous stabilizers

	Melt Point or Melt Range
P-1	Viscous Liquid
P-2	40-70°C
P-3	75°C
P-4	180°C
P-5	160-175°C

Figure 5. Comparison of physical properties.

   Structural differences designed to improve physical properties
were bound to affect the performance of the phosphorous compound as
a stabilizer in the polymer. How well a particular compound performs
in a specific polymer depends on a variety of factors such as the
phosphorus content of the stabilizer, its compatibility with the
polymer, and its volatility.

   The complexity of the situation makes it impossible to accur-
ately predict how well a new compound will perform based solely on
its phosphorus content. Therefore we have compared these new stabilizers to the old ones in a variety of stabilization tests.

   The new phosphorus compounds were compared to distearyl pentaerythritol diphosphite (P-2) and tris-nonylphenyl phosphite (P-1)
in a processing stability test. The phosphorus content of each stabilizer is indicated next to each melt flow stability curve in
Figure 6. It can clearly be seen from this graph that melt flow
stabilizing activity does not correlate with phosphorus content.
The compatibility of the stabilizer molecule with the polymer matrix
and the availability of the phosphorus atom to participate in stabilization reactions also influence the effectiveness of a particular
compound.

   In Figure 7 the color stabilizing activity of the new phosphites
P-4 and P-5 are compared to distearyl pentaerythritol diphosphite
(P-2) and tris-nonylphenyl phosphite (P-1). Note that P-5 has activity comparable to P-2. [Compare the color after the first extruder pass 75 to 76; the color after the fifth pass 73 to 73 and after
oven aging 68 to 63 (a significant improvement in color stability
is provided by P-5).] P-1 and P-4 gave equivalent color stability
throughout the test. [Compare the L-b color values after the first

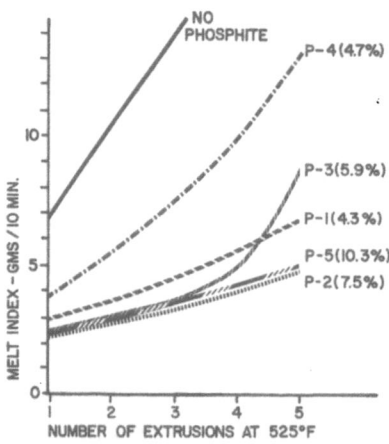

Figure 6. Polypropylene melt flow stability.

EXTRUDER

PASS	CONTROL	P-1	P-2	P-4	P-5
1	73	74	76	74	75
5	69	70	73	70	73

5 EXTRUSIONS + 400 HOURS AT 150°C

	56	59	63	59	68

FORMULATIONS: .08 AO + .05 CALCIUM STEARATE

Figure 7. Polypropylene color stability.

and fifth passes and after oven aging.] Phosphorus stabilizers were
used at .1 phr concentration level.

In the new types of linear low density polyethylene significant
improvements in color stability can be obtained with phosphorous
stabilizers as demonstrated in Figure 8. In addition to the phosphorous stabilizers listed here, all formulations contained 150 ppm
of Irganox[R] 1076. The bars on this chart represent the amount of
discoloration which occurred during seven extrusions. The left side
of the bar represents the color after the first extrusion; the right
side represents the color after the seventh extrusion. Stabilizers
P-2, P-3, P-4, and P-5 all reduced the amount of discoloration which
occurred after one extrusion. When more severe processing conditions
are encountered, such as the seven extrusions in this experiment,
P-2 and P-5 provided the best stability.

Distearyl pentaerythritol diphosphite (P-2) is frequently used
to improve the light stability of polyolefin fibers and films. A
comparison of the light stabilizing activity of the new phosphorous
stabilizers to P-2 is shown in the next two tables. A typical
medium denier polypropylene multifilament stabilized only with a
phenolic antioxidant can be expected to last two and a half to four
months when exposed outdoors. A fifty percent loss in tensile
strength is considered to be the failure point in this test. Addition of a one-tenth of a part (0.1) of phosphorous stabilizer to the
phenolic antioxidant containing formulation results in the changes in

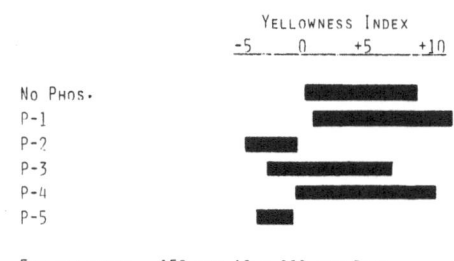

Figure 8. Color stability of LLDPE.

lifetimes of the fibers shown in Figure 9. P-5 is the only one of
the new phosphorous stabilizers which has the ultraviolet stabilizing
activity of distearyl pentaerythritol diphosphite (P-2).

The synergism which sometimes occurs between phosphites and ultraviolet light absorbers results in very effective and less expensive
U.V. stabilized formulations. P-2 is often used for this purpose.
The polypropylene fibers in Figure 10 were aged in the Caribbean.
The increase of the concentration of the U.V. absorber from .15 parts
to .60 parts increased the lifetime of these fibers from 4.8 to 7
months. The addition of .15 parts P-2 or P-5 to .15 parts of U.V.
absorber resulted in lifetimes over 9 months. The other phosphorus
stabilizers did not show this kind of synergism.

CONCLUSIONS

In summary we have found that the new phosphorous stabilizers
which have appeared in the last few years show improved handling advantages over the older commercial phosphites in certain types of
additive feeding equipment. Some of the new compounds, in particular

```
    ·  POLYPROPYLENE MULTIFILAMENT
    ·  FAILURE AT 50% LOSS OF TENSILE STRENGTH
    ·  OUTDOOR EXPOSURE

           LIFETIME COMPARED TO FORMULATION WITHOUT
           PHOSPHOROUS STABILIZER - 0.1 PTS.
    P-1              + 0.3 MONTHS
    P-2              + 2.5 MONTHS
    P-3              - 1.3 MONTHS
    P-4              + 0.2 MONTHS
    P-5              + 2.3 MONTHS
```

Figure 9. Light stability provided by phosphorous stabilizers.

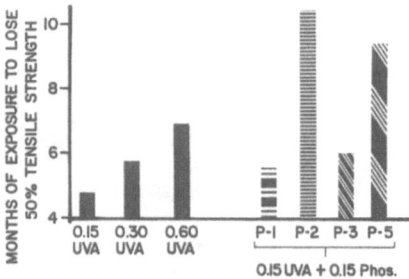

Figure 10. Phosphite-UV absorber synergism polypropylene multi-
            filament-outdoor exposure.

bis-(2,4-di-t-butylphenyl) pentaerythritol diphosphite, appear to be equivalent or even more effective stabilizers than the older compounds. Others of the new compounds which do not appear very effective in some tests are more effective in others. This demonstrates the need to screen a variety of stabilizers to select the most effective one for a particular application.

THERMOOXIDATION AND STABILIZATION OF URETHANE

AND URETHANE-UREA BLOCK COPOLYMERS

G. N. Mathur* and J. E. Kresta

Polymer Institute
University of Detroit
Detroit, MI 48221

INTRODUCTION

Urethane and urethane-urea block copolymers are subject to degradation under various environmental conditions. Among the different types of degradation, thermal, thermooxidation, UV degradation and hydrolysis are most important for the above classes of copolymers. The degradation of these polymers is quite complex due to the fact that they may contain, in addition to the urethane and urea groups, other functional groups such as ether, ester, allophanate, biuret, etc.

In non-oxidative thermal degradation the cleavage of the urethane linkages takes place, yielding either the starting components (isocyanate and polyol) or amine (urea), olefins and carbon dioxide, depending upon the nature of the substituents on the urethane linkages.

In contrast to the non-oxidative thermal degradation, it has been found that in thermooxidative degradation of urethane(urea)-polyether block copolymers, the urethane(urea) linkages are resistant to oxygen attack and the degradation starts in the polyether soft segments. Polyester urethanes are more resistant to thermooxidation than polyether urethanes. The factors influencing thermooxidation

---

*H. Butler Technological Institute, Kanpur

of urethane(urea) block copolymers include the effect of the structure of soft segments, effects of the structure of urethane and urea
groups, crosslink density and urethane catalysts.

Effect of the Structure of Soft Segments on the Thermooxidation
of Urethane-(Urea) Block Copolymers

The polyether soft segments are very sensitive to thermooxidation. The thermooxidative degradation studies carried out with
poly(oxyalkylene) glycols, such as poly(oxyethylene) diols (1),
poly(oxypropylene) diols (2) and poly(oxytetramethylene) diols (3)
showed that the oxidation proceeds by a radical chain process, forming hydroperoxides at the carbon atoms adjacent to the ether linkages. The decomposition of the hydroperoxides lead to the formation
of carbon dioxide, water and aldehydes, with smaller amounts of ace-
tates and formates. In the case of poly(oxytetramethylene) diols,
the oxidation is auto-catalytic (3) between 90 to 120°C.

The rate of oxygen absorption was found to be similar to the
rate of formation of hydroperoxides in the initial stages. However,
in the later stages, decomposition products had a significant effect
on the rate of oxygen absorption. The effect of temperature on the
concentration of hydroperoxides can be explained by the subsequent
rate of decomposition of these hydroperoxides to give chain branching, which is temperature dependent. The steps in the oxidation are
as follows:

$$\sim(CH_2)_4-O-(CH_2)_4\sim \xrightarrow{\text{oxidation}} \sim(CH_2)_3-CH-O-(CH_2)_4\sim$$
$$I$$

$$I + O_2 \longrightarrow \sim(CH_2)_3-\underset{\underset{O-O\cdot}{|}}{CH}-O-(CH_2)_4\sim$$
$$II$$

$$II + RH \longrightarrow \sim(CH_2)_3-\underset{\underset{OOH}{|}}{CH}-O-(CH_2)_4\sim + R\cdot$$
$$III$$

$$III \longrightarrow \sim(CH_2)_3-\underset{\underset{O\cdot}{|}}{CH}-O-(CH_2)_4\sim + HO\cdot$$
$$IV$$

$$\overset{O}{\overset{\|}{H C}}-(CH_2)_3-OCH_2\text{\textasciitilde} \quad + \quad \text{\textasciitilde}\cdot O-(CH_2)_4-O^\bullet$$

$$IV \xrightarrow{\text{cleavage}}$$

$$\text{\textasciitilde}(CH_2)_4-O-(CH_2)_2-CH_2^\bullet \quad + \quad \overset{O}{\overset{\|}{H C}}-O-(CH_2)_4\text{\textasciitilde}$$

In poly(oxypropylene) diols it was observed[4,5,6] that the thermooxidation starts around 80°C and that the acetylation of the hydroxyl groups did not improve the thermooxidation stability. In general poly(oxypropylene) chain segments are more easily attacked by oxygen than poly(oxyethylene) or poly(1,4-oxybutylene) chains.

Effect of Structure of Urethane and Urea Segments on the Thermooxidation of Urethane-(Urea) Block Copolymers

The effect of isocyanate structure on the oxidation of copolymers is less significant in the case of polyether urethanes as compared to polyester urethanes because polyether segments oxidize prior to urethane groups.

In the case of polyester urethanes, the rate of thermooxidation as a function of diisocyanate decreases in the following order:[2,7] hexamethylene diisocyanate > 4,4'-diphenylmethane diisocyanate > tolylene diisocyanate > 4,4'-diphenyl ether diisocyanate > 1,5-naphthylene diisocyanate. The increased aromaticity of these diisocyanates results in increased resistance to thermooxidation. (It is opposite to the results of UV initiated oxidation and non-oxidative thermo degradation.) Benzyl type diisocyanates, e.g., XDI, have a lower resistance to oxidation than aromatic diisocyanates. Oxidation of aromatic polyurethanes leads to the formation of quinoid, chromophore structures giving rise to discoloration of the polymers.[8,9,10,11]

It was established that the presence of urethane groups at the end of polyether chains increased thermooxidation stability of polyether segments. In order to investigate the effect of urethane groups,[4,12,13] polyether urethanes have been prepared from poly(oxypropylene) diol and phenyl isocyanate with different NCO/OH ratios. It was observed that the polyurethanes containing higher concentration of urethane groups had a lower rate of thermooxidation. Incorporation of small amounts of urea in polyether urethanes increased also the induction period for degradation and decreased the loss of weight at higher temperatures. Similar retardation of oxidation of poly(oxypropylene) diols were observed in the presence of added urethane and urea compounds.[14] It must be mentioned that the retarda-

tion effect was observed only with aromatic urethane (urea) groups
and not with aliphatic ones.[15]

The Effect of Crosslink Density on Thermooxidation of
Urethane-(Urea) Block Copolymers

The increased resistance to thermooxidation was observed for
polyether urethane films with increasing crosslink density.[16] This
study was carried out with polyurethanes in which the diisocyanate
structure was varied as well as the polyether polyol functionality,
which ranged from three to eight. Similar results were obtained with
polyester urethanes.[12,17]

The Effect of Urethane Catalysts on Thermooxidation
of Urethane-(Urea) Block Copolymers

The effect of 1,4-diazabicyclo[2,2,2] octane (Dabco), stannous
octoate and diethyltin dicaprylate (DETDC) on the thermooxidation of
polyether urethanes based on tolylene diisocyanate and poly(oxypropylene) diol was studied.[18,19] The oxygen absorption in the presence
of DABCO started around 130-140°C, while in the presence of DETDC it
started around 170-180°C. Stannous octoate was also found to have a
stabilizing effect on polyether urethanes in the presence of air.

Stabilization of Urethanes

Many excellent reviews are to be found in literature on the stabilization of polymers.[15,20-24] The most commonly used stabilizers
may be classified as radical chain terminators and peroxide decomposers. Most of the work regarding the effectiveness of stabilizers
has been done on polyolefins and rubbers, while relatively little
information has been available regarding the effectiveness of these
stabilizers in polyurethane systems. The effectiveness of stabilizers depends on the chemical structure, their diffusion mobility,
solubility in the polymeric materials and volatility. The structure
of the polymer also plays an important role: in the case of micro
nonhomogeneity of polymers (as can be expected in urethane (urea)
block copolymers), the diffusion (oxygen, radical centers, stabilizers, etc.) and kinetics (reaction of alkyl macroradicals with oxygen,
cleavage of macromadicals, recombination of macroradicals, etc.) can
have a polychromatic character.[24]

In this paper the stabilization of urethane and urethane-urea
block copolymers by combinations of various stabilizers will be
discussed.

EXPERIMENTAL

Chemicals

All chemicals used in this study are listed in Table I.

Poly(oxytetramethylene) glycol of molecular weight 1000 (Polymeg 1000) was dried by applying a vacuum of 1 mm mercury at 60°C for 6-8 hours until there was no more evolution of gas bubbles. Urethane grade 1,4-butanediol was also degassed in the same manner. Trimethylolpropane, 4,4'-methylene bis (o-chloroaniline) (MOCA) and all the stabilizers were used as received from the supplier. Toluene was dried over activated molecular sieves (Linde 5A).

Preparation of Films

Prepolymer Formation. Tolylene diisocyanate (174 g) and dried toluene (150 g) were added to a one-liter reaction kettle fitted with a stirring assembly. The degassed Polymeg 1000 (504 g) was added slowly from a dropping funnel under a nitrogen blanket to give a final NCO/OH ratio of 2/1. The temperature of the reaction mixture was maintained at 70°C. The progress and completion of the reaction was determined by means of NCO group analysis using the di-n-butylamine titration method.[25] The prepolymer was stored in predried air-tight cans under a nitrogen blanket.

Chain Extension with 4,4'-Methylene bis (o-chloroaniline) (MOCA). After determining the isocyanate concentration of the prepolymer, an equivalent amount of MOCA was dissolved in dry toluene, added to the prepolymer and mixed thoroughly to give a final NCO/OH ratio of 1.0. The stabilizers were also dissolved in toluene and added at this stage. Films of approximately 0.002" thickness were then drawn on glass plates. The oxygen absorption for films of 0.002" thickness was not diffusion controlled, as some authors[4] have found. The glass plates were kept in the oven under nitrogen at 80°C for 16 hours, and then at 120°C for the final two hours of curing. The cured films were removed from the glass plates by immersing them in water for 2 to 3 hours.

Chain Extension with 1,4-Butanediol and Trimethylolpropane. After determining the isocyanate concentration of the prepolymer, dried and degassed 1,4-butanediol and trimethylolpropane were added in the ratio of 5.5:1 to give a final NCO/OH ratio of 1.05.

In order to accelerate the cure of these films, 0.05% by weight of dibutyltin dilaurate (T-12 catalyst) was also added. The solution of stabilizer(s) in toluene was added and thoroughly mixed before drawing films of approximately 0.002" thickness by means of a film applicator. The films were drawn on glass plates and cured under nitrogen in the oven at 80°C for 16 hours and at 120°C for

Table I. Chemicals

Chemical Identification	Supplier	Designation
80-20% Mixture of 2,4- and 2,6-isomers of tolylene diisocyanate	BASF Wyandotte	TDI
Toluene, reagent grade	Eastman Chemical Products, Inc.	
Poly(oxytetramethylene) glycol	Quaker Oats Co.	Polymeg 1000
4,4'-Methylene bis (o-chloroaniline)	E. I. duPont de Nemours & Co.	MOCA
1,4-Butanediol	GAF Corp.	
Trimethylolpropane	Celanese Chemical Co.	TMP
N,N'-Di-beta-naphthyl-p-phenylene diamine	R. T. Vanderbilt Co., Inc.	Agerite White
Aldol-alpha-naphthylamine	R. T. Vanderbilt Co., Inc.	Agerite Resin
Polymerized 2,2,4-trimethyl-1,2-dihydroquinoline	R. T. Vanderbilt Co., Inc.	Agerite Resin D
2,2'-Methylene-bis (4-methyl-6-tertiary butyl phenol)	American Cyanamid Co.	Antioxidant 2246
Ditridecyl thiodipropionate	Evans Chemetics	DTDTDP
Dilauryl thiodipropionate	Evans Chemetics	DLTDP
Distearyl thiodipropionate	Evans Chemetics	DSTDP

Chemical	Manufacturer	Trade Name
2,6-Di-tertiary-butyl-4-methyl phenol	Shell Chemical Co.	Ionol
Octadecyl 3-(3', 5',-di-tertiary-butyl-4'-hydroxy phenyl) propionate	Ciba-Geigy Corp.	Irganox 1076
Tetrakis[methylene 3-(3', 5'-di-t-butyl-4' hydroxy phenyl) propionate] methane	Ciba-Geigy Corp.	Irganox 1010
Tri (mixed mono and dinonylphenyl) phosphite	Naugatuck Chemicals Div., Uniroyal, Inc.	Polygard HR
2,5-Di-tertiary amylhydroquinone	Monsanto Chemical Co.	Santovar A
Polymeric carbodiimide	Bayer AG	Staboxol PCD
2,(2'-Hydroxy-3', 5'-di-tert-amylphenyl) benzotriazole	Ciba-Geigy Corp.	Tinuvin 328
Tri (mixed mono and dinonylphenyl) phosphite	Naugatuck Chemicals Div., Uniroyal, Inc.	Polygard

two hours. They were removed from the glass plates by immersion in
water for 1-2 days.

Methods Used for Evaluation of Thermooxidation

Oxygen Uptake. A constant volume type of apparatus was designed
and fabricated with a pressure transducer as the sensing element to
measure the induction period and the initial rate of oxygen absorption. This apparatus consisted of an aluminum sample tube which was
kept in a constant temperature AL block, connected to the pressure
transducer and recorder. This particular oxygen uptake apparatus
has significant advantages over other similar assemblies as the measurements are more precise and the rate of oxygen absorption as well
as the change in mechanical properties can be measured on the same
film sample.

All the samples were preconditioned by heating them under vacuum
at 100°C for 4 hours in order to avoid the effect of absorbed impurities such as moisture and solvents. The samples were cut in the form
of dog bones from films of uniform thickness, in order to have approximately the same weight ($\sim$ .03 g), surface area and thickness
(.002"-.003").

The samples were suspended in an aluminum sample tube. The
tube was then closed tightly and placed in the heating block and connected to the vacuum pump. When the temperature had equilibrated,
the system was twice purged with oxygen and finally filled up with
pure dry oxygen at atmospheric pressure. The pressure in the system
was continuously monitored by a pressure transducer and was recorded.
The induction period and rate of oxygen uptake were measured. Temperatures were chosen so as to measure thermooxidative degradation
without the effect of thermal degradation within a reasonable period
of time. The samples were removed during the course of the run and
tested for change in tensile strength.

Oxygen Bomb Test. This is a relatively rapid test to determine
the effect of thermooxidative degradation on the mechanical properties
of the films. The films, cut in the form of dog bones, were attached
to the sides of a glass beaker. The beaker was kept in a stainless
steel bomb at 100°C and 100 psi of pure oxygen pressure. The optimum
time of exposure was found to be seven days following a set of probe
experiments. All samples were tested for tensile strength and elongation after exposure, to measure the loss in mechanical properties,
due to thermooxidation.

Thermogravimetric Analysis. A duPont 950 Thermogravimetric
Analyzer, along with a duPont Differential Thermal Analyzer, was
used in this study. The sample weight used was generally 10 mg and
a platinum crucible was used for keeping the sample suspended from

the quartz rod. The atmosphere was dry air and the flow rate was
maintained at 1.5 SCF/hr. The heating rate was constant at 15°C
per minute.

Stress-Strain Measurements. An Instron tester (Model 1130) was
used for testing the samples for tensile strength and elongation at
break, according to ASTM D-1708. The films were cut in the form of
dog bones with the help of a micro die (3/16" x 3/4") and tested
on the Instron tester with a 10 lb. load cell and a crosshead speed
of 2 inches per minute at room temperature.

RESULTS AND DISCUSSION

Structure of Polyurethane and Poly(urethane-urea) Block Copolymers

     The mechanical properties of urethane (urea) block copolymers
depend on the type and degree of crosslinking and phase separation
of hard and soft segments. The crosslinking could be either chemical or physical, the latter being primarily due to the phase separation and hydrogen bonding. For crosslinked polyurethanes the
strength properties are a function of the degree of crosslinking
and pass through an optimum with an increasing number of crosslinks.[26] In the case of crosslinked polyurethanes where the ratio
of diol to triole was 5.5:1, the molecular weight per crosslink was
3729 and was close to the optimum strength properties as suggested
by Smith and Magnusson.[27]

     The good strength properties of linear urethane-urea block
copolymers were mainly due to the physical crosslinks formed between the urea groups because of their high cohesive energy density. Fujiko and Goto[28] have shown that an increase in urea group
concentration also increases the tensile strength properties of
urethane-urea block copolymers due to hydrogen bonding. The concentrations of urethane and urea groups in the poly(urethane-urea)
used in this study were 7.25% and 7.13%, respectively.

Thermooxidative Degradation of Urethane and (Urethane-urea)
Block Copolymers

     The thermooxidative degradation of urethane and (urethane-urea)
block copolymers was measured by means of the oxygen uptake at
155°C. At this temperature, as was established by measurements
under nitrogen, the thermal degradation of the above-mentioned polymers did not take place to any significant degree. During the induction period the mechanical properties did not change. In the
latter stage of degradation it was observed that the retention of
tensile strength was higher for stabilized (urethane-urea) copolymers than for the stabilized polyurethane for the same amount of
oxygen absorbed (Fig. 1). In the case of urethane-urea block co-

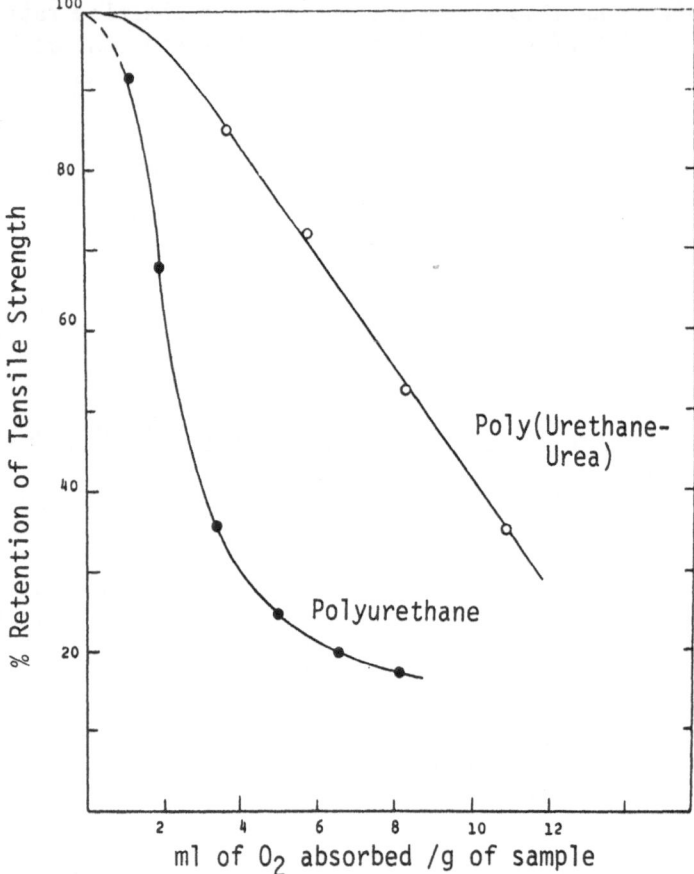

Figure 1.   Correlation of retention of tensile strength of poly
            (urethane-urea) and polyurethane block copolymers with
            the amount of absorbed oxygen at 155°C.

polymer with 1% of antioxidant 2246, the loss of tensile strength
per milliliter of absorbed oxygen per gram of polymer was 6.5% as
compared to 20% for polyurethanes with 1% of Irganox 1076.   The
structural difference between these two copolymers was in the hard
segment.   In the case of (urethane-urea) block copolymers the hard
segments (urea-urethane segments of the chain) form strong physical
crosslinks, which are less vulnerable to cleavage than polyether
chemical crosslinks.   For this reason the overall properties of
urethanes (crosslinked chemically) decrease faster with advancing
degradation.   Similar physical phenomena were observed with oriented
polymers, which showed, with increasing orientation of polymer chains,
higher retention of properties per volume of absorbed oxygen than
nonoriented polymers.[24]

Correlation of Testing Methods for Measurement of Thermooxidation

The thermooxidation of stabilized block copolymers was studied by using oxygen bomb method and thermogravimetric analysis (TGA). These methods measure thermooxidative degradation, but at different stages of the degradation process. In the case of oxygen bomb aging at 100°C, the acceleration of oxidation was achieved by increasing the oxygen pressure to 100 psi (the acceleration factor was 35 times).[22]

In dynamic TGA, the weight loss starts only at a very advanced stage of oxidation with the formation of volatile products. A linear regression analysis was carried out for TGA data (temperature at 25% weight loss) and oxygen bomb data (percent loss in tensile strength after 7 days of exposure) for the stabilized samples (Fig. 2). The linear regression can be expressed by the following equation:

$$X = 1.617Y + 249.75$$

where Y is the temperature at which the sample loses 25% of its weight in TGA and X is the percent retention in tensile strength after oxygen bomb aging. The correlation coefficient was 0.87. The above-mentioned results showed that the TGA method can be used (with certain precautions) for prescreening of stabilizers.

Evaluation of Effectiveness of Stabilizers in Urethane Copolymer Systems

A number of commercially available stabilizers were evaluated for their effectiveness in the stabilization of urethane and (ure-

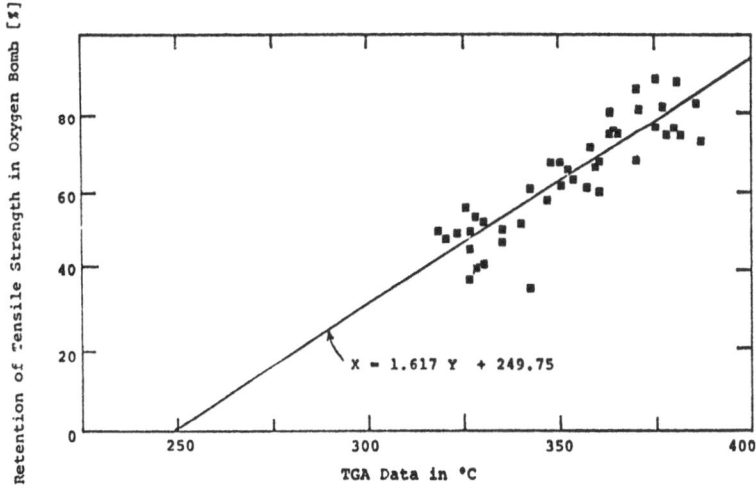

Figure 2. Correlation between oxygen bomb aging and TGA data.

thane-urea) block copolymers. Five substituted phenols, three amines, two phosphites, three thioethers and polycarbodiimide were used in the initial screening experiments. Dynamic TGA in air was used for screening the stabilizers by measuring the weight loss of stabilized block copolymers with temperature. This method exhibited good correlation with the oxygen bomb method and was much faster. From the TGA results, the best stabilizer from each group was chosen for further investigation. (Tables II, III.)

The effect of concentration of stabilizer in the range of 0.167% to 1%, based on the weight of the polymer, on the thermooxidation of polyurethanes and (urethane-urea) block copolymers was evaluated by means of oxygen bomb aging for 7 days. The tensile strength losses for both of these polymers were not linear with a decrease in the stabilizer concentration.

Employing Antioxidant 2246 and Agerite White as stabilizers in the poly(urethane-urea) system, the tensile strengths were "S" shaped as a function of stabilizer concentration. The critical concentration for poly(urethane-urea) block copolymers at 100°C and 100 psi of oxygen pressure have been obtained for the following stabilizers: Antioxidant 2246 - 0.0038 moles/kg; Agerite White - 0.0039 moles/kg. Similar "S" shaped curves were obtained for Irganox 1076 and Agerite White in the polyurethane system. The extrapolated critical concentrations for polyurethanes at 100°C and 100 psi oxygen pressure were as follows: Irganox 1076 - 0.00279 moles/kg; Agerite White - 0.00466 moles/kg.

In order to investigate the effectiveness of mixtures of the stabilizers, the five stabilizers chosen from the screening experiments were blended to a set pattern. The design of these experiments was based on a partial factorial design with five variables at five levels. The total percentage of the stabilizers for all mixtures was kept constant at 1%. The individual percentages of the components varied from 1% to 0.2% as the number of components increased from one to five. The results of the oxygen bomb tests are summarized in Tables IV and V. The loss in tensile strength decreased with an increase in the number of components of stabilizers, as was observed in the two and three component systems of (urethane-urea) block copolymers.

The three component systems gave the best results with only 11.0% loss in tensile strength for a combination of Antioxidant 2246, Agerite White and Staboxol (ABF). The combination Antioxidant 2246, Agerite White and DTDTDP also had only a 11.7% loss in tensile strength. On further increasing the number of components to four (1/4% of each stabilizer), there was actually an increase in the loss in tensile strength and the same was the case with five stabilizers.

STABILIZATION OF URETHANE BLOCK COPOLYMERS 147

Table II. Thermogravimetric Analysis Data for Poly(urethane-urea) System 1-Component Stabilizer

	Temperature (°C) of Weight Loss				
Additives	1%	10%	25%	50%	
1. No additives	245	302	342	385	
2. Irganox 1076	230	295	340	375	
3. Irganox 1010	225	311	352	388	substituted phenols
4. Ionol	253	300	343	375	
5. Santovar A	225	308	342	377	
6. Antioxidant 2246	255	310	354	392	
7. Agerite White	265	309	350	385	
8. Agerite Resin D	245	305	350	382	amines
9. Agerite Resin	232	282	328	370	
10. Polygard HR	238	300	347	380	phosphites
11. Polygard	230	300	342	380	
12. Distearyl thiodipropionate	245	300	343	377	
13. Dilauryl thiodipropionate	250	305	346	380	thioethers
14. Ditridecyl thiodipropionate (DTDTDP)	235	320	360	397	
15. Staboxol	232	290	328	364	scavenger

Table III. Thermogravimetric Analysis Data of Polyurethane System 1-Component Antioxidant

Additives	Temperature (°C) of Weight Loss				
	1%	10%	25%	50%	
1. No additives	245	301	332	365	
2. Irganox 1076	245	300	335	380	substituted phenols
3. Irganox 1010	230	295	328	370	
4. Ionol	225	285	315	356	
5. Santovar A	232	282	320	364	
6. Antioxidant 2246	245	297	332	370	
7. Agerite White	252	300	336	373	amines
8. Agerite Resin D	244	297	332	395	
9. Agerite Resin	232	284	325	370	
10. Polygard HR	240	299	333	362	phosphites
11. Polygard	237	295	325	348	
12. Distearyl thiodipropionate	231	285	318	358	thioethers
13. Dilauryl thiodipropionate	230	292	325	358	
14. Ditridecyl thiodipropionate (DTDTDP)	232	307	342	375	

Table IV. Loss in Physical Properties After 7 Days at 100°C and 100 psi Oxygen for Poly(urethane-urea) System Using a Combination of Stabilizers

Stabilizers	None	Antioxidant 2246 A		Agerite White B		Polygard HR C		DTDTDP D		Staboxol F
Loss in Tensile Strength, %	64.6	32.7		39.3		32.9		40.2		60.0
Stabilizers	AB	AC	AD	AF	BC	BD	BF	CD	CF	DF
Loss in Tensile Strength, %	25.3	32.2	24.8	39.2	38.3	33.9	17.2	41.0	46.3	50.3
Stabilizers	ABC	ABD	ABF	ACD	ACF	ADF	BCD	BCF	BDF	CDF
Loss in Tensile Strength, %	18.7	11.7	11.0	35.5	25.0	27.2	34.4	23.2	19.0	32.6
Stabilizers	ABCD	ABDF	ABCF	ABDF	BCDF					
Loss in Tensile Strength, %	49.4	39.5	60.0	51.5	46.3					
Stabilizer	ABCDF									
Loss in Tensile Strength, %	46.2									

Table V. Loss in Physical Properties After 7 Days at 100°C and 100 psi Oxygen for Polyurethane System Using a Combination of Stabilizers

Stabilizers	None	Irganox 1076 A		Agerite White B		Polygard HR C		DTDTDP D		Staboxol F
Loss in Tensile Strength, %	85.1	76.2		74.2		84.3		79.5		84.7
Stabilizers	AB	AC	AD	AF	BC	BD	BF	CD	CF	DF
Loss in Tensile Strength, %	75.8	79.6	59.9	77.5	61.3	60.9	60.9	55.4	55.7	65.1
Stabilizers	ABC	ABD	ABF	ACD	ACF	ADF	BCD	BCF	BDF	CDF
Loss in Tensile Strength, %	42.4	36.2	34.2	50.2	34.3	44.6	46.3	51.3	36.6	34.2
Stabilizers	ABCD	ABDF	ABCF	ABDF	BCDF					
Loss in Tensile Strength, %	71.0	66.5	75.1	77.3	74.5					
Stabilizer	ABCDF									
Loss in Tensile Strength, %	71.5									

The polyurethane systems showed a similar trend of decreasing tensile losses in the oxygen aging as the number of components in the stabilizer system was increased up to three components beyond which the tensile losses increased again. The minimal percentage loss in tensile strength was obtained with Irganox 1076, Agerite White and Staboxol (34.2%). Another good combination was found to consist of ditridecylthiodipropionate, Polygard HR and Staboxol at a concentration of 1/3% each, with a 34.2% loss in tensile strength.

The synergism between radical chain terminators and peroxide decomposers was previously well established.[20],[22] Another synergism between phenolic and amine stabilizers is due to the regeneration mechanism of the amine stabilizer by a phenolic one.[24] Both stabilizers react with peroxyradicals according to equation:

$$HA + PO_2^\bullet \longrightarrow A^\bullet + POOH$$

The reaction constant for the amine is about one order higher than for the phenolic stabilizer. The amine radical in comparison to phenoxy is more active in the chain transfer reaction:

$$A_{am}^\bullet + PH \longrightarrow HA_m + P^\bullet$$

and therefore amine stabilizers are weaker inhibitors. In the presence of both stabilizers in the system, amine radicals react preferentially with phenolic stabilizers, regenerating amine stabilizer:

$$A_{am}^\bullet + HA_{ph} \longrightarrow HA_{am} + A_{ph}^\bullet$$

This regeneration is a key to the above-mentioned synergism.

In our study carbodiimide (Staboxol) did not show any stabilization effect alone, but in combination with other stabilizers, especially with amine stabilizer. it showed remarkable effect. The action of carbodiimide may be somehow associated with the regeneration of antioxidants.

SUMMARY

In this paper thermooxidation and stabilization of urethane and (urethane-urea) block copolymers were discussed. It was found that mechanical properties of (urethane-urea) block copolymers decreased more slowly during the oxidation in comparison to urethanes under similar conditions. This was attributed to higher stability of the physical crosslinks (in hard segments) to thermooxidation.

The stabilization of block copolymers by various combinations of stabilizers (radical chain terminators, peroxide decomposers) was studied. It was found that the best protection against thermodegradation (measured by the oxygen bomb technique) was provided by the synergistic mixture of amine and phenolic stabilizers and carbodiimide.

REFERENCES

1. R. S. Goglev and M. B. Neiman, Vysokomol. Soed. A9, No. 10, 2083 (1967).
2. A. B. Blyumenfel'd, M. B. Neiman and B. M. Kovarskaya, Vysokomol. Soed. A9, No. 7, 1587 (1967).
3. M. B. Neiman, B. M. Kovarskaya, I. I. Levantovskaya and M. P. Yazrikova, Soviet Plastics, (1) 43 (1967).
4. O. G. Tarakanov, V. A. Orlov and V. K. Beljakov, J. Polymer Sci., Part C 23, 193 (1968).
5. O. G. Tarakanov, L. H. Kondrateva and L. V. Nevskii, Soviet Plastics (6), 36 (1970).
6. P. A. Okunev and O. G. Tarakanov, Vysokomol. Soed. A10, No. 1, 173 (1968).
7. L. V. Nevskii, O. G. Tarakanov and V. K. Belyakov, Soviet Plastics (7) 45 (1966).
8. C. S. Schollenberger and K. Dinsbergs, SPE Trans. 1, (1), 31 (1961).
9. V. A. Orlov and O. G. Tarakanov, Soviet Plastics (5), 18 (1966).
10. H. Schultze, Makromol. Chem. 172, 57 (1973).
11. H. C. Beachell and I. L. Chang, J. Polymer Sci., A10, 503, (1972).
12. V. A. Orlov and O. G. Tarakanov, Soviet Plastics (6) 44 (1967).
13. O. G. Tarakanov and L. N. Kondrateva, Vysokomol. Soed., A13, 565 (1971).
14. O. G. Tarakanov and L. N. Kondrateva, Vysokomol. Soed. A14, 806, (1972).
15. K. B. Piotrovskii and Z. N. Tarasova, Aging and Stabilization of Synthetic Rubbers and Vulcanizates. Khimia (1980).
16. S. L. Reegen and K. C. Frisch, J. Polymer Sci., Part C, 16, 2733, (1967).
17. J. H. Engel, S. L. Reegen and P. Weiss, J. Appl. Polymer Sci., 7, 1679, (1963).
18. V. A. Orlov and O. G. Tarakanov, Soviet Plastics (11), 49, (1967).
19. V. A. Orlov and O. G. Tarakanov, Vysokomol. Soed. 8, No. 6, 1139 (1966).
20. G. Scott, "Atmospheric Oxidation and Antioxidants," Elsevier Publishing Co., Chap. 4, (1963).
21. F. H. Winslow, "Polymer Degradation and Stabilization," Polymer Conference Series, Univ. of Detroit (1970).

22. W. L. Hawkins, "Polymer Stabilization," Interscience Publishers, New York, 1972.
23. L. Reich and S. S. Stivala, "Autooxidation of Hydrocarbons and Polyolefins," Marcel Dekker, Inc., New York, 1969.
24. N. M. Emanuel and A. L. Buchachenko, Chemical Physics of Aging and Stabilization of Polymers, Nauka, (1982).
25. D. J. David and H. B. Staley, "Analytical Chemistry of the Polyurethanes," Part III, Wiley-Interscience, New York, 1969.
26. J. M. Buist and H. Gudgeon, "Advances in Polyurethane Technology," Wiley-Interscience, New York, 1969.
27. T. L. Smith and A. B. Magnusson, J. Polymer Sci. 42, 391 (1960).
28. K. Fujiko and J. Goto, J. Rubber Ind. Japan, 37, 773 (1964).

MODEL COMPOUNDS STUDIES TO ESTABLISH THE BASIC

MECHANISM FOR TIN STABILIZERS IN POLYVINYLCHLORIDE

Alain Guyot, Alain Michel and Tran Van Hoang

CNRS - Laboratoire des Materiaux Organiques

BP 24, 69390 Vernaison, France

INTRODUCTION

Among PVC stabilizers the organotin compounds, mainly thiogly-colate and maleate derivatives, are enjoying a growing importance. In addition, the best recipe includes a mixture of mono- and dialkyl derivatives, which shows a synergistic effect. It was attempted to compare these systems with the metal soap recipe, which includes a synergistic mixture of barium and cadmium, or calcium and zinc car-boxylates. Such a comparison was recently suggested by Burley and Hutton,[1] who showed from infrared evidences that exchange of chlorine with thioglycolate groups occurred between the mono- and dialkyl compounds; so, the formation of dibutyltin dichloride is favored instead of butyltin trichloride, which is more strongly prodegradant than dibutyltin dichloride. The situation may be compared to that of the calcium-zinc system, where the exchange reaction favors the formation of calcium chloride instead of the strongly prodegradant zinc chloride. Other aspects of the mechanisms involved in the zinc calcium recipe have been extensively treated in our previous works.[2] They deal mostly with the catalytic activity of zinc chloride versus the dehydrochlorination reaction, as well as the substitution reaction and other side reactions.

In this paper we would like to report briefly the present state of our work on the basic reaction where tin chloride compounds act as catalysts. We will limit it to the studies carried out using chlorohexene (a mixture of 4-chloro-hex-2-ene and 2-chloro-hex-3-ene) as a model compound for unstable allylic chlorine atoms in the PVC chain. The results already published[3-4] are reviewed and new results are presented. Most of the experimental methods have already been described.[2,3,4]

RESULTS AND DISCUSSION

It must first be recalled that the thermal degradation of the chlorohexene (CH) model compound in chlorinated solvent is a reaction equilibrated with an autocatalysis through a strongly colored transfer complex between HCl and the hexadiene products (mainly 2-4 cis and trans isomers).[5]

$$CH_3 - CH = CH - \underset{\underset{Cl}{|}}{CH} - CH_2CH_3 \rightleftharpoons HCl + CH_3 - CH = CH - CH = CH - CH_3 \quad (1)$$

Some kinetic data are reported in Table I. It has been noted that the rate of the reaction increases with the dielectric constant of the solvent, in agreement with an ionic mechanism. It is to be noted that the reaction is much quicker in 1,2-dichlorethane (DCE) than in 1,4-chlorobutane (DCB).

The catalytic effect of organotin chloride on that reaction has been studied. In DCE solution, the tributyltin chloride shows a powerful stabilizing effect, since no reaction is observed at 60°C. The dibutyltin dichloride also shows a stabilizing effect, although some slow reaction is observed. Butyltin trichloride, on the other hand, is a catalyst for dehydrochlorination, the reaction being first order versus tin compound, as well as versus chlorohexene. At 60°C, the rate constant is $40.10^{-3} mole^{-1}.1.sec^{-1}$, a value similar to the one found using $ZnCl_2$ as a catalyst.

Table I. Kinetic Data for Reaction (1)

Solvent	Temperature (°C)	Rate Constants Direct (s^{-1})	Reverse ($mole.1^{-1}.s^{-1}$)	Activation Energy (kJ/mole) Direct	Reverse
1-2 DCE	40		0.68		
	60	$0.6. 10^{-5}$		38	10
	70	$0.87.10^{-5}$			
	80		1.11		
1-4 DCB	30		$0.4.10^{-3}$		80
	42.5		$1.4.10^{-3}$		
	60	$0.08.10^{-5}$		88	
	80	$0.48.10^{-5}$			

In DCB solution, the same kind of kinetic law was observed in the presence of R_2SnCl_2 compounds, R being an alkyl, aryl or ester-alkyl group. Again, an E2 mechanism is obeyed, with the rate constant listed in Table II.

Both sets of data pointed out the determining effects of the Lewis acidity of metal chloride in catalysis. This statement is strongly supported by the fact that the rate constants can be quite well correlated with the Mossbauer isomeric shifts of tin chloride compound; the literature values of these isomeric shifts are also reported in Table II.

The values for rate constants in the case of estertin compounds (carbomethoxyethyl) are low as compared with the values that should be expected, considering their isomeric shift. The reason for that is obviously the internal coordination of the ester group onto the tin atom. This points out the fact that the rate-determining step in catalysis is probably the coordination of the labile chlorine atom of the model compound to the tin atom, prior to the formation of

Table II. Rate Constants in E2 Dehydrochlorination of Chlorohexene in DCB Solution at 80°C in the Presence of Various Organotin Chloride Compounds

Organotin Compound	Rate Constant.10^{-5} ($mole^{-1}.1.sec^{-1}$)	Isomeric Shift (mm/s)
Dioctyltin dichloride	12	1.731
Dibutyltin dichloride	18	1.624
Dimethyltin dichloride	30	1.55
Diphenyltin dichloride	40	1.387
Di-β-carbomethoxyethyltin dichloride	15	1.45
Butyltin trichloride	283	1.321
β-Carbomethoxyethyltin trichloride	100	1.11

pentacoordinated counter ion of the allylic carbocation involved in dehydrochlorination.

$$CH_3-CH=CH-CH-CH_2CH_3 \quad CH_3-CH=CH-CH-CH_2-CH_3 \quad CH_3-CH-CH-CH-CH_2-CH_3 \quad (2)$$

$$
\underset{C_4H_9SnCl_3}{\overset{Cl}{\underset{+}{|}}} \quad \rightleftarrows \quad \underset{C_4H_9SnCl_3}{\overset{Cl}{\underset{\vdots}{|}}} \quad \rightleftarrows \quad \left[C_4H_9SnCl_4\right]^{\ominus}
$$

Another point in favor of the mechanism is the fact that butyltintrichloride causes a change in the activation energy of dehydrochlorination, which drops from 88 kJ/m to 41 kJ/m.

The reverse reaction (addition of HCl onto hexadiene) is also catalyzed by tin chloride compounds. Again, the catalytic effect is related to the Lewis acidity of chloride: dioctyltin dichloride causes a decrease in the reaction rate, and a strong catalytic effect is observed with the butyltin trichloride. The activation energy for that reverse reaction in the presence of that compound is only 22 kJ instead of 80 in 1,4 DCB. A few data are reported in Table III.

Some side reactions are observed: in a few cases[6] (phenyltin compounds) the stability of carbon-tin bond is low, so that the bond is cleaved in the presence of HCl. It finally results in the formation of $SnCl_4$, which is the strongest Lewis acid; its formation is, however, delayed since the exchange reaction is very rapid. But when

$$\phi_2SnCl_2 \quad + \quad SnCl_4 \quad \rightarrow \quad 2 \ \phi \ SnCl_3 \qquad\qquad (3)$$

Table III. Initial Rates for the Addition of HCL (260 m mole.1^{-1}) to Hexadiene (260 m mole.1^{-1}) in the Presence of Organotin Chlorides in DCB at 80°C

Organotin Chloride	Concentration (m mole.1^{-1})	Initial Rate (m mole.1^{-1}.sec^{-1})
Dioctyltin dichloride	90	14
Dimethyltin dichloride	90	26
Dibutyltin dichloride	90	21
Butyltin trichloride	18	36

the extent of the tin-carbon cleavage is high enough, a very strong
acceleration of the reaction is observed. For ϕ_2SnCl_2, the cleavage
takes place at the reaction temperature used for model compounds
studies, as well as at the PVC processing temperature. For β-car-
boxyethyl compounds, the cleavage is observed only at the processing
temperature.[4]

A second reaction is the cationic oligomerization of hexadiene:
such reaction is a possible model reaction for crosslinking of PVC.
That reaction has been shown to be strongly catalyzed by $ZnCl_2$ in
the presence of HCl. It is also strongly catalyzed by $SnCl_4$, so that
when such a compound may be formed, for instance, after cleavage of
the carbon-tin bonds by HCl in the case of ϕ_2SnCl_2, the equilibrium
between dehydrochlorination and the reverse addition reaction is
shifted in favor of dehydrochlorination. As shown in Figure 1, as
the consumption of the model compound is increasing, the amount of
hexadiene found goes through a maximum, the difference being attributed to the oligomerization. The catalytic activity of the tin trichlorides is much lower than that of $ZnCl_2$, however. (See Figure
2). There is almost no activity for the di- and monochloride derivatives.

A final side effect is the possible coordination of HCl with the
tin atom which may compete with the allylic chloride for occupying
the vacant coordination site of the tin. Such a competition causes
a limitation of the catalysis, and then of the dehydrochlorination

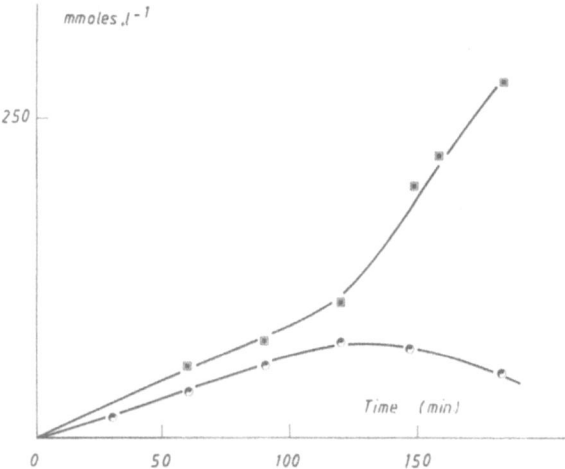

Figure 1.   Degradation of chlorohexene (380 m moles.1^{-1}) in 1,4-
            dichlorobutane at 80°C in the presence of ϕ_2SnCl_2 (50
            m moles.1^{-1}) (■) chlorohexene disparition; (●) hexadiene
            formation.

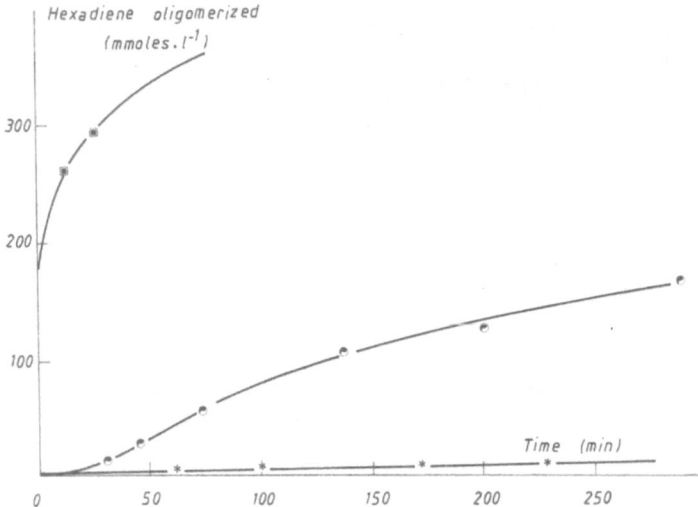

Figure 2.   Hexadiene oligomerization during the dehydrochlorination
of chlorohexene (400 m moles.1^{-1}) in dichlorobutanone at
80°C in the presence of 30 m moles.1^{-1} of ZnCl$_2$ (■);
BuSnCl$_3$ (◐); and Bu$_2$SnCl$_2$ (*).

extent. Evidences for this phenomenon were described in the case of
SnCl$_4$[4] where a limitation of the reaction is observed, although the
oligomerization of hexadiene is so rapid that the reverse reaction
cannot take place. Actually, the formation of hexadiene is not
observed directly. In the case of Bu$_2$SnCl$_2$, the complexation with
HCl causes a rather large decrease of the isomeric shift in the Mössbauer spectrum.

Interaction of tetra- or tri- or dialkyltin compounds with HCl
is the only way to explain the stabilizing efficiency of these compounds of very low Lewis acidity. Tetra- and trialkyl tin compound
react with HCl to give dichloride.[12] We suggest that these compounds
are able to destroy or displace or inhibit the formation of the charge
transfer complex between HCl and hexadiene which is the catalyst for
the dehydrochlorination.[5] The decrease of the isomeric shift in the
Mössbauer spectrum of Bu$_2$SnCl$_2$ upon complexation with HCl can explain
that the coordination site for the allylic chlorine atom of CH is no
longer available.

Because of its Lewis acidity, ZnCl$_2$ is a strong catalyst, not
only for dehydrochlorination (degradation effect in the case of PVC)
but also for the substitution reaction of carboxylate and other nucleophilic organic reagents to the labile allylic chloride[2] (stabilization effect in the case of PVC). The reason for that is because
both reactions involve the same rate determining reaction, namely,

formation of allylic carbocation. In the presence of nucleophilic
reagent, there is a competition between the elimination reaction and
the substitution reaction. Because some tin compounds, and mostly
the strong Lewis acids, were shown to catalyze formation of that carbocation, they are expected to show a catalytic effect versus the
substitution reaction.

Among the organotin stabilizers, dialkyltin dicarboxylates were
first used, but they were found to be less efficient than dialkyltin
diisothioglycolates or maleate derivatives. The reasons for that
rather poor efficiency are not known at the moment. According to
Klemchuk,[7] the reaction of dibutyltin dilaurate (Bu_2SnL_2) with chloro-
pentene as model compound follows a complex autocatalytic law and it
is suggested that the monoreacted species Bu_2SnLCl should be the
active one. A further study by Parker and Carman[8] was not conclusive
on that point but stated from NMR evidence that there is an equilibrium exchange reaction between the dilaurate and the dichloride; in
the case of the thioglycolate derivative, the exchange reaction is
easily completed towards the formation of the monochloride molecule.

At 80°C in DCB there is no reaction between Bu_2SnL_2 and the model
compound chlorohexene. Infrared spectroscopy also shows no trace of
interaction or complex formation between these two products. The
reaction takes place in the presence of organotin chlorides, with an
autocatalytic character, whatever tin chloride compound is used.
Substitution (Figure 3) and elimination (Figure 4) reactions are both
observed with their yield ratio of about 2:1 for the dichloride of
dibutyl,dioctyl and dimethyl derivatives. The dimethyl derivative is
slightly more reactive and the diphenyl derivative is much more active
as is the butyltin trichloride. So the order of the catalytic activity of the chlorides follows again their order of Lewis acidity. The
initial rates can be described by the following kinetic law:

$$R = K \{CH\}^a \{RSnCl_x\}^b \{Bu_2SnL_2\}^{-c} \tag{4}$$

where a, b and c depend on the nature of the chloride. These data
suggest again that the rate-determining step in both the substitution
reaction and the elimination reaction involves the intermediate formation of a carbocation through a reaction between the Lewis acid and
the allylic chlorine atom after coordination of the latter with the
tin atom. The substitution reaction involves further the nucleophilic
attack of the carbocation by the dilaurate but that reaction is not
expected to be rate-determining. So the contribution of the dilaurate concentration to the kinetic law needs further explanation, which
probably involves the contribution of the exchange reaction. The
latter can be studied by IR using the shift of the 1610 cm^{-1} carbonyl
band of the laurate group, or preferably by using ^{1}H NMR chemical
shifts of protons of the methylene (or methyl) group directly attached
to the tin atom (Table IV). The extent of the exchange reaction

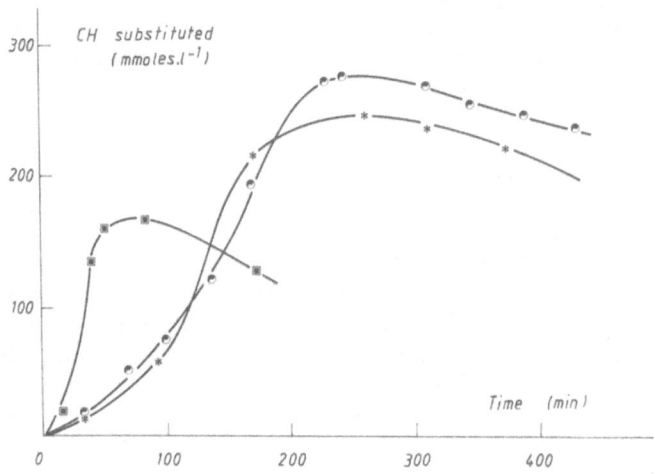

Figure 3.   Esterification of chlorohexene in DCB at 80°C by dibutyl-
            tin dilaurate in the presence of organotin chlorides.
            (*) CH: 843; Bu_2SnCl_2: 180; Bu_2SnL_2: 150 m moles.1^{-1};
            (◉) CH: 528; Oct_2SnCl_2: 192; Bu_2SnL_2: 150 m moles.1^{-1};
            (■) CH: 380; $BuSnCl_3$: 48; Bu_2SnL_2: 120 m moles.1^{-1}.

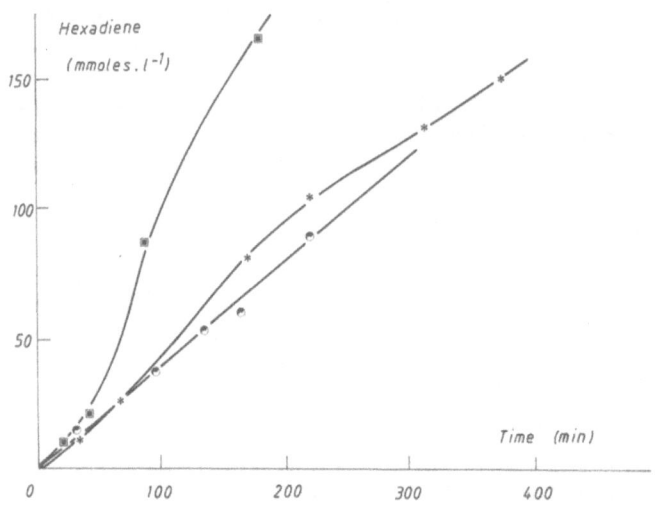

Figure 4.   Degradation of chlorohexene in DCB at 85°C by dibutyltin
            dilaurate in the presence of organotin chlorides.   (*)
            CH: 483; Bu_2SnCl_2: 180; Bu_2SnL_2: 150 m moles.1^{-1}; (◉) CH:
            528; Oct_2SnCl_2; 192; Bu_2SnL_2: 150 m moles.1^{-1}; (■) CH:
            380; $BuSnCl_3$: 48; Bu_2SnL_2: 120 m moles.1^{-1}.

$$Bu_2SnL_2 + R_XSnCl_Y \rightleftarrows Bu_2SnClL + R_XSnLCl_{Y-1}$$

$$\uparrow\downarrow$$

$$Bu_2SnCl_2 + R_XSnL_2Cl_{Y-2}$$

can be estimated from the value of the chemical shift. When X = 3 evidences are conclusive only in the case of Me_3SnCl and when for X = 2 the shifts are observed in all cases and their extent follows the order of the Lewis acidity, the larger shifts are observed for the trichloride derivatives. In all cases when enough chloride is present, the formation of Bu_2SnCl_2 can be proved using the gas chromatographic analysis. If $ZnCl_2$ is used instead of the organotin chloride, the exchange reaction is quantitative and all the laurate groups become attached to the zinc atoms. Most probably the exchange reaction involves formation of an intermediate bimolecular complex which makes the exchange of the ligands easier.

In the presence of the model compound CH, a competition between the CH and Bu_2SnL_2 for complexation with the organotin chloride may be taking place. Such a competition might explain the negative exponent for Bu_2SnL_2 in Equation 4. Further, the nucleophilic character of Bu_2SnL_2 is weak and might be weaker again upon complexation. For that reason the ratio of substitution/elimination is low and as a result a rather large amount of hexadiene is produced in all cases (Figure 4) unless it is oligomerized, as in the case of the strongest Lewis acid organotin trichloride.

In order to check the initial suggestion of Klemchuk about the catalytic efficiency of the monochloride Bu_2SnLCl, the compound which is impossible to isolate, it was produced through the exchange reaction under conditions such that the extent of the reaction was limited and the formation of Bu_2SnCl_2 was avoided. This was in the case when a stoichiometric mixture of Bu_2SnL_2 and R_3SnCl was used. If CH was added to such a mixture in DCB solution at 80°C, there was no consumption of CH, except in the case of the trimethyltin chloride. where a very small amount of hexadiene was produced. It may be concluded that the monochloride is not an efficient catalyst; the stabilization involves mainly the catalytic-contribution of a di- (or tri-) chloride with enough Lewis acidity. Further, the monochlorocarboxylate Bu_2SnLCl is not an active species.

As shown in Figure 3, a maximum in the substitution reaction is observed when a strong catalyst is used and when the amount of allylic chloride is higher than the amount of the laurate. Under these conditions HCl can be formed and reacted with the substitution product, giving lauric acid and CH which will lead to hexadiene only. Such an effect has been well documented in the case of the zinc-calcium systems.[2]

Table IV.   Chemical Shift of the Protons, CH-Sn-Cl Adjacent
to Tin Atom in Organotin Chlorides

| Compounds | Chemical Shifts in ppm, Referred to TMS | |
	Alone	In the Equimolar Mixture with Dibutyltin Dilaurate
Trimethyltin chloride	0.68	0.60
Dimethyltin dichloride	1.21	1.06
Dibutyltin dichloride	1.80	1.75
Dioctyltin dichloride	1.86	1.75
Diphenyltin dichloride	7.75	8.05
Methyltin trichloride	1.2	1.3
Butyltin trichloride	2.5	1.85
Octyltin trichloride	2.38	1.80

Another possibility for the limitation of the substitution
reaction is described in Equation 5:

$$- CH{=}CH{-}CH{-} + RSnCl_3 \rightleftarrows -CH{=}CH{-}CH{-} + RSn(OCOR)Cl_2 \qquad (5)$$
$$\underset{OCOR}{|} \qquad\qquad\qquad \underset{Cl}{|}$$

The possibility of that exchange reaction is raised by the fact
that the maximum of the substitution product shown in Figure 3 occurred long before all the laurate groups had been used. In that
case, the CH produced is easily reacted to give hexadiene.

We have also obtained a few data about tin maleates; this
study, however, is still in progress and will be completed soon.
There are two kinds of tin maleate compounds: the alkyltin alkyl-
maleate such as the tetrabutyl derivative (DBTDBM):

$$Bu_2Sn \{OCO{-}CH{=}CH{-}CO{-}OBu\}_2$$

and the alkyltin maleate such as the dibutyl derivative (DBTM):

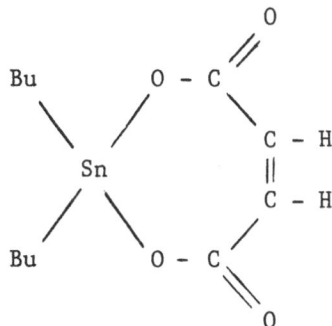

In both cases reaction with chlorohexene takes place in DCE at 60 or at 80°C in DCB, being first order versus both CH and tin concentration, and producing mainly the substitution product with a very small amount of hexadiene. The DBTM (rate constant 80°C, DCB-$0.37.10^{-3}$ mole.1^{-1}.s^{-1}) is slightly more reactive than DBTDBM $(0.25.10^{-3})$.

In the presence of Bu_2SnCl_2, the behavior of DBTDBM (DCE 60°C) is similar to that of the dilaurate Bu_2SnL_2: autoacceleration for both substitution and elimination but the ratio is more in favor of substitution (5 instead of 2). However, Bu_2SnCl_2 does accelerate the reaction of DBTM and CH and changes the kinetic law, which becomes first order versus Bu_2SnCl_2 but zero order versus DBTM. Here the mechanism becomes similar to that observed in the presence of zinc chloride, the rate-determining step being the formation of the intermediate carbocation. The dioctyltindichloride seems to follow the same mechanism, but with a catalytic activity at a very low level, hardly perceptible. $BuSnCl_3$ is much more active but the kinetic laws are not simple, probably due to the secondary reactions: hexadiene oligomerization and possible Diels-Alder condensation of the diene and the maleic anhydride. HCl is formed when the maleate groups have been consumed and will lead to the formation of maleic acid through reaction with the substitution product. $BuSnCl_3$ is able to dehydrate the maleic acid into maleic anhydride (as well as Bu_2SnCl_2, but more rapidly); the latter reacts readily with hexadiene. When the hexadiene produced is consumed in secondary processes, it is no longer possible to study the kinetics of the substitution reaction, because the difference between the consumption of the chlorohexene and the production of hexadiene is no more representative of the substitution.

The basic ionic nature of the mechanism of the reaction of dialkyltin dithioglycolates compounds with allylic chlorides has been recognized as early as 1970 by Ayrey, Poller and Siddiqui[9] using chlorobutenes (3,1 and 1,2 isomers) as models. These authors also have shown evidence for isomerization through a carbocationic intermediate. The reaction was almost complete after 45 min. at 180°C. In our laboratory, Nölle[3] has shown that the reaction with chloro-

hexene isomers does not take place at room temperature but becomes
rapid at 100°C. The substitution and isomerization takes place and
the reaction has an autocatalytic character. The reaction can be
initiated at room temperature upon addition of either HCl or the
monochloride tin derivative $Bu_2Sn(Cl)-S-CH_2COOR$ or also the thioglycolic ester $HSCH_2COOCH_3$. The latter ester is able to react with the
chlorohexene, not at room temperature but at 90°C, giving the substitution product and HCl. So its catalytic effect is probably the
result of an interaction with the tin thioglycolate derivative.

According to Poller,[10] the enhanced reactivity of the bis thioglycolate tin derivatives can be explained by the internal coordination of the carbonyl group to the tin atom, which leads to a nucleophilic assistance to weaken the Sn-S bond, which is then prone to
react with the allylic chloride. Possibly an external coordination
of the free thioglycolic ester plays the same role.

Another specific character of the tin thioglycolic derivatives
is seen in the exchange reaction between the thioglycolate group and
the chlorine atom. Parker and Carman[8] have shown that the exchange
reaction is readily complete:

$$Bu_2SnCl_2 + Bu_2Sn(S-CH_2-COOR)_2 \rightarrow 2\ Bu_2Sn(Cl)(S-CH_2COOR) \qquad (6)$$

while the reaction with the carboxylate derivatives reaches an equilibrium. The findings of Parker and Carman were confirmed by Nölle[3]
for a series of methyl-derivatives. More recently Burley and Hutton[1]
made a systematic study of the exchange reaction for isooctyl thioglycolate compounds concluding that the exchange reaction favors the
situation where the chlorine atoms are attached to a maximum of tin
molecules, so the formation of highly acidic $RSnCl_3$ compounds is not
favored. It may be then concluded that in PVC processing, that compound may appear only at very high degrees of degradation.

Although the organotin compounds are often used without other
stabilizers, the similarity of the ionic carbocation mechanism observed with these compounds and the zinc-calcium recipe suggests a
study of the effect of a combination of Lewis acid tin chloride and
nucleophic agents in the substitution reaction on the allylic chlorine of the model compound CH.

Calcium stearate was tried as a nucleophilic agent. No reaction
takes place at 60°C in DCE without tin compound, or in the presence
of Bu_3SnCl or Bu_2SnCl_2. Both substitution and elimination reaction
were observed in the presence of $BuSnCl_3$. The kinetic laws for substitution and elimination (60°C, DCE) are as follows:

Substitution: $R_S = 12.10^{-3}$ (CH)(Sn)

   Elimination: $R_E = 21.10^{-5}$ (CH)(Sn)(Ca)$^{-1}$

The mechanism seems to be similar to the one observed with $ZnCl_2$, but reactions are slower. Most probably, the first steps of these reactions are exchange reaction, such as:

$$Cl_3SnBu + Ca(St)_2 \rightarrow CaCl_2 + BuSnCl_a(St)_b$$

with a + b = 3 and a being 0, 1 or 2.

The actual nucleophile is probably the tin carboxylate compound but its formation involves a decrease in the actual catalyst concentration. Thus, minus 1 order versus calcium concentration for elimination reaction can be explained.

In addition, the use of organotin chloride in combination with epoxy compounds was tried. Then, three main reactions were observed: elimination and substitution reactions, but also epoxy polymerization. The situation is again similar to the one observed using $ZnCl_2$, but the kinetics were much more complex. Two epoxy compounds were used: epoxy-butane (BO) and epoxy-cyclohexane (HO). A simple SN2 mechanism for the substitution (etherification) reaction was observed with Bu_2SnCl_2 and BO. However, the reaction was, in addition, first order versus epoxy compound, if HO was used in the presence of Bu_2SnCl_2 or if BO was used in the presence of $BuSnCl_3$. The last combination (HO and $BuSnCl_3$) leads chiefly to the epoxy polymerization. Similar kinetics were obtained for elimination reaction : E2 mechanism with the BO-Bu_2SnCl_2 system, but reaction of an order minus 1 for the two other systems was considered. Then, it seems that the epoxy compounds are involved in the rate-determining step. From NMR studies some evidence for coordination of the epoxy group to the tin atoms can be found. Possibly there is a competition between oxirane and allylic chlorine atoms for coordination to the tin atom. This may explain the negative order observed for the elimination reaction, because the actual concentration of the catalyst (uncomplexed tin compound) available for the formation of the allylic carbocation is lowered. However, other possibilities are open by the coordination of the epoxy compound to the tin atom. First, sixth coordination position remains available for the chlorine atom of the allylic model compound. Then a concerted mechanism may occur inside the coordination sphere of the tin which may lead to the formation of the β chloroether, provided both oxirane and CH are coordinated at a suitable vicinal coordination site (cis). But most probably, after coordination, the oxirane becomes polarized:

Then, a charge transfer to the tin will lead to the initiation
of the oxirane polymerization, but also, the chlorine atom of the
allylic CH may attack the carbocation of the oxirane, leaving an
allylic carbocation which undergoes the nucleophilic attack of the
polarized oxygen; again, this mechanism will lead to the formation
of the β chloroether. Both mechanisms may account for the observed
kinetic laws. More details will be published elsewhere.[11]

CONCLUSION

SN2 or E2 mechanisms for stabilization and degradation reactions
of PVC model compounds are valid for organotin compounds, as well as
for metal soaps stabilizing systems. However, owing to various exchange reactions, and because of the easy coordination of tin atoms
with different species, the situations are more complex. The very
high efficiency of the organotin system in the thermal stabilization
of PVC can be partly explained by at least two factors: the fast
exchange reactions, which delay the formation of strong prodegradant
Lewis acid chlorides (trichlorides, mainly) and the high efficiency
in the catalysis of the reverse addition of HCl on dienes. But a
clear explanation is still lacking for the stabilizing power of
weakly acidic compounds, such as dioctyltin dichloride or trialkyl-
tin chlorides, except for their interaction with HCl.

The conclusion of model compounds studies at 60 - 80°C cannot
be applied to the actual PVC processing without caution, owing to
the difference of temperature which may cause further effects of
radical processes, and also of the morphology of solid PVC which
makes difficult a rapid and complete mixing of the stabilizers.

REFERENCES

1. J. W. Burley and R. P. Hutton, Polym. Degr. and Stab., 3, 285
      (1981).
2. A. Guyot and A. Michel, in "Developments in Polymer Stabiliza-
      tion-2, Chap. 3," Ed., G. Scott, Applied Science Pub.,
      London (1980), pp. 89-124.
3. A. Michel, A. Guyot and D. Nölle, Polym. Degr. and Stab., 2,
      277 (1980).
4. Tran Van Hoang, A. Michel and A. Guyot, Polym. Degr. and Stab.,
      3, 137 (1980-81); 4, 213, 365 (1982).

5. Tran Van Hoang, A. Michel, C. Pichot and A. Guyot, Europ. Polym. J., <u>11</u>, 469 (1975).
6. G. Ayrey, F. P. Man and R. C. Poller, Europ. Polym. J., <u>17</u>, 45 (1981).
7. P. P. Klemchuk, in A.C.S. Chemistry Series n° 85 "Stabilization of Polymers and Stabilization Processes," Ed., R. F. Gould, 1968, pp. 1-17.
8. R. G. Parker and C. J. Carman, in A.C.S. Chem. Ser. n° 169, "Stabilization and Degradation of Polymers," Eds., D. L. Allara and W. L. Hawkins, 1978, pp. 363-373.
9. G. Ayrey, R. C. Poller and I. H. Siddiqui, J. Polym. Sci. Part B, <u>8</u>, 1 (1970), and J. Polym. Sci. Part A1, <u>10</u>, 725 (1972).
10. R. C. Poller, private communication.
11. Tran Van Hoang, A. Michel and A. Guyot, Polym. Deg. and Stab. (in press).
12. R. K. Ingham, S. D. Rosenberg and H. Gilman, Organotin Compounds - Chemical Reviews, <u>60</u>, Oct., 1960, n° 5, p. 483.

REACTIONS OF PVC WITH ORGANOTIN STABILIZERS

UNDER CONTROLLED CONDITIONS

G. Ayrey, S. Y. Hsu and R. C. Poller

Queen Elizabeth College
Campden Hill Road
London W8 7AH, England

INTRODUCTION

Research and development work in this area often concentrates on attempts to reproduce in the laboratory conditions which approximate to those in the actual processing equipment where degradation and stabilization occur. Such experiments are successful in evaluating and development stabilizers but are less effective in clarifying the chemistry involved. Our own work, at first with model compounds[1,2] and then with PVC itself,[3-5] has been designed to show how the polymer reacts under carefully controlled conditions so that the various reaction parameters can be isolated.

To eliminate morphological factors and effects due to incomplete mixing, all experiments described in this paper were carried out in solution. Air was excluded from all reactions which were carried out with polymer from a single commercial batch of PVC, suspension polymerized at 51°C (Corvic D65/02X I.C.I. Plastics Division), $\overline{M}_n$ = 5.46 x 10^4, $\overline{M}_w$ = 1.20 x 10^5.

REACTIONS OF PVC WITH LABELLED STABILIZERS

The following radio-labelled stabilizers were prepared; all had acceptable elemental analyses and the radiochemical purities were 100% within experimental error. These compounds were chosen as being representative of the main classes of stabilizers in current commercial use: DBTL = dibutyltin di([1-^{14}C]laurate); DBTM = dibutyltin bis(methyl[1-^{14}C]maleate); DBTT = dibutyltin bis(2-ethylhexyl mercapto[1-^{14}C]acetate).

171

The methods of synthesis are summarized in the following equations:

DBTL:

$$Bu_2SnO + 2CH_3(CH_2)_{10}{}^{14}COOH \xrightarrow[\text{(-2H}_2\text{O)}]{\overset{\text{Heat}}{C_6H_6}} Bu_2Sn(O^{14}COC_{11}H_{23})_2 \tag{1}$$

Specific activity of product 5.297×10^{10} dpm mol^{-1}

DBTM:

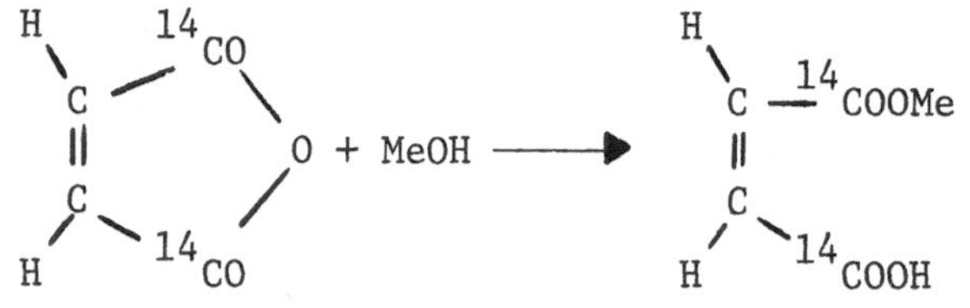

(2)

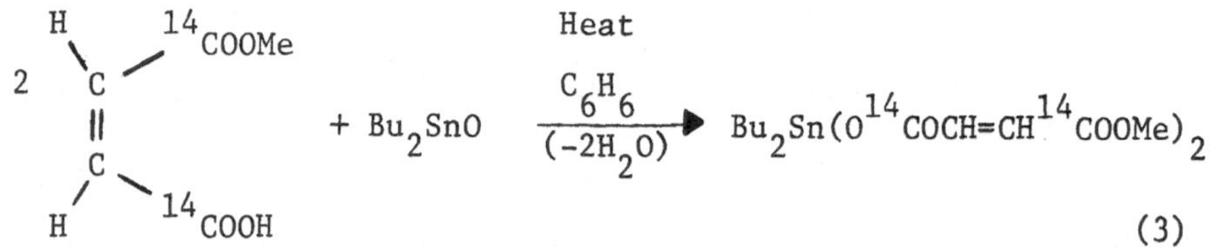

(3)

Specific activity of product 7.480×10^9 dpm mol^{-1}

DBTT:

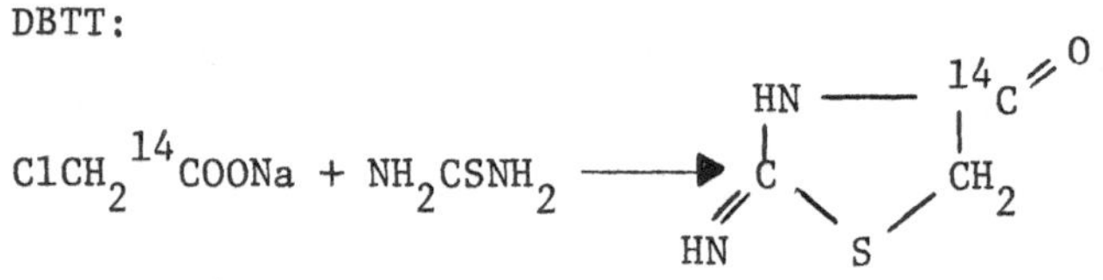

$$ClCH_2{}^{14}COONa + NH_2CSNH_2 \longrightarrow$$

(4)

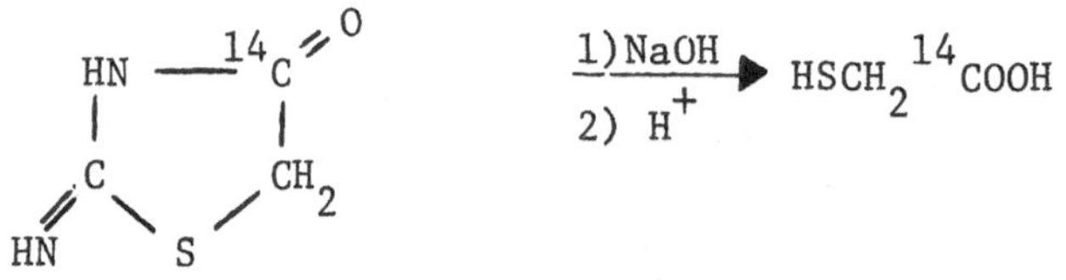

$\xrightarrow[\text{2) H}^+]{\text{1) NaOH}}$ $HSCH_2{}^{14}COOH$ (5)

$$HSCH_2^{14}COOH + CH_3CH_2CH_2CH_2\underset{\overset{|}{CH_2CH_3}}{CH}CH_2OH \xrightarrow{H^+} HSCH_2^{14}COOC_8H_{17} \quad (6)$$

$$2HSCH_2^{14}COOC_8H_{17} + Bu_2SnO \xrightarrow[C_6H_6(-2H_2O)]{Heat} Bu_2Sn(SCH_2^{14}COOC_8H_{17})_2$$

$$(7)$$

Specific activity of product 1.745×10^{10} dpm mol^{-1}

A series of tubes each containing a degassed solution of PVC (0.200 g) and stabilizer (1.583×10^{-5} mol) in chlorobenzene was sealed under high vacuum and heated at 180°C for various periods of time. The polymer was precipitated with methanol and purified by two cycles of dissolution in tetrahydrofuran and precipitation with methanol. The incorporation of the radioactive labels into the polymer were determined by liquid scintillation counting and it was demonstrated that a further purification cycle had little effect on the specific activities. The results are shown in Figure 1.

It is clear that the rate of uptake of ^{14}C from DBTL is much lower than from the other two stabilizers. Even after 15 h heating only ~ 0.1 laurate group is incorporated per polymer molecule. This low rate is because the only mechanism for incorporation of the label is by exchange between labile chlorine atoms in the polymer and laurate residues on the stabilizer (Equation 8).

$$Bu_2Sn(OCOC_{11}H_{23})_2 + 2 \;\sim\!CH=CHCHClCH_2\!\sim\; \longrightarrow Bu_2SnCl_2 +$$

$$2 \;\underset{\overset{|}{OCOC_{11}H_{23}}}{\sim\!CH=CHCHCH_2}\!\sim\; (or \;\underset{\overset{|}{OCOC_{11}H_{23}}}{\sim\!CHCH=CHCH_2}\!\sim\;) \quad (8)$$

It was established previously that organotin carboxylates undergo exchange reactions with alkylic chlorine atoms much more slowly than organotin thiolates.[1] Uptake of ^{14}C from DBTM is much more rapid and increases linearly up to 15 h when, on average, each polymer molecule has taken up one maleate residue. Although exchange reactions will occur, the dominant mechanism for incorporation of the label is probably by Diels-Alder reactions between the DBTM, or of maleic acid residues derived from it, and conjugated diene units in the degrading polymer. For processors of PVC it is the reactions which occur at the earliest stages of heating which are of most immediate interest and we see that the initial rate of uptake of ^{14}C from the thioglycolate (DBTT) is very high. After about one hour the rate begins to fall and from 2 - 15 h incorporation of the label occurs more slowly at a fairly steady rate. From earlier work with other organotin thiolates[3] it is likely the reaction responsible for the slower

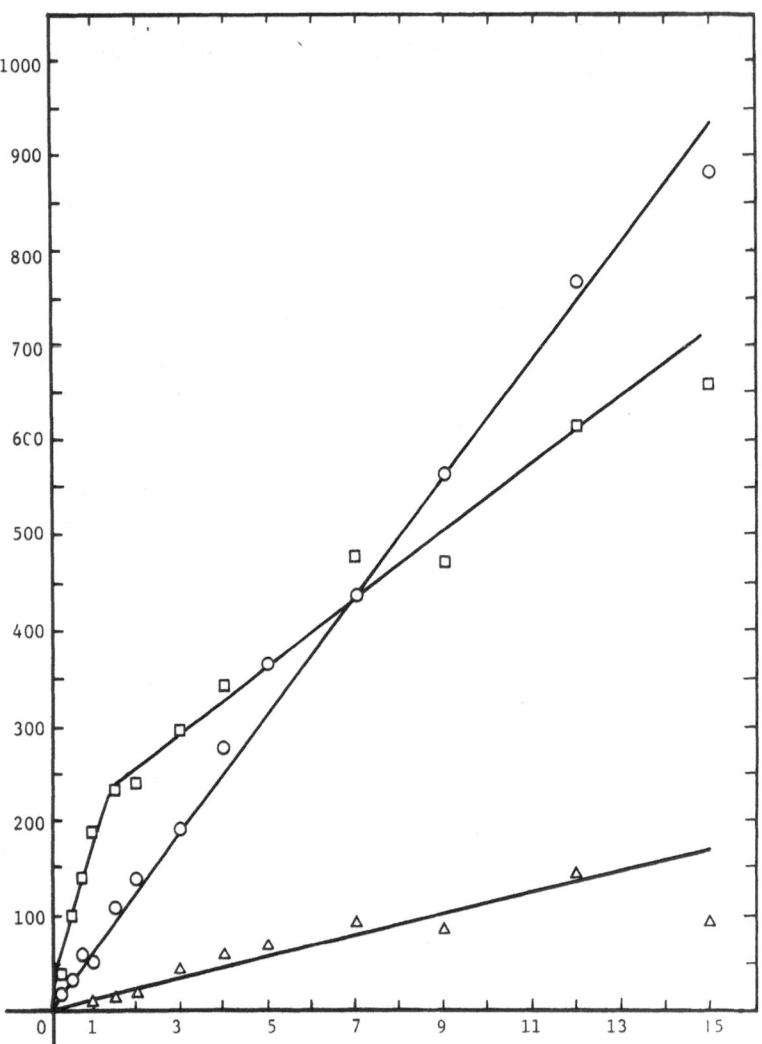

Figure 1.   Rate of uptake of radiolabelled stabilizers by PVC on
            heating at 180°C for various periods. DBTL (-△-), DBTM
            (-○-), DBTT (-□-). Ordinate: uptake of stabilizers
            (mol g^{-1} PVC x 10^8). Abscissa: heating time (h).

uptake of ^{14}C is addition to alkene units in the degrading polymer
of 2-ethylhexyl thioglycolate (Equation 9). This is obtained from
Sn - S bond cleavage of the stabilizer by hydrogen chloride produced
in the degradation (Equation 10).

$$\sim CH = CHCHCl\sim + HSCH_2COOC_8H_{17} \longrightarrow \sim \underset{\underset{SCH_2COOC_8H_{17}}{|}}{CHCH_2CHCl} \sim \quad (9)$$

$$Bu_2Sn(SCH_2COOC_8H_{17})_2 + 2HCl \longrightarrow Bu_2SnCl_2 + 2HSCH_2COOC_8H_{17} \quad (10)$$

The initial rapid reaction is presumably exchange between the thio-glycolate residues bound to tin and labile chlorine atoms in the poly-mer.

   The PVC samples obtained in the above experiments were dissolved in THF and their UV and visible spectra recorded. Polymer heated with DBTT remained colorless for 9 h and Figure 2 shows that the con-centration of longer chain conjugated polyene sequences responsible for the absorption at the longer wavelength end of the spectrum in-

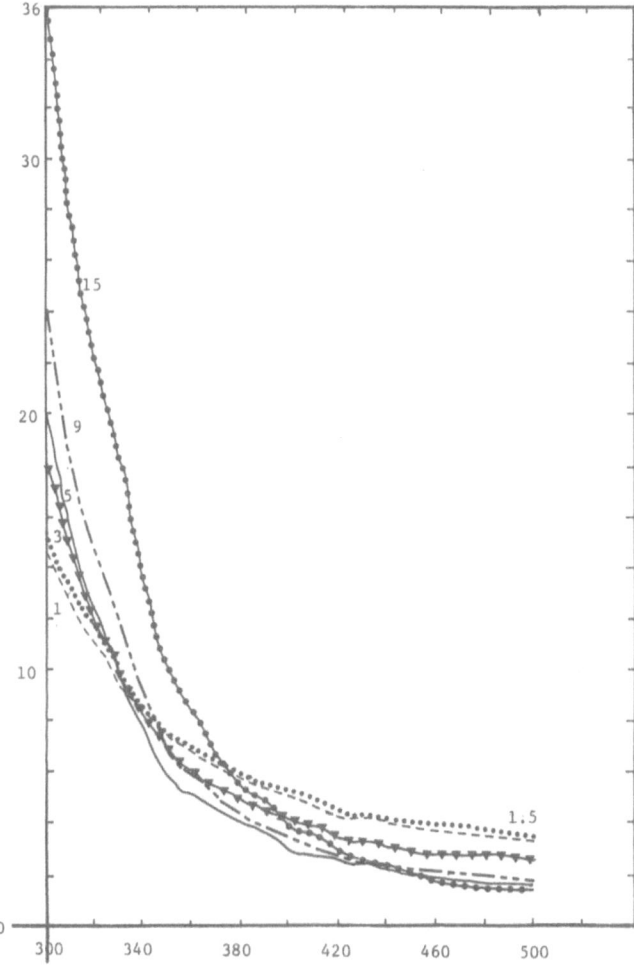

Figure 2. UV absorption spectra (in THF) of PVC stabilized with DBTT and heated at 180°C. (Numbers on curves indicate time of heating in hours). Ordinate: $E_{1\ cm}^{1\%}$ x 10^{-2}. Abscissa: λ (nm).

creases slightly up to 1.5 h but hererafter is decreased presumably by 2-ethylhexyl thioglycolate additions. However, these additions are unable to suppress the development of the shorter chain sequences which cause a steady increase in the absorption in the 300 nm region of the UV spectrum. Dibutyltin dilaurate has no power to decolorize degraded polymer since it cannot add to double bonds. DBTL prevented discoloration for only ~ 0.75 h and its poor stabilizing action is seen in Figure 3. The rapid increase in absorption intensity throughout the spectrum as heating is continued is striking with $E_1^{1\%}$cm values of above 250 x 10^2 being recorded. The superior sta-

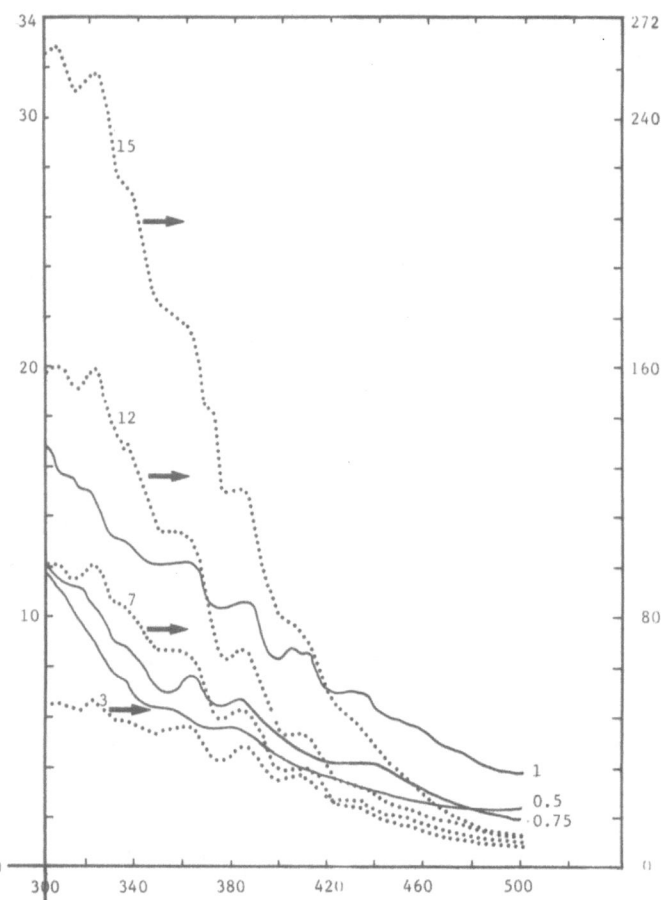

Figure 3. UV absorption spectra (in THF) of PVC stabilized with DBTL and heated at 180°C. (Numbers on curves indicate time of heating in hours.) Left hand and right hand ordinates: $E_{1\ cm}^{1\%}$ x 10^{-2}. Abscissa: λ (nm).

bilizing power of DBTM which, like DBTT, can both prevent and cure
degradation is seen in Figure 4, which resembles Figure 2. PVC heated
with this stabilizer had not developed any color even after 15 h when
the experiment was terminated.

When virgin PVC was heated from 1 h at 180°C with the experimental conditions described earlier, the solution became orange-
colored. After removal of all hydrogen chloride the labelled thioglycolate stabilizer DBTT was added under anearobic conditions at
the same concentration as had been used previously. Heating at 180°C
was continued, the color slowly faded and the solution was colorless

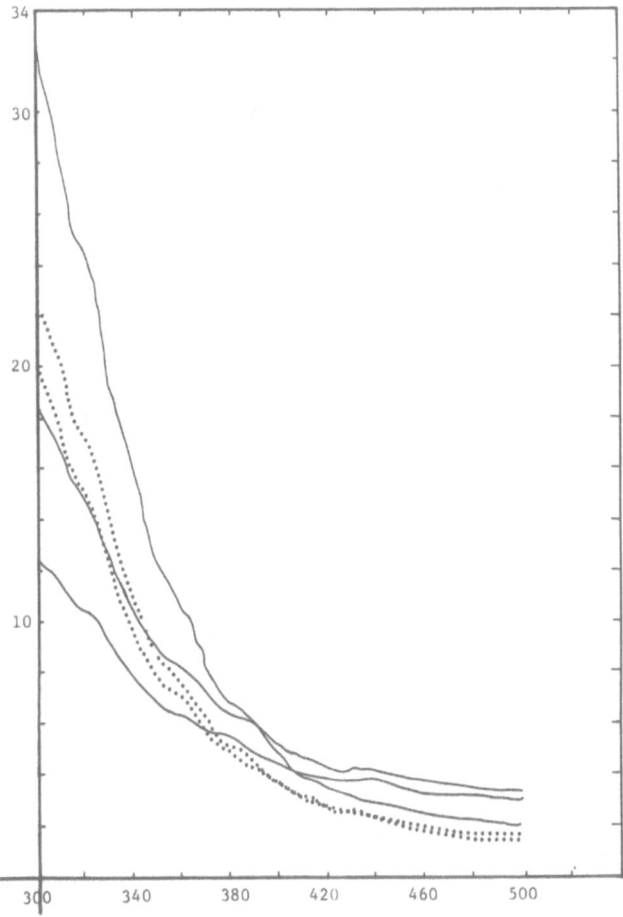

Figure 4.   UV absorption spectra (in THF) of PVC stabilized with
            DBTM and heated at 180°C. (Numbers on curves indicate
            time of heating in hours.) Ordinate: $E^{1\%}_{1\,cm}$ x 10^{-2}.
            Abscissa: λ (nm).

after 4 h. As shown in Figure 5, uptake of ^{14}C by degraded and un-
degraded polymer was identical within experimental error. This implies that the number of labile chlorine atoms in degraded PVC is
the same as in virgin polymer. Thus, in the early stages of degradation initiator sites are neither created nor destroyed. The face
that the rate of addition of free thioglycolate residues to the double

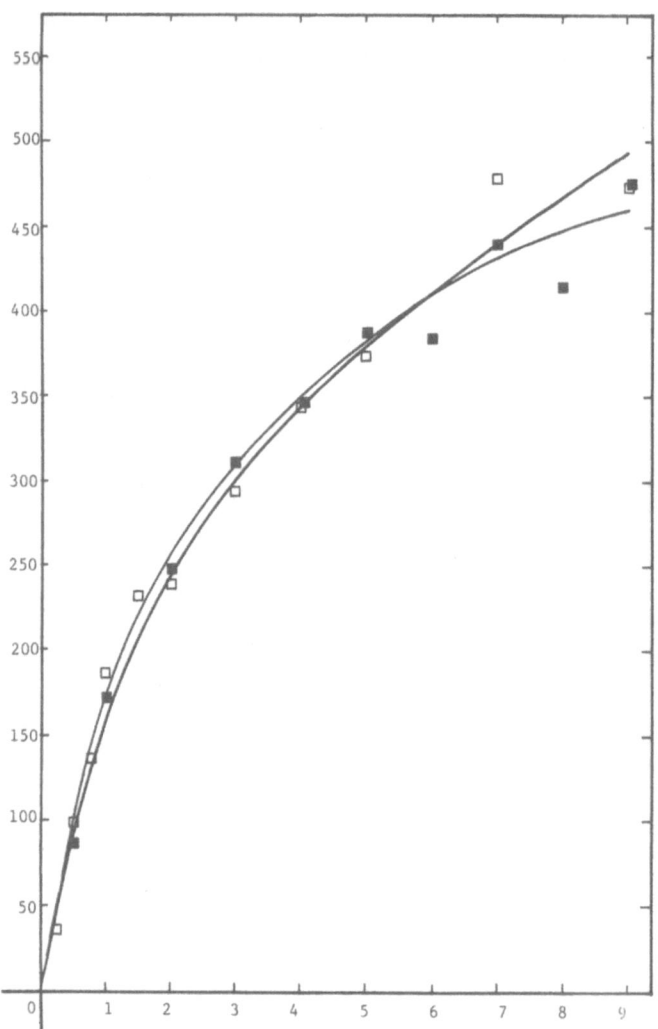

Figure 5.   Rate of uptake of radiolabelled DBTT by PVC which has
            been degraded at 180°C for 1 h (-■-) and by virgin PVC
            (-□-) on heating at 180°C for various periods. Ordinate:
            uptake of stabilizer (mol g^{-1} PVC x 10^8). Abscissa:
            heating time (h).

bonds is unchanged, even though the concentration of double bonds is increased, suggests that some other process may be rate determining. The most likely possibility is scission of Sn-S bonds by hydrogen chloride to liberate 2-ethylhexyl thioglycolate.

Similar experiments in which degraded PVC was treated with DBTM and DBTL were carried out. With DBTM the color was completely discharged in 3 h and the rate of uptake of ^{14}C was marginally increased compared with virgin PVC (Figure 6). Absence of a marked increase in rate strongly suggests that it is not the intact stabilizer which undergoes the Diels-Adler reactions but a maleate residue which has to be removed from tin in a prior rate-determining process. In the DBTL experiment the color persisted and became deeper as heating was continued confirming that, since addition reactions do not occur, this stabilizer has no healing properties. As expected, there is little difference between virgin and degraded polymer with respect to ^{14}C incorporation from DBTL (Figure 6).

MEASUREMENT OF RATES OF DEHYDROCHLORINATION OF PVC

Solutions of PVC (2.000 g) in 1,2,4-trichlorobenzene (100 cm^3) either without additive or in the presence of non-radioactive stabilizer (1.583 x 10^{-4} mol, i.e., same proportion as in previous set of experiments), were heated at 180°C in a stream of oxygen-free nitrogen using the apparatus shown in Figure 7. This allowed volatile acid to be continuously estimated to automatic titration with standard KOH solution, the pH in the titration vessel being maintained at 7.6.

In Figure 8 the expected, non-autocatalytic, rapid evolution of hydrogen chloride from PVC alone is shown and under these conditions DBTL inhibits dehydrochlorination for ~ 16 h while in the presence of DBTT there was no rapid evolution of acid, even after 36 h. When DBTM was added an unexpected increase in the rate of evolution of volatile acid relative to PVC alone was observed (Figure 9).

The apparatus was then modified (Figure 10) so that a solution of PVC (2,000 g) without additive in 1,2,4-trichlorobenzene (100 cm^3) was degraded at 180°C in a nitrogen stream and the effluent gas passed into a second flask in the same heating bath. The second flask contained the same quantity of stabilizer (1.583 x 10^{-4} mol) as used previously but now dissolved separately in an equal volume of 1,2,4-trichlorobenzene. The gas stream from the flask was then led into the titration cell as before. Figure 11 shows that when the stabilizer was DBTL or DBTT no acid was evolved for 7 h. The amount of HCl evolved from PVC alone in 7 h corresponds exactly to 2.0 mol HCl per mol of stabilizer. Thus for these two stabilizers we can not separate the HCl scavenging reactions (10) and (11) from other stabilizing reactions.

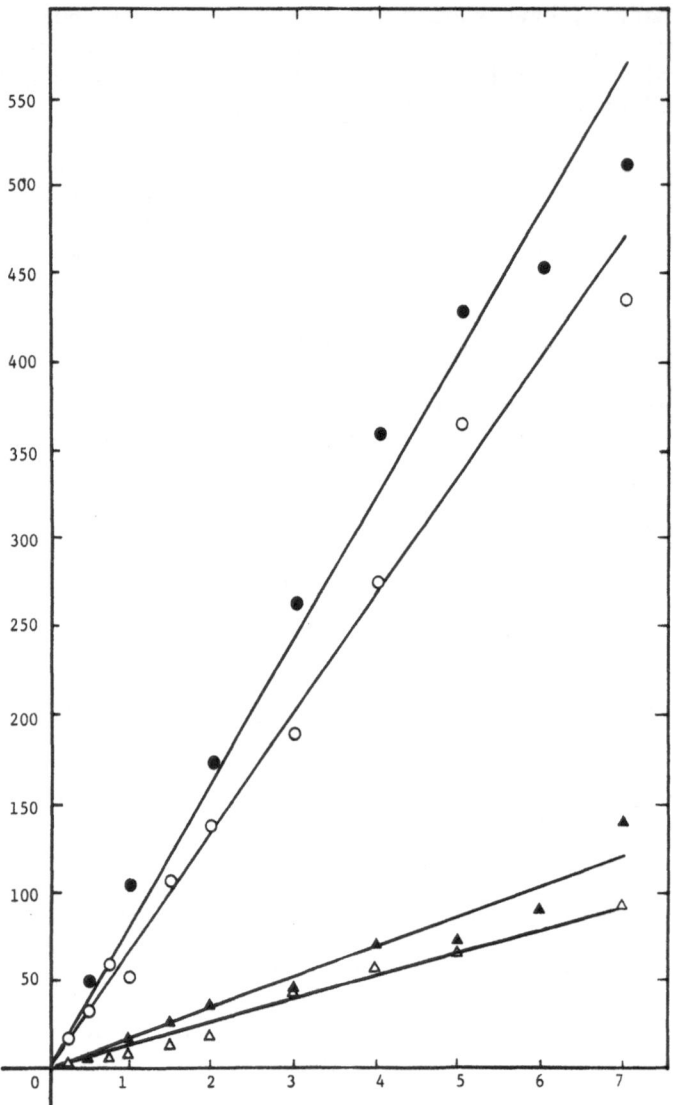

Figure 6. Rates of uptake of radiolabelled DBTL and DBTM by PVC
which has been degraded at 180°C for 1 h and by virgin
PVC on heating at 180°C for various periods (DBTL - virgin
PVC -△-, DBTL - degraded PVC -▲-, DBTM - virgin PVC -○-,
DBTM - degraded PVC -●-). Ordinate: uptake of stabilizer
(mol g^{-1} PVC x 10^8). Abscissa: heating time (h).

$$Bu_2Sn(OCOC_{11}H_{23})_2 + 2HCl \longrightarrow Bu_2SnCl_2 + 2C_{11}H_{23}COOH \qquad (11)$$

It was seen (Figure 8) that when these stabilizers were in the

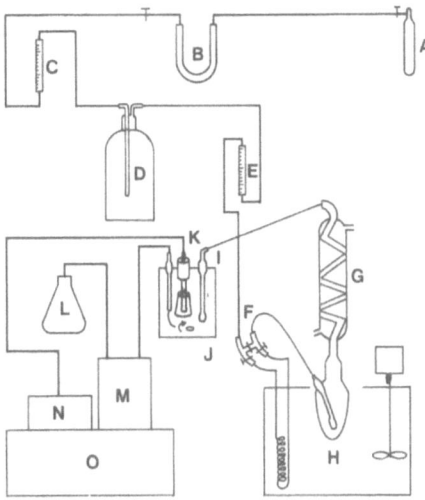

Figure 7. Apparatus used to measure the evolution of volatile acids.
Key: A - oxygen free nitrogen supply; B - "Anhydrone"
drying tube; C - flow rate meter; D - three liter ballast
tank; E - flow rate meter; F - alternate flow stopcocks
to select gas pre-heating; G - condenser; H - reaction
vessel; I - Sintred glass gas disperser; J - titration
vessel; K - pH electrodes; L - standardized NaOH solution;
M - micro dispenser; N - pH meter; O - automatic control
and printer.

presence of PVC, then evolution of HCl was suppressed for 16 h by
DBTL and for > 36 h by DBTT. In the case of DBTL, experiments with
the labelled stabilizer had shown that limited exchange occurred between labile chlorine atoms and laurate groups. It appears that this
exchange is marginally more effective in the inhibition of HCl evolution than direct reaction of HCl with the stabilizer. For DBTT
the HCl scavenging reaction is only of minor importance so the major
stabilizing processes must be exchange reactions and removal of allylic chlorine atoms by addition of $HSCH_2COOC_8H_{17}$ to double bonds.[9]

It is seen from Figure 11 that when DBTM is placed in the second
flask the flow of volatile acid to the titration cell is not impeded,
in fact, the rate of evolution of acid is more rapid than when PVC
and DBTM were in the same vessel. This suggested that DBTM itself
was decomposing to produce a volatile acidic product and this was
confirmed by passing nitrogen through a flask containing only DBTM
dissolved in 1,2,4-trichlorobenzene and titrating volatile acid as
before. Results obtained from the three different experiments with
DBTM are shown in Figure 12 and it is clear that, while the initial
rates of evolution of acid are all very similar, in the absence of
PVC or HCl the rate of evolution falls off more rapidly. To under-

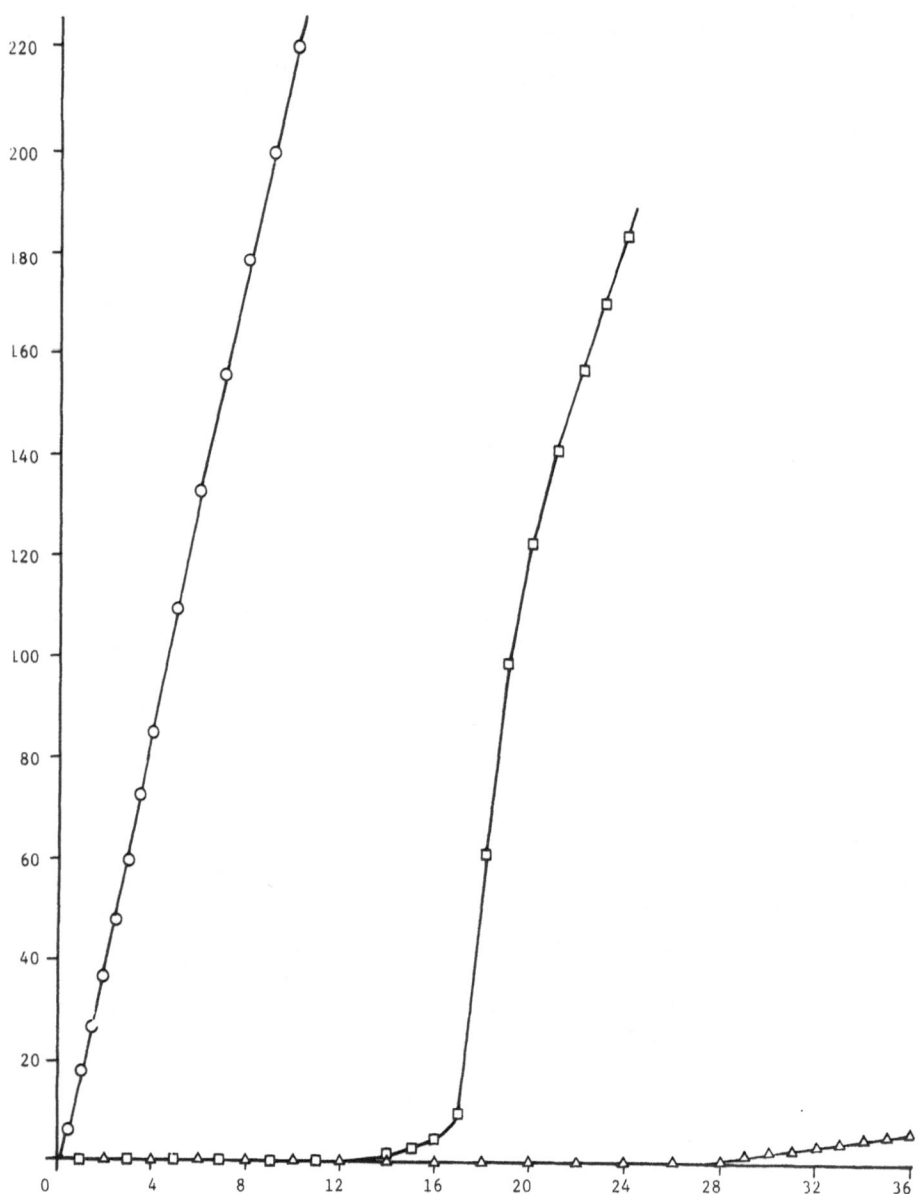

Figure 8. Evolution of volatile acid at 180°C under nitrogen from
solution of PVC alone (-O-), in the presence of DBTL
(-□-), and in the presence of DBTT (-Δ-). Ordinate:
volatile acid titrated (mol KOH g^{-1} PVC x 10^6).
Abscissa: time (h).

stand what is happening we need to know the identity of the volatile
acid produced in the thermal decomposition of DBTM and also when PVC
is involved; is this acid being titrated along or together with HCl?

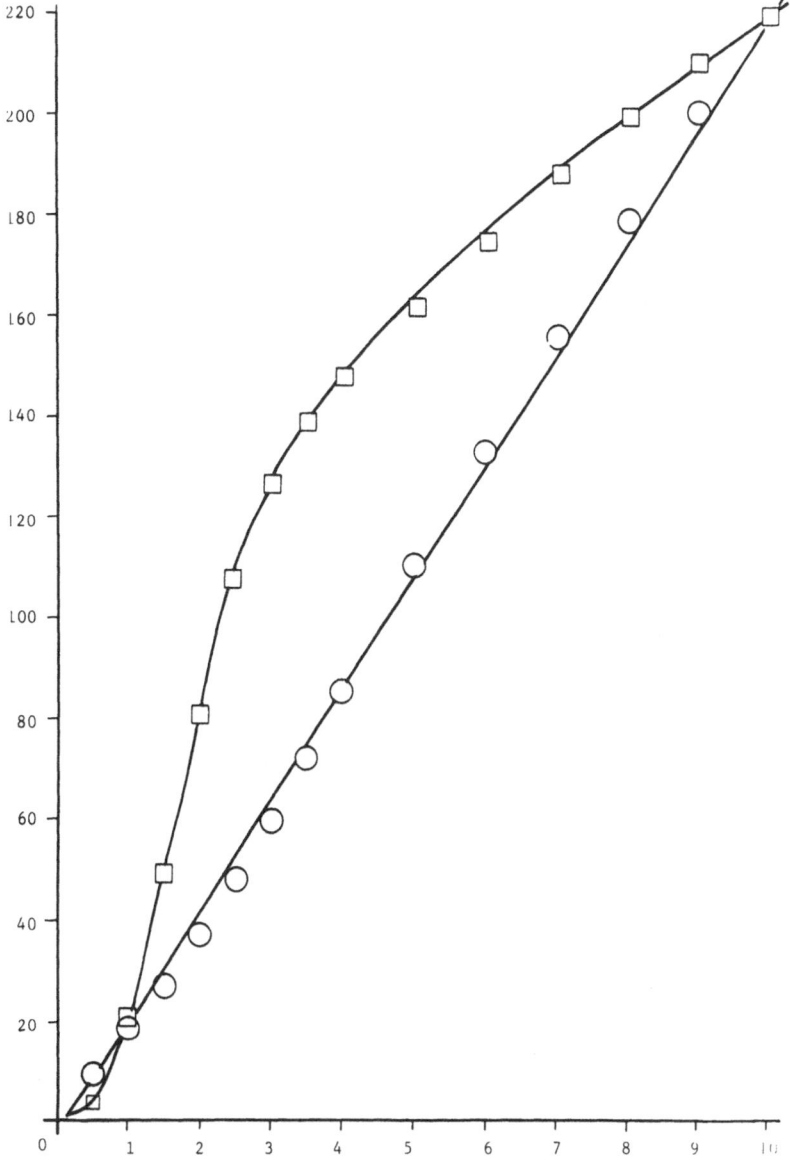

Figure 9. Evolution of volatile acid at 180°C under nitrogen from
           solutions of PVC alone (-O-) and in the presence of DBTM
           (-□-). Ordinate: volatile acid titrated (mol KOH g^{-1}
           PVC x 10^6). Abscissa: time (h).

        Examination of the three stabilizers by the DSC technique showed
the following approximate initial decomposition temperatures, DBTL
275°C, DBTT 252°C, DBTM 180°C, confirming that only DBTM would decompose thermally at the temperature of the dehydrochlorination experiments.

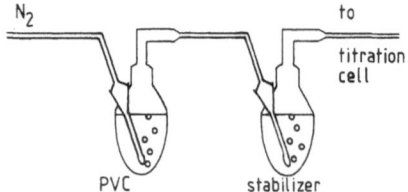

Figure 10. Modification of apparatus used to measure the evolution
            of volatile acids.

To determine the volatile products obtained in the purely thermal
decomposition, DBTM in trichlorobenzene was heated at 180°C in a
stream of nitrogen. The gas stream was passed through coils maintained at -70°C before reaching the titration cell from where it was
passed over a column of I_2O_5 at 110°C and finally through starch solution. The experiment was allowed to run for 10 h, when the following was observed: (a) no iodine was liberated from the I_2O_5 indicating no CO evolved; (b) carbon dioxide was not liberated since
no KOH was consumed in the titration cell; (c) one major product
(mixed with trichlorobenzene) was collected in the cold trap. The
mass and IR spectra together with elemental analysis showed that the
product was maleic anhydride. When the volatile acid from thermally
decomposed DBTM was titrated more than 2 mols of KOH per mol of stabilizer were consumed. Thus a reasonable possibility for the thermal
decomposition reaction is as follows:

$$
Bu_2Sn(OCOCH = CHCOOMe)_2 \longrightarrow Bu_2Sn(OMe)_2 + 2 \;
\begin{array}{c}
\text{HC} \\ \| \\ \text{HC}
\end{array}
\begin{array}{c}
\nearrow CO \searrow \\ O \\ \searrow CO \nearrow
\end{array}
\qquad (12)
$$

We are currently examining the effect of HCl on the thermal decomposition of DBTM. It appears that substantial amounts of maleic
anhydride are produced but work on this reaction is not yet complete.

SUMMARY AND CONCLUSIONS

The main results obtained can be summarized as follows:

(a) The initial rate of uptake of radiolabelled groups Y from
Bu_2SnY_2 decreases in the order DBTT > DBTM > DBTL.

(b) The relative effectiveness of the stabilizers both in de-

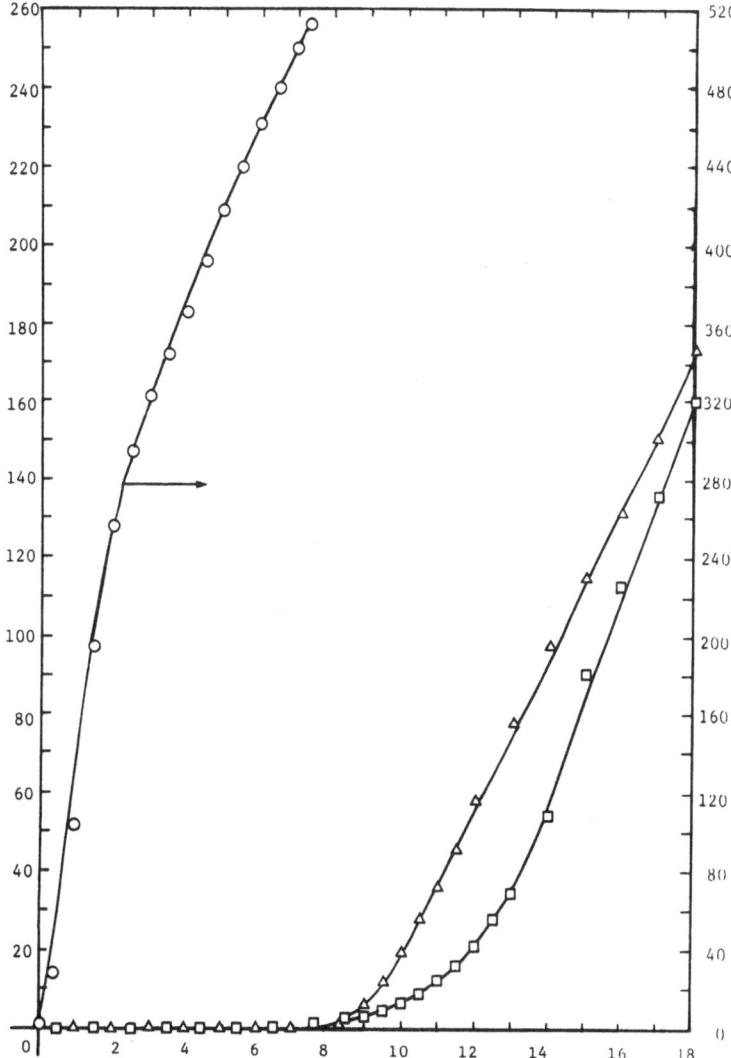

Figure 11.   Evolution of volatile acid at 180°C under nitrogen from
             solutions of PVC and stabilizer in separate flasks
             (DBTL -△-, DBTM -O-, DBTT -□-).   Left hand ordinate:
             volatile acid titrated (mol KOH g^{-1} PVC x 10^6).   Right
             hand ordinate: volatile acid titrated (mol KOH x 10^6).
             Abscissa: time (h).

laying color formation and in discharging color in previously degraded PVC is DBTM > DBTT >> DBTL.

     (c) The relative effectiveness of the stabilizers in delaying
evolution of HCl from PVC is DBTT > DBTL. DBTM was also effective,

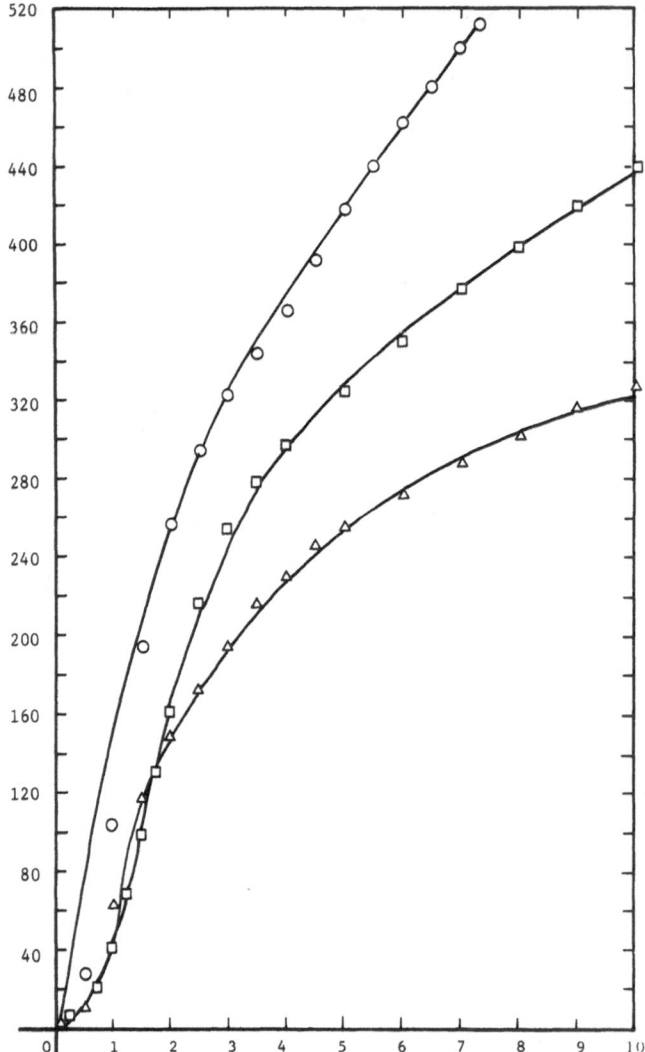

Figure 12. Comparison of volatile acid evolution at 180°C under
nitrogen (PVC + DBTM in trichlorobenzene (TCB) in the
same flask -□-, PVC in TCB and DBTM in TCB in separate
flasks -O-, DBTM alone in TCB -Δ-). Ordinate: volatile
acid titrated (mol KOH x 10^6). Abscissa: time (h).

but quantitative comparison is not yet possible.

(d) When placed in separate vessels both DBTL and DBTT quantitatively absorbed 2.0 mols of HCl from degrading PVC (this took 7 h
under our experimental conditions). When placed in the same vessel
as the PVC these stabilizers prevented HCl evolution for much longer
periods.

(e)   DBTM decomposes at ~ 180°C to give maleic anhydride.

It is concluded that exchange of Y groups with labile chlorine atoms in the polymer is important for all 3 stabilizers:

$$Bu_2SnY_2 + 2 \ \sim\!\!\sim\!CH\!=\!CHCHCl\!\sim\!\!\sim \ \longrightarrow \ Bu_2SnCl_2 \ =$$

$$2 \ \sim\!\!\sim\!CH\!=\!CHCHY \sim\!\!\sim (or \ \sim\!\!\sim\!CHYCH\!=\!CH\!\sim\!\!\sim) \tag{13}$$

A second major stabilizing reaction for all the stabilizers is scavenging of hydrogen chloride and for DBTT and DBTL this can be expressed by equation 14. The reaction of DBTM with HCl is still being investigated.

$$Bu_2SnY_2 + 2HCl \ \longrightarrow \ Bu_2SnCl_2 + 2HY \tag{14}$$

DBTL has no healing effect on degraded polymer and utilizes only reactions 13 and 14 for stabilization, these being of approximately equal importance.

For DBTT there is a third major reaction involved in stabilization, namely, the addition of thiol liberated in reaction 14 to double bonds (Equation 9). DBTM similarly utilizes a third major reaction, the addition of maleic anhydride to conjugated double bonds. The high efficiency of this last stabilizer may be due to its ability to release maleic anhydride, which can diffuse rapidly through the polymer by thermal decomposition.

REFERENCES

1. G. Ayrey, R. C. Poller and I. H. Siddiqui, J. Polymer Sci. A-1,
    (1972) 10, 725.
2. G. Ayrey, F. P. Man and R. C. Poller, J. Organomet. Chem., (1979),
    173, 171.
3. F. Alavi-Moghadam, G. Ayrey and R. C. Poller, Europ. Polymer J.,
    (1975), 11, 649.
4. F. Alavi-Moghadam, G. Ayrey and R. C. Poller, Polymer (1975),
    16, 833.
5. R. C. Poller, J. Macromol. Sci.-Chem. (1978) A12, 373.

NEW MERCAPTO ESTERS AS INTERMEDIATES FOR PVC STABILIZERS

Gerry P. Mack

Germack Chemical Corporation

Jackson Heights, NY   11372

INTRODUCTION

The recent introduction of antimony mercaptides as PVC heat stabilizers is most timely in view of recent developments in the PVC field.

First, the use of twin and multiple screw extruders for PVC pipes and other rigid forms permits the use of low levels of stabilizers on the order of 0.4% and this low level is favorable for antimony stabilizers which otherwise would have many drawbacks at high concentration levels (203%).

Secondly, the erratic price of tin which recently reached approximately $8.00/lb. with forecasts of ever higher prices made antimony an attractive metal as its price of approximately $2.00/lb. has been very stable.

Third, the final approval by N.S.F. of antimony mercaptide stabilizers at a limit of 0.4% stimulated their use for PVC pipes.

One myst remember that the antimony stabilizers on the market are basically the same as first introduced in 1954 by Weinberg et al.[1] At the time they had a short shelf life and usually in a few weeks they precipitated. This phenomenon is caused by their sensitivity to air and moisture. Antimony stabilizers are also sensitive to light and transfer to black liquid quite rapidly. Probably the following scheme summarizes this change:

(1) $Sb(10TG)_3$ $\xrightarrow{\text{UV + O}_2}$ Sb_2O_3 (white precipitate)

(2) $Sb(10TG)_3$ $\xrightarrow{\text{UV + N}_2}$ Sb_2S_3 (black sulfide)
$$Sb_2S_3$$

Recently Dieckman[2] in a U.S. patent claimed that the early
color heat performance of antimony organic sulfur containing compounds was significantly improved if they were combined with ortho-
dihydric phenols. Improvements in the long term heat stability also
are achievable, according to the patent, and in addition, the compositions are asserted to be liquids which are shelf-stable at ambient
temperatures. Dieckman points out that liquid antimony stabilizer
compositions tend to deteriorate on standing, as observed by the
formation and/or precipitation of solids in the liquid compounds
forming heterogeneous liquids, which increase the problems of measuring and mixing the antimony compounds into vinyl halide resins
for stabilization. This problem, it is asserted, is overcome by
incorporating the ortho-dihydric phenols with the liquid antimony
stabilizer. In these combinations, metal carboxylates, and particularly calcium stearate, can also be incorporated to achieve the
advantages of the previously issued Dieckman patent.

A number of patents have suggested the use of antimony compounds, particularly sulfur-containing compounds such as the antimony mercaptides. These include U.S. Patents 2,680,726; 2,684,956;
3,340,285; 3,999,220; 3,466,261; and 3,530,158. These patents dis-
close various types of organic sulfur-containing antimony compounds,
but none have been adequate in inhibiting the development of an
early yellow discoloration.

German Patent 1,114,808 to Deutsche Advance proposed antimony
compounds of the formula:

$$(XS)_2SbS(CH_2)_xCOO-A-COO(CH_2)_xSSb(SX)_2$$

where x is an integer from 1 to 4, A an alkylene residue of up to
ten carbon atoms, with or without OH groups, or merely -a bond, and
SX is the residue (having from eight to eighteen carbon atoms) of
an aliphatic of aromatic mercaptan, or of an ester of a thioalcohol
of thio acid, as stabilizers for polyvinyl halide resins.

Chemische Werke Barlocher British Patent 1,194,414, published
June 10, 1970, suggested antimony compounds of the formula:

$$\begin{matrix} R^1 \\ (R^2)_2 \end{matrix} Sn-S-CH_2-COO-CH_2-CH_2-OOC-CH_2-S-Sb \begin{matrix} S-R^3 \\ S-R^4 \end{matrix}$$

East German Patent No. 71,360 patented Feb. 20, 1970, suggested antimony mercaptocarboxylate esters such as Sb-tris (2-ethylhexyl-thioglycolate), used together with the corresponding organotin mercapto carboxylic ester.

As the references cited indicate, though there has been considerable patent activity on antimony stabilizers, they are mostly a variation of an antimony mercapto-structure described in the Weinberg[1] patent.

In view of the above, it was decided to try to develop a new antimony mercapto-structure. As the author in the past had developed and used oxathiolanes as synergists for organotin stabilizers to lower the cost of such stabilizers, it was decided that perhaps such structures would lead to a new type of antimony stabilizer. Incidentally, these structures were also useful in preparing new types of organotin stabilizers which are described in a patent issued to Mack.[3]

These authors showed that aldehydes would condense with mercapto-acids to obtain the 1,3-oxathiolan-5-one derivative and a bis-thio glycollic acid derivative and described several of these compounds. They proposed the following reaction:

$$RCHO + HSCH_2COOH \longrightarrow RCH \begin{array}{c} S-CH_2 \\ | \\ O-CO \end{array} + RCH \begin{array}{c} SCH_2COOH \\ \\ SCH_2COOH \end{array}$$

Such derivatives have been made from various aldehydes, but as they had a strong and in some cases a foul odor, the stabilizers, although they had good heat stability, were not acceptable.

In view of the odoriferous nature of the aldehyde compounds, synthesis based on ketones was carried out and these thiolanes had a much more pleasant odor. Work was concentrated on the cyclohexanone condensation with thioglycollic acid.

This research led to discovery of a new reaction of substituted cyclic thiolactones with mercaptides which results in the opening of the lactone ring, and the formation of the corresponding mercaptocarboxylic acid. The organic radical of the mercaptide becomes attached to the keto carbon atom of the starting aldehyde or ketone. This reaction is unique and literature or patent references have not been found.

As will be discussed later through a further new synthesis, these percusor mercaptocarboxylic acids are esterified with glycol

or higher polyglycols to give th polyol monoesters. By further esterification of the free alcohol group of the polyol monoester with
a mercaptocarboxylic acid, the corresponding ester with a free mercapto group is formed.

The properties and performance of these new stabilizers in PVC
systems will be discussed in this paper.

EXPERIMENTAL

The new Sb stabilizers were synthesized in five reaction steps:

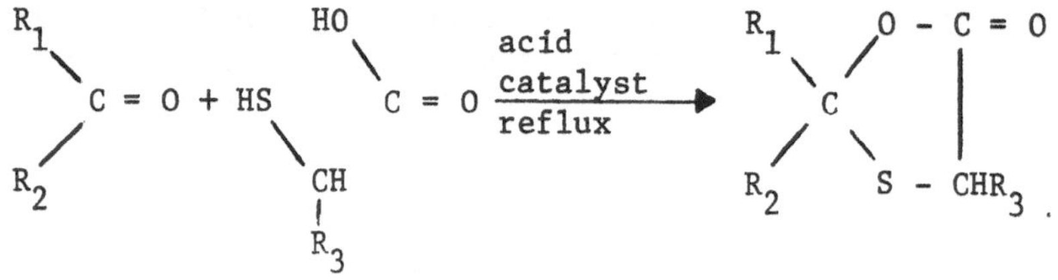

R_1, R_2, and R_3 = hydrocarbon group, and R_1 and R_3 can also be hydrogen.

Next, there is added one mole of a mercaptide and an acid catalyst (having a minimum ionization constant k_{ion} in water at least
10^{-4}) if an acid catalyst is not used in the first reaction, and
which can be an α- or β-mercaptocarboxylic acid ester, mercaptoalcohol ester, or aryl mercaptan R_4SH, and reaction is continued
under reflux for several hours to form the corresponding acid.
This reaction proceeded with opening of the lactone ring, and the
formation of the corresponding mercaptocarboxylic acid, with the
organic radical of the mercaptide becoming attached to the keto
carbon atom of the starting aldehyde or ketone:

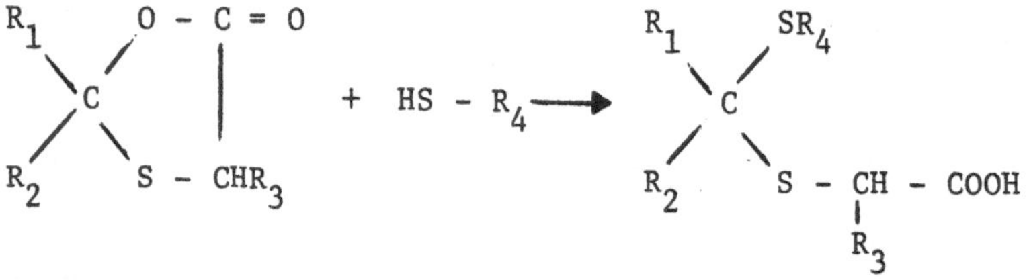

R_4 = residue of the mercaptide. The mercaptocarboxylic acid was
esterified with a glycol or higher polyol to give the polyol monoester:

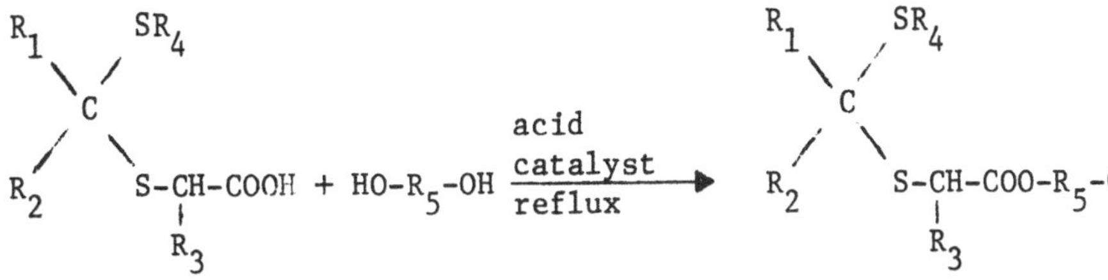

$$R_1, R_2 > C < SR_4, S-CH(R_3)-COOH + HO-R_5-OH \xrightarrow[reflux]{acid\ catalyst} R_1, R_2 > C < SR_4, S-CH(R_3)-COO-R_5-OH$$

3

where R_5 = divalent aliphatic or cycloaliphatic residue from polyol.

The synthesis of the mercaptocarboxylic acid ester is completed by the esterification of the polyol monoester with the same or another mercaptocarboxylic acid:

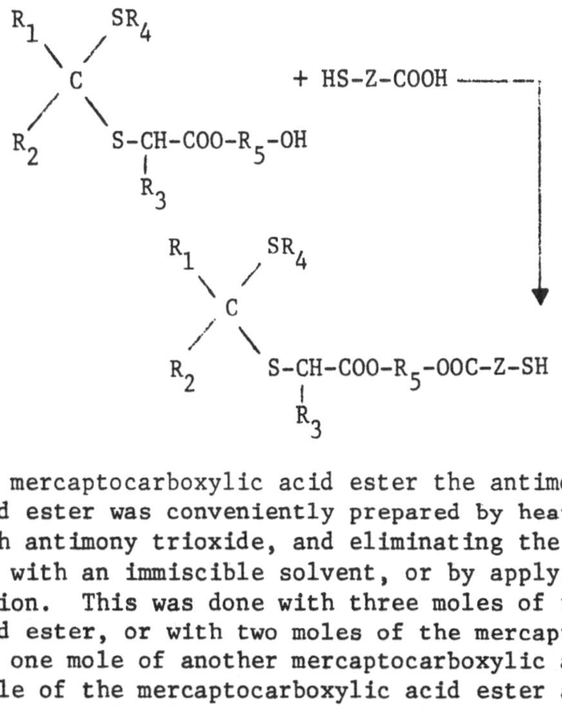

$$R_1, R_2 > C < SR_4, S-CH(R_3)-COO-R_5-OH \quad + HS-Z-COOH \dashrightarrow$$

4

$$R_1, R_2 > C < SR_4, S-CH(R_3)-COO-R_5-OOC-Z-SH$$

From this mercaptocarboxylic acid ester the antimony mercapto-carboxylic acid ester was conveniently prepared by heating the mercaptoester with antimony trioxide, and eliminating the reaction water by azeotroping with an immiscible solvent, or by applying vacuum to the warm solution. This was done with three moles of the mercapto-carboxylic acid ester, or with two moles of the mercaptocarboxylic acid ester and one mole of another mercaptocarboxylic acid ester, or with one mole of the mercaptocarboxylic acid ester and two moles of another mercaptocarboxylic acid ester. The three reactions form-ing these three kinds of antimony compounds are as follows:

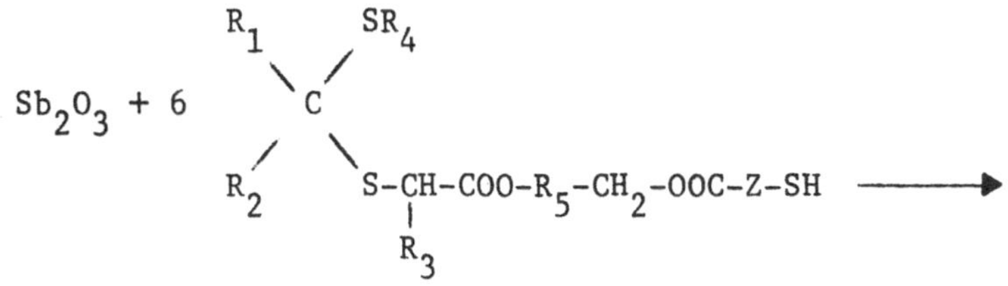

$$Sb_2O_3 + 6 \quad R_1, R_2 > C < SR_4, S-CH(R_3)-COO-R_5-CH_2-OOC-Z-SH \longrightarrow$$

$$\longrightarrow 3H_2O + 2Sb-\left[S-Z-COO-R_5-OOC-CH-S\underset{R_3}{\overset{}{|}}\text{---}C\overset{R_1}{\underset{R_2}{\diagdown}}\overset{SR_4}{\diagup}\right]_3 \quad 5a$$

$$Sb_2O_3 + 4\ \ R_2\overset{R_1}{\underset{S-CH-COO-R_5-OOC-Z-SH}{\diagdown}}C\overset{SR_4}{\diagup}\quad + 2\ HS-Z-COOR_6 \longrightarrow$$
$$\underset{R_3}{}$$

$$\longrightarrow 3H_2O + 2\ R_6OOC-Z-S-Sb-\left[S-Z-COO-R_5-OOC-\underset{R_3}{\overset{}{C}H}-S\text{---}C\overset{R_1}{\underset{R_2}{\diagdown}}\overset{SR_4}{\diagup}\right]_2 \quad 5b$$

$$Sb_2O_3 + 2\ \ R_2\overset{R_1}{\underset{S-CH-COO-R_5-OOC-Z-SH}{\diagdown}}C\overset{SR_4}{\diagup}\quad + 4\ HS-Z-COOR_6 \longrightarrow$$
$$\underset{R_3}{}$$

$$\longrightarrow 3H_2O + 2\left[R_6OOC-Z-S\right]_2-Sb-S-Z-COO-R_5-OOC-\underset{R_3}{\overset{}{C}H}-S\text{---}C\overset{R_1}{\underset{R_2}{\diagdown}}\overset{SR_4}{\diagup} \quad 5c$$

In reaction (1), the mercaptocarboxylic acid must have the mercapto group in the α-position to the carboxylic acid group. β-Mercaptocarboxylic acids do not yield the desired reaction product.

Reaction (2) is believed to be new, and so is the synthesis including it in combination with reactions (1) and (3) to (5).

No reference to reaction (2) has been found in the literature. However, there is significant evidence that the reaction proceeds in this way, inasmuch as the titratable acidity of the reaction mixture increases in the course of the reaction, indicating that the reactants which are nonacidic react to form a titratable acid. At the same time, the disappearance of the titratable mercapto function -SH of the mercaptocarboxylic acid ester starting material is also confirmed by analysis. Infra-red and proton magnetic resonance spectra are consistent with the structure indicated of the reaction product.

In reaction (3) any polyol can be used, and in reaction (4) any mercaptocarboxylic acid, including not only mercaptocarboxylic acids but also mercaptocarboxylic acid esters, and mercaptoalcohol carboxylic acids and esters having a free mercapto group.

The reaction products from steps (2), (3) and (4) of this synthesis are predecessor intermediates not only to the antimony mercaptocarboxylic acid esters but also to other monovalent and polyvalent metal mercaptocarboxylic acid ester salts, such as the Sn, Ca, Zn, Cd, Mg, Ba, Pb, and Li compounds, and are further considered to be new compounds.

RESULTS AND DISCUSSION

The following new Sb stabilizers were prepared according to procedures described above and their performance in the PVC systems was evaluated:

$$Sb \left[S-CH_2-COO-CH_2CH_2-OOC-CH-S \begin{array}{c} iso-C_8H_{17}-OOC-CH_2-S \\ \\ CH_3 \end{array} \bigcirc S \right]_3 \qquad I$$

$$\left[\bigcirc S \begin{array}{c} S-CH_2-COO-C_8H_{17}-iso \\ \\ S-CH_2-COO-CH_2CH_2-OOC-CH_2S \end{array} \right]_2 -Sb-S-CH_2-COO-C_8H_{17}-iso \qquad II$$

$$
\underset{\text{S-CH}_2\text{-COO-CH}_2\text{CH}_2\text{-OOC-CH}_2\text{-S-Sb}[\text{-S-CH}_2\text{-COO-C}_8\text{H}_{17}\text{-iso}]_2}{\overset{\text{S-CH}_2\text{-COO-C}_8\text{H}_{17}\text{-iso}}{\bigcirc\hspace{-1.5em}\times}} \qquad \text{III}
$$

$$
\text{Sb-}\left[\begin{array}{l} \text{iso-C}_8\text{H}_{17}\text{-OOC-CH}_2\text{- S} \\ \text{S-CH}_2\text{-COO-CH}_2\text{CH}_2\text{-OOC-CH}_2\text{-S} \end{array}\right]_3 \qquad \text{IV}
$$

$$
\text{Sb-}\left[\begin{array}{l} \text{iso-C}_8\text{H}_{17}\text{-OOC-CH}_2\text{-S}\diagdown \hspace{1em} \text{CH}_3 \\ \hspace{6em} \text{C} \\ \text{S-CH}_2\text{-COO-CH}_2\text{CH}_2\text{-OOC-CH}_2\text{-S}\diagup \hspace{0.5em} \text{C}_4\text{H}_9\text{-iso} \end{array}\right]_3
$$

An unpigmented rigid, i.e., nonplasticized, polyvinyl chloride resin formulation was prepared having the following composition:

Component	Parts by Weight Example I
Polyvinyl chloride resin polymer (Diamond 40)	100
Calcium stearate	0.6
Wax 160 (160°F m.p. paraffin)	0.1
Low molecular weight polyethylene	0.1
Stabilizers (as shown in Table I)	(as in Table I)

The stabilizer was mixed in the resin in the proportion indicated in Table I on a two-roll mill to form a homogeneous sheet and sheeted off. Strips were cut from the sheet and heated in an oven at 175°C for up to ninety minutes. Pieces of each strip were removed at fifteen minute intervals, and affixed to cards, to show the progressive development of the discoloration. During the first fifteen to thirty minutes of heating early discoloration manifests itself. After thirty minutes of heating, long term heat stability can be observed.

Color of each sample was rated in accordance with color scale "A" as follows:

Table I

Example No.	Stabilizer	% Sb	Parts per 100 parts resin	Initial unexposed color	After (minutes)					
					15	30	45	60	75	90
Control 1	Antimony tris (iso-octyl thio-glycolate)	16.7	0.4	0	1	2	3	4	5	7
1)	I	14.5	0.4	0	1	1	2	2	3	3
2)	II	11.8	0.4	0	1	1	1	2	2	2
3)	II	11.8	0.3	0	1	1	2	2	3	3
Control 2	Antimony tris (iso-octyl thio-glycolate) t-butyl catechol	16.7	0.4	0	1	1	2	3	4	6
4)	II + t-butyl-catechol	11.8	0.3 / 0.02	0	1	1	1	2	2	5
5)	I + t-butyl-catechol	14.5	0.4 / 0.02	0	1	1	1	2	2	5

	Scale A
0 - Clear and colorless	5 - Light amber
1 - Trace of color	6 - Medium dark yellow
2 - Very light tan	7 - Dark amber
3 - Light tan	8 - Green-brown
4 - Very light amber	9 - Green-black

The stabilizers used in each resin composition and the results obtained are shown in Table I.

Despite a lower antimony content by weight, each of the stabilizers imparts better early color and long term stability than the Control antimony stabilizer, with or without added t-butyl-catechol.

Rigid pigmented polyvinyl chloride resin formulations were prepared having the following composition:

Component	Parts by Weight
Polyvinyl chloride resin homopolymer (Diamond 40)	100
Titanium dioxide (pigment)	1
Calcium carbonate	0.6
Calcium stearate	1
Acrylic processing aid	1
Wax 160 (160°F m.p. paraffin)	1
Low molecular weight polyethylene	0.2
Stabilizer (as shown in Table II)	0.4

The stabilizer was mixed in the resin in the proportion indicated in Table on a two-roll mill to form a homogeneous sheet, and sheeted off. Strips were cut from the sheet and heated in an oven at 175°C to determine the onset of early discoloration during the first stages of heating. Pieces of each strip were removed at fifteen minute intervals, and affixed to cards, to show the progressive development of the discoloration for the first fifteen to thirty minutes. The effect on long term heat stability was determined by continuing the test for 120 minutes.

The development of early discoloration is evaluated by the intensity of tint formed. The observed discoloration of each sample is described in Table II. The following abbreviations are used: w = white; ow = off white; c = cream; t = tan; b = brown; cc, tt and bb indicate darker shades of cream, tan and brown, respectively.

Table II

Example No.	Stabilizer	% Sb	Initial unexposed color	After (minutes) 15	30	45	60	75	90	105	120
Control 3	Antimony tris (iso-octyl thioglycolate)	16.7	w	ow	ow	t	tt	tt	b	b	b
6)	II	11.8	w	w	w	c	cc	t	t	tt	tt
7)	III	12.64	w	w	w	c	cc	t	t	tt	tt
8)	IV	7.59	w	w	w	c	cc	t	t	tt	tt
9)	V	10.90	w	w	ow	c	cc	t	t	tt	tt

The results show that each of the new antimony stabilizers imparts better heat stability, both in preventing early discoloration and in lessening the severity of discoloration in the long term, than antimony tris (isooctyl thioglycolate). Note that the product #2 stabilizer which is a methyl isobutyl ketone derivative with only 10.90% antimony had superior early color retention to the other stabilizers with higher antimony content.

Finally, rigid pigmented polyvinyl chloride resin formulations were prepared having the following composition:

Component	Parts by Weight
Polyvinyl chloride resin homopolymer (Diamond 40)	100
Titanium dioxide (pigment)	1
Calcium carbonate	0.6
Calcium stearate	1
Acrylic processing aid	1
Wax 160 (160°F m.p. paraffin)	1
Low molecular weight polyethylene	0.2
Catechol	0.02
Stabilizer (as in Table III)	0.4

The preparation and testing of samples was carried out as described above. The results are summarized in Table III.

SUMMARY

The results show that each of the new antimony stabilizers imparts better heat stability, both in preventing early discoloration and in lessening the severity of discoloration in the long term, than antimony tris (isooctyl thioglycolate). The response to catechol addition also differs. With antimony tris (isooctyl thioclycolate) the improved resistance to early discoloration resulting from catechol addition is compromised by more severe discoloration later, while with the new Sb stabilizers catechol addition is helpful both to early color control and to long term heat stability.

ACKNOWLEDGEMENTS

The author thanks Argus Chemical Corp., a subsidiary of Witco Chemical Corp., for sponsoring this research. Special thanks go to Mrs. Dina Sneddon for her assistance in preparing this paper.

Table III

Example No.	Stabilizer	% Sb	Initial unexposed color	After (minutes)							
				15	30	45	60	75	90	105	120
Control A	Antimony tris(iso-octyl thio-glycolate)	16.7	w	w	w	c	t	tt	b	bb	bb
10)	II	11.8	w	w	w	ow	c	cc	cc	t	t
11)	III	12.64	w	w	w	ow	c	cc	cc	t	t
12)	IV	7.59	w	w	w	ow	c	cc	t	t	
13)	V	10.90	w	w	w	c	cc	cc	t	t	t

REFERENCES

1. G. P. Mack, U.S. Patent 4,269,731 (May 26, 1981).
2. D. Dieckman, U.S. Patent 3,887,508 (June 3, 1975).
3. E. Weinberg, U.S. Patent 2,680,727 (June, 1956).

THE EFFECTS OF BROMINE AND ANTIMONY-CONTAINING FLAME RETARDANTS

ON THE BURNING CHARACTERISTICS OF POLYETHYLENE

Y. -L. Hsieh and K. Yeh*

Division of Textiles & Clothing
University of California
Davis, CA
*Dept. of Textiles & Consumer Economics
University of Maryland
College Park, MD

INTRODUCTION

The combinations of antimony trioxide with a halogen source are often described as being synergistic and have been an effective and widely used flame retardant system in polymeric materials.[6] In the Sb_2O_3-halogen systems, the primary site of retardant action was thought to be in the gas phase. A major function of the halogen has been thought to transport the relatively non-volatile antimony to the gas phase. The volatile species produced in these systems have been generally agreed to be antimony halides (Sb_2X_3 or probably as SbX when insufficient halogen is used)[5,6,9,10,11] where antimony halides decompose and inhibit the flame.

Sb_2O_3 in combination with a chlorine source in various polymers was found to be more effective than either component alone and greater than the sum of effects of the two components.[2,4,7,8] Based on reported thermo-analytical data, interaction between Sb_2O_3 and various chlorinated paraffins,[5,13] polyvinyl chloride[5] and perchloropentacyclodecane[5,13] were found to be in the condensed phase, resulting in volatile halogen and antimony species. A 3 to 1 atomic ratio, or a 2 to 1 weight ratio, has been found to be most efficient for polyethylene,[1,7,11] polystyrene and ABS[11] based on oxygen index results.

The research reported herein deals with the effects of antimony trioxide and decabromodiphenyl oxide on the burning characteristics of polyethylene films. Decabromodiphenyl oxide (DBDPO) has wide

application as flame retardant in textiles (polyester) and thermoplastic resins (polyester, epoxy and phenolics).[11] A new technique
has been developed for the evaluation of the burning characteristics
of polymers in a controlled environment. The additive effects, both
individual and combined effects of these two components, are studied.
Retardation mechanisms of these systems are also investigated and
compared.

EXPERIMENTAL

A new technique developed for the evaluation of the burning
behavior of polymeric materials was employed for this study.[3] A
horizontal apparatus was designed and constructed to carry out burning experiments in a controlled environment (Figure 1). Fundamental
burning characteristics such as total heat release and carbon monoxide and carbon dioxide generation were selected for investigation
with this technique due to their significance as creteria for evaluating combustion mechanisms of polymeric materials.

The temperature measurement for total heat release was detected
with a four-junction chromel-constantan thermocouple. Propane gas
of 99% purity was burned as the heat source for heat calibration.
A previously calibrated flowmeter was used to monitor the supply of
propane. Assuming complete combustion of propane gas, the heat rates
and total heat produced were calculated.

Total heat in the range of 1 to 19 kilo-calories was produced

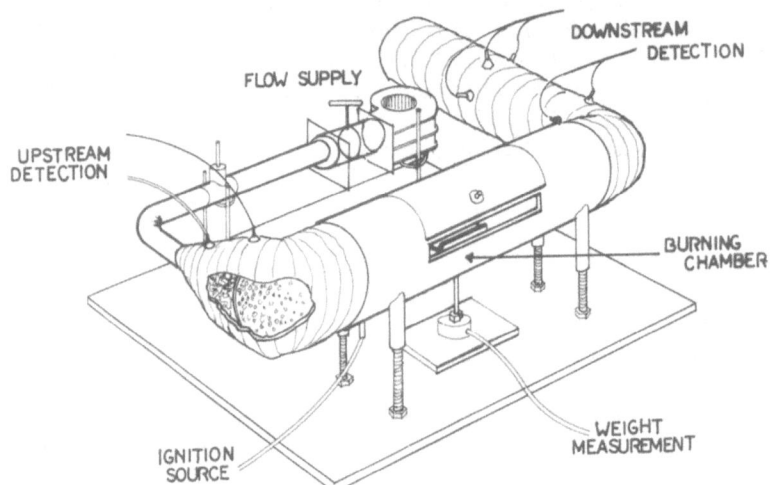

Figure 1.   Horizontal apparatus designed and constructed for
            burning experimentation.

from various burning times at heat rates of 33, 45 and 58 calories per second. Linear regression of the total heat data versus the total areas under the temperature-time curves resulted in the heat conversion as:

$$\text{Heat (cal)} = 4.64 \ (\pm \ 2.9\%) \ \text{x Area (mv} \cdot \text{s)} + 1649.1 \qquad (1)$$

Carbon monoxide and carbon dioxide generations were continuously measured by two Beckman infrared analyzers. Four standard $CO/CO_2/N_2$ mixtures of known concentrations were used for both concentration and volume calibration. The conversions for volumes from the integration of the potentiometric outputs were obtained from the linear regression of calibration data as follows:

$$\text{CO volume (cc)} = 205.98 \ \text{x Integration} \qquad (2)$$

$$CO_2 \text{ volume (cc)} = 415.52 \ \text{x Integration} \qquad (3)$$

All potentiometric outputs from the detection of parameter measurements were processed and recorded by a multi-channel Data Acquisition System. The data were read, manipulated and analyzed on a computer. All error terms for calibration conversion equations in this apparatus were approximately 3.5% or less. The error in the entire measurement system was estimated to be approximately 6 to 8 per cent.

Materials studied were 100% polyethylene and three groups of additive incorporated polyethylene. Decabromodiphenyl oxide (DBDPO) and antimony trioxide (Sb_2O_3) were added both individually and in combination in polyethylene at several levels. In the DBDPO and Sb_2O_3 combined system, $DBDPO/Sb_2O_3$ weight ratio was held at a 2 to 1 (or a 3 to 1 atomic ratio) ratio (Table I). Polymer specimens were prepared by a high-pressure melt molding technique to sheets

Table I. Burning Characteristics of Additive Polyethylene

SPECIMEN	HEAT RELEASE (CAL/G)	CARBON MONOXIDE (CC)	CARBON DIOXIDE (CC)
POLYETHYLENE	8834 (4.4%)	52 (4.4%)	2084 (7.2%)
Sb_2O_3 - 2.5%	8582 (2.6%)	51 (11.2%)	2014 (17.7%)
4.6%	8653 (1.7%)	44 (22.0%)	1759 (3.3%)
6.5%	8504 (3.1%)	48 (17.1%)	2041 (7.7%)
DBDPO - 5.1%	7562 (4.0%)	104 (7.8%)	1967 (14.4%)
9.3%	7255 (2.3%)	137 (9.2%)	1790 (4.1%)
12.9%	6524 (4.4%)	174 (14.7%)	1686 (16.7%)
14.8%	6116 (3.9%)	173 (15.2%)	1422 (12.9%)
Sb_2O_3/DBDPO 0.6%/1.3%	7115 (11.2%)	261 (12.8%)	1804 (3.9%)
0.98%/1.97%	6433 (11.7%)	539 (10.8%)	1794 (3.7%)
1.5%/3.0%	3238 (19.9%)	1811 (29.8%)	1481 (17.7%)

of 1 mm in thickness and 2.5 x 10 cm in dimension. Depending upon
the composition, sample weight ranged from about 1.3 to 2.1 grams.

Ignition was provided with 99% purity propane at a heat rate
of 107 cal/s. Triplicate burning was performed for each specimen.
The complete heat of combustion for the original specimens and their
burned residues was measured by an adibatic bomb calorimeter. Elemental analysis was performed on residue from selected additive incorporated specimens.

RESULTS AND DISCUSSION

The calibration of measurement parameter and the material burning studies were all conducted at room atmosphere and temperature
with a systematic gaseous flow of 125 liters per minute. Calibration
equations were used to convert all potentiometric output into parameter measurement.

Heat Release

Total heat release from burning polymeric materials was obtained
by converting the total areas under the temperature-time curves using
Equation 1. Ignition heat was deducted from the heat calculated.
The net heat data were then normalized by the total sample weight
consumed (Table I) and plotted versus the additive contents (Figure
2).

Heat release data from samples treated with Sb_2O_3 at all three
levels were slightly lower than that of the pure polyethylene, yet
within the standard deviation of the polyethylene control. Antimony
trioxide was considered not to have any effect on the heat release
of polyethylene. Decabromodiphenyl oxide, on the other hand, intro-

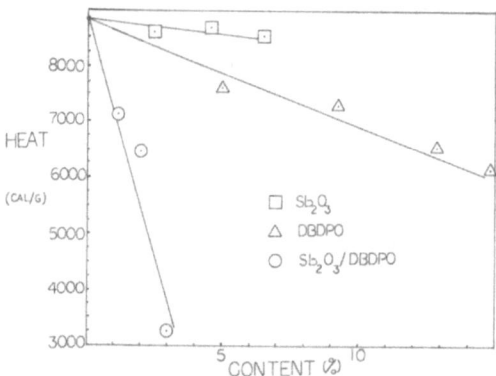

Figure 2. Heat generation.

duced a linear reduction on the heat release of polyethylene as the additive contents increased. In the Sb_2O_3 and DBDPO combined system, tremendous reduction of heat release was observed at significantly lower contents of either components. A more than additive effect of these two components became apparent, and the synergistic effect from the heat reduction data seems to be obvious.

Carbon Dioxide Generation

Total carbon dioxide volumes generated from burning were derived from the integration values using the conversion Equation 3. Correction of carbon dioxide generated from the ignition source was made, and net CO_2 volume per gram sample consumed was calculated (Table I).

Carbon dioxide formation from the addition of Sb_2O_3 was expected to be the same as that of the polyethylene control, since no effect was observed on the heat release data of the same specimens. The much lower CO_2 generation at 4.6% Sb_2O_3 content was suspected to be an experimental error. For DBDPO, a gradual decrease in CO_2 generation was observed as the additive contents increased (Figure 3). The patterns of decreasing CO_2 formation with increasing DBDPO contents were found parallel to those observed in the heat release data. A more profound decrease in the CO_2 generation was shown in the Sb_2O_3/DBDPO combined system.

Carbon Monoxide Generation

Total volumes of CO generation during burning were converted from the integration values using Equation 2. No correction for CO volume from the ignition was necessary. Antimony trioxide did not exert any effect, whereas DBDPO increased CO formation linearly with increased additive contents. A very drastic increase in the amount of CO was also observed with the combined system (Figure 4).

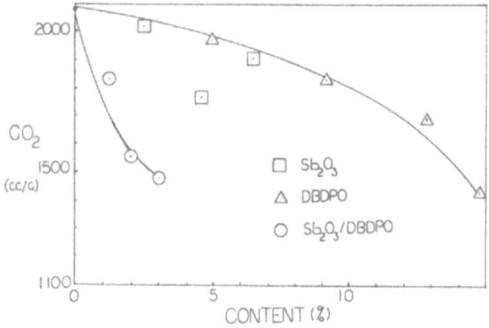

Figure 3. CO_2 generation.

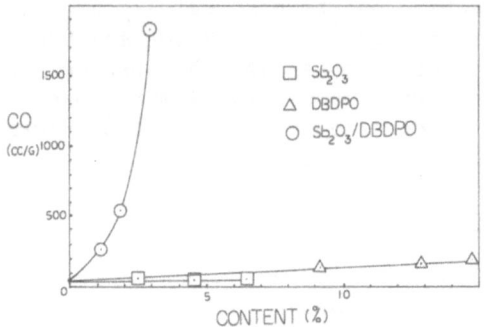

Figure 4.   CO generation.

Correlation and Retardant Mechanism

The increase in the incomplete combustion of polyethylene in the DBDPO and DBDPO/Sb_2O_3 systems was demonstrated both by the deduction of heat release and the increases in the CO generation with increasing additive contents. Assumption was made that the heat release reduction resulted from the increased incomplete combustion. Combustibles generated from the incomplete combustion of polyethylene included CO, fragments of the polymer chains and the additives, interaction of these fragments and/or additives and smoke. The modes of promoting incomplete combustion by DBDPO and DBDPO/Sb_2O_3 systems could be elucidated partially from the proportion of CO formation in the total incomplete combustion.

Values based on per gram "polyethylene" in the consumed sample were used for comparison. The net heat reduction, $\delta(\Delta H_1)$, was calculated for each content level in these two systems as (Table II):

$$\delta(\Delta H_1) = (\Delta H_1)_{polyethylene} - (\Delta H_1)_{additive} \qquad (4)$$

The increase in CO generation was also calcualated as:

$$\delta(Volume_{CO}) = (V_{CO})_{additive} - (V_{CO})_{control} \qquad (5)$$

Carbon monoxide, with a heat of combustion of 67.636 kcal/mole and a density of 1.25 g/1, is calculated to have a heat of combustion of 3.02 cal/cc. The heat release reduction contributed by the increase of CO generation was thus calculated to be:

$$\delta(\Delta H_{CO}) = 3.02 \text{ (cal/cc)} \times \delta(Volume_{CO}) \qquad (6)$$

The proportion of carbon monoxide in the incomplete combustion products was described in per cent as: $\delta(\Delta H_{CO})/\delta(\Delta H_1)$.

Table II.   Heat Release Reduction and CO Generation

SPECIMEN	$\delta(\Delta H_l)^a$ (cal/g)	$\delta(V_{CO})^b$ (cc/g)	$\delta(\Delta H_{CO})^c$ (cal/g)	$\delta(\Delta H_{CO})/\delta(\Delta H_l)$ (%)
DBDPO - 5.1 %	920	58	174	18.9
9.3 %	943	99	299	31.7
12.9 %	1492	147	444	29.7
14.7 %	1831	151	455	24.9
Sb_2O_3/DBDPO				
0.6%/1.3%	1595	160	641	40.2
0.98%/1.97%	2243	501	1512	67.4
1.5%/3.0%	5475	1784	5547	100.0

a calculated from equation 4
b calculated form equation 5
c calculated from equation 6

Earlier data indicated increased CO formation with increased DBDPO contents in the DBDPO system. The proportion data further showed that higher DBDPO contents did not necessarily increase the relative amount of CO in the total increase of incomplete combustibles. A large portion of heat release reduction was probably contributed by the increased formation of combustibles other than CO.

However, the Sb_2O_3/DBDPO combination demonstrated a definite effect on increasing CO generation both quantitatively and proportionately in the total amount of combustibles. These data indicated that the promotion of incomplete combustion by the combined system resulted in CO as the major incomplete combustible, whereas DBDPO alone encouraged the formation of combustibles other than CO. This would tend to suggest that these two systems may impose flame retardation through different mechanisms.

The complete heat of combustion for both the original polymer specimens, $(\Delta H_C^0)_{sample}$, and the burned residues from these samples, $(\Delta H_C^0)_{residue}$, was measured by an adibatic bomb calorimeter. Theoretical values for heat of combustion of the same materials were estimated to confirm the postulated mechanisms.

The theoretical heat of combustion for Sb_2O_3 treated specimen was calculated as:

$$(\Delta H_C^0)_{th} = (1-X)(\Delta H_C^0)_{polyethylene} \tag{7}$$

where X is the fraction of Sb_2O_3 in the samples. Both the measured and theoretical values listed in Table III confirmed that antimony trioxide, being in its highest oxidative form, contributed nothing to the heat of combustion of its treated polyethylene.

Table III. Heat of Complete Combustion (Calories/Gram Sample)

SPECIMEN (% content)	ORIGINAL POLYMER experimental	ORIGINAL POLYMER theoretical	RESIDUE experimental
Polyethylene	11106	-----	10978
Sb_2O_3- 2.5%	10860	10828	9907
4.6%	10579	10595	8686
6.5%	10543	10384	7631
DBDPO- 5.1%	10650	10592	10716
9.4%	10283	10173	10623
12.9%	9741	9801	10457
14.8%	9809	9611	10376
17.4%	9390	9308	10484
Sb_2O_3/DBDPO 0.6%/1.3%	10938	10906	10905
0.98%/1.97%	10836	10797	10935
1.5%/3.0%	10700	10634	10868

Heat of combustion of DBDPO was estimated under the assumption that total heat of combustion of a mixture was the summation of heat from each of the components. Thus, the total heat of combustion for DBDPO treated polyethylene can be calculated as Equation 8:

$$(\Delta H_C^0)_{additive} = (1-X)(\Delta H_C^0)_{polyethylene} + X(\Delta H_C^0)_{DBDPO}$$

$$= (\Delta H_C^0)_{PE} + [(\Delta H_C^0)_{DBDPO} - (\Delta H_C^0)_{PE}] \cdot X \qquad (8)$$

where X is the fraction of DBDPO in the specimen. Theoretical values for DBDPO treated as well as the Sb_2O_3/DBDPO treated specimens were calculated according to their compositions (Table III). Excellent agreement between the calculated and experimental values was found.

In terms of the heat of combustion for the residues, the combustion heat values in the Sb_2O_3 system were considerably lowered as the amount of additive increased. For the DBDPO system, the combustion heat values for the burned residues were slightly lower than that of the residues from pure polyethylene, whereas residues from the Sb_2O_3/DBDPO treated polyethylene was essentially the same combustion heat values as residue of the pure polyethylene.

The above observation seemed to suggest that both Sb_2O_3 and DBDPO in the combined system were removed from the polymer substrate during burning. The argument that DBDPO provides means of transport for Sb_2O_3 to escape from solid substrate during burning was thus supported for the same reason. In the DBDPO system, it seemed that the majority of the additive was removed from the polymer during burning, whereas all or most of Sb_2O_3 in the Sb_2O_3 system seemed to remain in the residues.

Three residue samples, each from the highest additive level in the three treated polymer systems, were analyzed for Sb and Br element contents. The measured elemental contents in the residues and the calculated element contents in the original samples are both listed in Table IV for comparisons.

The measured contents of these elements generally supported the previous postulation derived from the heat of combustion data for these residues. However, the experimental values were still higher than expected in the Sb_2O_3 and the combined system. Since only limited amounts of residues were available for testing in either the determination of the heat of combustion or the elemental analysis, small differences between the experimental results and postulation were attributed by both insufficient sampling and experimental errors.

From the heat of combustion data of both the polymers and their residues, the amount of fuel available per gram sample consumed (ΔH_2) can be calculated as:

$$\Delta H_2 = \frac{(\Delta H_C^O)_{additive} \times W_t - (\Delta H_C^O)_{residue} \times W_r}{\text{Weight Consumed}} \qquad (9)$$

where W_t and W_r are the weight of the original sample and that of the residue. For the combined system, $(\Delta H_C^O)_{additive}$ values were the same as ΔH_2 since the combustion heat data for the original and residue were considered to be the same.

Table V summarizes the values of total fuel available (ΔH_2), their normalized data $[\Delta H_2/(\Delta H_C^O)_s]$ and the ratios of $\Delta H_1/\Delta H_2$. The ratio of $\Delta H_1/\Delta H_2$ describes the combustion efficiency as the fraction of total available fuel actually combusted.

The combustion efficiency for the DBDPO treated samples stayed at an almost constant fraction of 0.70 despite various amounts of additive. Furthermore, the normalized fuel values, $\Delta H_2/(\Delta H_C^O)_{sample}$, for these samples decrease slightly with increasing DBDPO content.

Table IV. Element Contents (%)

SPECIMEN	ORIGINAL		RESIDUE	
(% CONTENT)	(CALCULATED)		(MEASURED)	
	Sb	Br	Sb	Br
6.5 % Sb_2O_3	5.4	0	33.5	0
14.8 % DBDPO	0	12.4	0	4.7
1.5% Sb_2O_3/3.0% DBDPO	1.3	2.5	0.7	0.3

Table V.   Combustion Efficiency

SPECIMEN	$(\Delta H_c^0)_{ORIG.}$	$(\Delta H_c^0)_{RES.}$	ΔH_2*	$\dfrac{\Delta H_2}{(\Delta H_c^0)_s}$	$\dfrac{\Delta H_1}{(\Delta H_c^0)_s}$	$\dfrac{\Delta H_1}{\Delta H_2}$
POLYETHYLENE	11106	10978	11106	1.00	0.80	0.80
DBDPO						
5.1%	10650	10716	10631	1.00	0.71	0.71
9.3%	10293	10623	10152	0.99	0.71	0.72
12.9%	9741	10457	9310	0.96	0.67	0.70
14.8%	9809	10376	8853	0.96	0.62	0.69
Sb_2O_3/DBDPO						
0.6%/1.3%	10938	10905	10938	1.00	0.65	0.65
0.98%/1.97%	10836	10935	10836	1.00	0.59	0.59
1.5%/3.0%	10700	10868	10700	1.00	0.30	0.30

*CALCULATED ACCORDING TO EQUATION 9

This tends to suggest that the system is a solid-phase system, based on Yeh's model.[14] However, considering the error involved in the ΔH_2 values, the observed difference may not be of significance to justify the interpretation. On the other hand, analyzed results in the combined system clearly indicated that the additive action is a vapor-phase system.

SUMMARY

A previously developed technique was applied in studying the additive effects of antimony- and bromine-containing compounds on the combustion of polyethylene. These additives included decabromodiphenyl oxide (DBDPO), antimony trioxide (Sb_2O_3), and the combination of the two at a 3 to 1 Br/Sb ratio. Fundamental burning characteristics, such as total heat release, CO generation, and CO_2 generation, were evaluated for the several levels of add-ons in each additive system.

Antimony trioxide alone was not found to affect the burning of polyethylene. The chemical remained in the burn residue after burning. DBDPO, by itself, exerted a gradual retardation effect on the burning of polyethylene. With add-on of about 15%, the heat release of polyethylene was reduced by about 20%. Relatively gradual decrease in CO_2 generation and increase in CO generation were also observed.

The combination of DBDPO and Sb_2O_3 at a 2 to 1 weight ratio was found to greatly increase the flame retardation effect of either component alone, and even more than the summation of the two. Results strongly supported the vapor phase retardation mode for the Sb_2O_3/DBDPO combined systems. These further supported the theories proposed in the literature that bromine and antimony react in the solid phase to form Sb_2Br_3 which subsequently imposes flame retardation in the vapor phase.

The mode of action for the DBDPO system could not be fully established from the results of present study. However, visual observation indicated that DBDPO promoted much more smoke formation than the combination of Sb_2O_3 and DBDPO. Carbon monoxide formation was found to account for only 25 to 32 per cent of the incomplete combustion in the DBDPO systems. In the combined system, however, the amount of carbon monoxide generation not only increased with the increases of additive but also was the major product of the incomplete combustion. Analyses of the results also suggested that the retardation mechanism may be different in the DBDPO system than in the Sb_2O_3/DBDPO combined system.

REFERENCES

1. S. K. Brauman and A. S. Brolly, J. Flame Retardant Chem., 3, 66 (1976).
2. R. M. Fristrom, NBS Special Publication, No. 357, 131 (1972).
3. Y. -L. Hsieh, Ph.D. Dissertation, University of Maryland, 1981.
4. G. S. Learmonth, A. Nesbitt and D. G. Thwaite, Br. Polym. J., 1, 149 (1969).
5. G. S. Learmonth and D. G. Thwaite, Br. Polym. J., 2, 104 (1970).
6. J. W. Lyons, The Chemistry and Uses of Fire Retardants, Wiley-Int., 1970.
7. J. J. Pitts, J. Fire & Flammability, 3, 51 (1972).
8. J. J. Pitts, P. H. Scott and D. G. Powell, J. Cellular Plastics, 6, 35 (1970).
9. J. A. Phys, Chemistry & Industry (London), 187 (1969).
10. W. G. Schmidt, Trans. J. Plastics Institute, 247 (1965).
11. T. E. Tabor and S. Bergman, 1974 Proceedings of International Symposium on Flammability & Fire Retardants, sponsored by Alena Enterprise of Canada.
12. G. C. Tesoro, Paper presented at Polymer Conference Series on Flammability Characteristics of Polymeric Materials, University of Utah, June, 1970.
13. Z. Touval, J. Fire & Flammability, 3, 130 (1970).
14. K. Yeh, M. J. Drews and R. H. Barker, J. Fire Retardant Chemistry, 7, 99 (1980).

MOLYBDENUM SMOKE SUPPRESSANTS IN POLYVINYL CHLORIDE*

Fred W. Moore, George A. Tsigdinos and Thomas R. Weber

Climax Molybdenum Company of Michigan
A Division of AMAX of Michigan, Inc.
Ann Arbor, Michigan, 48106

INTRODUCTION

The necessity for preventing fires and the evolution of smoke and toxic gases from burning synthetic materials cannot be overemphasized. A report on a study entitled America Burning[1,2] estimates the occurrence of 2.5 million fires a year. More than 50% of the victims succumbing to a fire die of smoke and toxic gas evolvement and not of direct fires and burns.[2] Similar fire statistics from the United Kingdom for the period of 1955-71 clearly demonstrate the increasing importance of smoke and toxic gas effects.[3,4] Total casualties rose from 4,000 to 6,000 per year in the period under study so that the percentage attributable to gas or smoke rose from 5 to nearly 20%. Fatalities attributable to gas or smoke increased to above 50%. Recent fires such as the one that occurred in November, 1980, at the MGM Grand Hotel in Las Vegas, Nevada, resulting in 84 deaths and 300 injuries, point to the ever-increasing need for smoke and toxic gas control from burning materials. Our involvement in the investigation of molybdenum compounds as flame retardants and smoke suppressants in various plastics systems was prompted by the growing number of regulations for flame and smoke reduction and the necessity for es-

---

*The terms "flame retardant" and "smoke suppressant," as used herein, are relative terms and are not intended to indicate hazards presented by plastics containing these products or any other material, under actual fire conditions. The data presented in this paper represent results from controlled laboratory testing and are not intended to reflect hazards presented by this or any other material under actual fire conditions.

tablishing and optimizing flame and smoke formulations using Mo compounds in various plastics formulations, as different formulations
even within the same plastics system behave differently.

Previous work by us has shown that molybdenum-containing additives are effective flame/smoke suppressants in polyvinyl chloride
(PVC) formulations[5-7] and also in other plastics systems.[8] The present work is an extension and up-date of a paper presented at the American Chemical Society Meeting held in Las Vegas, Nevada, March 25-
April 2, 1982.[9] It includes the evaluation of molybdenum trioxide
and of molybdenum compounds extended on inert cores in rigid and plasticized PVC formulations. Included is the evaluation of smoke suppression by three smoke chambers: namely, the NBS, the Arapahoe and
the Ohio State Release Rate (OSRR) apparatus. Data are also presented on the use of another smoke chamber, the Fumenometer, utilized
in Europe.[4] An additional objective of this work has centered on
evaluating, by the OSRR apparatus, the effect of molybdenum trioxide
and extended molybdenum compounds on the flammability of rigid and
plasticized PVC formulations. A discussion is also included on the
mechanistic aspects of the role of molybdenum in the smoke suppression
of PVC and polyesters.

EXPERIMENTAL METHODS

The composition of the materials used as smoke suppressants is
described below. The molybdenum, trioxide, Climax M Grade, was air-
micronized prior to use. The particle size found by sedimentation
is 0.5-3.2 (50% 1.42) microns. Electron microscopy gave a range of
0.5-10 microns. Ammonium octamolybdate, $(NH_4)_4Mo_8O_{26}$, had an average
size of 10 microns. The extended materials used here are white powders that contain less than 10% Mo in the form of metal molybdates.
Additives A and D contain calcium, zinc and molybdenum. Their average particle size is 2.50 microns, as determined by the Joyce Loebl
method. Additive B contains zinc and molybdenum, and Additive C
contains zinc, phosphorus and molybdenum as the active components.
These materials have an average particle size of less than 5 microns.
The composition of Additives B and C is based on a recent patent.[10]

The amount of smoke evolved on burning was measured using the
Arapahoe Smoke Chamber,[11] the NBS Smoke Chamber[12,13] and the Ohio
State Release Rate (OSRR) apparatus.[14] The OSRR test was run with
vertical 1/8-inch (3.2 mm) thick samples with a small pilot flame at
the bottom center of the sample. The samples were evaluated at various heat fluxes. Sample thicknesses of 75 mils (1.9 mm) were used
for the flexible samples and 125 mils (3.2 mm) for the rigid samples
in the Arapahoe test. Samples were tested in the NBS Chamber at a
thickness of 40 mils (1.0 mm). In addition, data are also presented
employing the Fumenometer Smoke Chamber. It has now been introduced

in Europe by Eternit S.A. It is a relatively new, simple and apparently reliable test.[4]

RESULTS AND DISCUSSION

Mechanistic Investigations

Early work at this laboratory[5-7] has shown that molybdenum acts as a flame/smoke suppressant in the solid phase in rigid and plasticized PVC and in polyesters. This behavior is unlike that obtained for Sb_2O_3 and, in part, for phosphorus. Antimony oxide forms volatile halide and oxyhalide species which act in the vapor phase as free radical scavengers[15-17] and phosphorus compounds act in both the solid and vapor phases depending on the substrate.[15,18] The contrasting behavior of Sb_2O_3 and MoO_3 in flame and smoke retardance is borne out by the reactivity studies of these oxides determined thermogravimetrically and shown in Figure 1, in a 20% HCl-80% N_2 gas stream. Antimony oxide reacts with gaseous HCl at room temperature, as ascertained by the weight increase to form volatile halides, followed by weight loss above 40°C. However, for MoO_3 the rate of formation of $MoO_2Cl_2 \cdot H_2O$ must nearly equal the rate of volatilization since no gain in weight is observed.[6] These reactions are represented by the equations given below.

$$Sb_2O_3 + 6HCl \xrightarrow{\sim 25°C} 2SbCl_3 + 3H_2O$$

$$MoO_3 + 2HCl \xrightarrow{\sim 100°C} MoO_2Cl_2 \cdot H_2O$$

$$MoO_2Cl_2 \cdot H_2O \xrightarrow{\sim 200°C} MoO_2Cl_2 + H_2O$$

The formation of volatile $MoO_2Cl_2 \cdot H_2O$[19,20] is indicated, however, by the slow weight loss above 85°C, the volatilization being complete at 265°C.

That volatile molybdenum species can act as flame inhibitors, like volatile antimony halides, has been shown in H_2-O_2-N_2 flame mechanistic studies where molybdenum catalyzes the recombination of H;, ·OH and O·radicals, the rate of such recombination being proportional to the concentration of Mo species in the vapor phase.[21-23] Such catalytic radical removal proceeds according to the steps

$$HMoO_3 + H· \rightarrow MoO_3 + H_2$$

$$MoO_3 + H_2O \rightarrow H_2MoO_4$$

$$H_2MoO_4 + H· \rightarrow HMoO_3 + H_2O$$

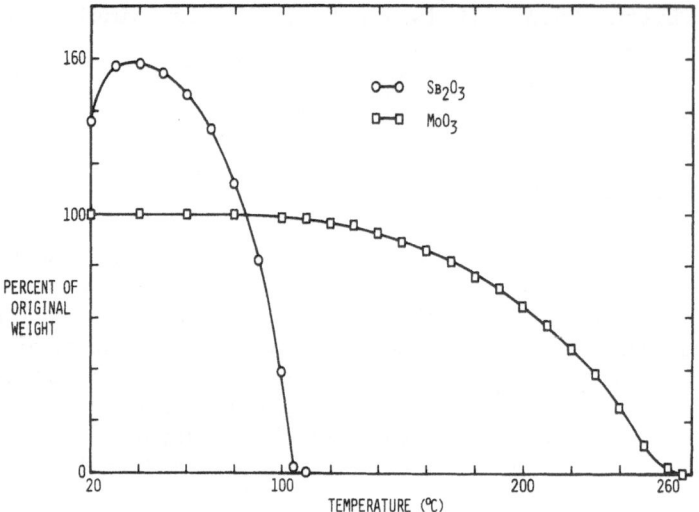

Figure 1. Volatilization of Sb_2O_3 and MoO_3 by HCl.

$$H^{\bullet} + H^{\bullet} \rightarrow H_2$$

the net effect being the recombination of the hydrogen atoms in hydrogen.

   To corroborate the solid state mode of action of molybdenum
flame/smoke suppression, a comparison was carried out of the flammability of plasticized PVC in oxygen and in nitrous oxide atmospheres.[7]
Comparisons of the flammability of a flame retarded material when
burned in an oxygen atmosphere with its behavior when nitrous oxide
is used as the oxidant can yield clues as to whether the flame retardant acts in the solid or vapor phase.[24,25] When oxygen is used as
the oxidant a large proportion is consumed in a free radical branching
step, whereas nitrous oxide is mainly consumed in a non-branching
step. Thus, if a flame retardant acts as a gas phase inhibitor, its
activity should be confined to combustion in an oxygen-containing
atmosphere and the activity of a solid state flame retardant should
be unaffected by a change in oxidant since its activity is specific
to pyrolysis reactions of the polymer. The oxygen index data obtained
for a 30 phr plasticizer formulation are plotted in Figure 2 and the
nitrous oxide data are given in Figure 3.[7] In oxygen, antimony oxide
is very effective in raising the oxygen index, but molybdenum trioxide
is also effective. Increasing the concentration of antimony oxide has
little effect on the nitrous oxide index, but increasing the concentration of molybdenum trioxide has a dramatic effect on the nitrous
oxide index. These results confirm that antimony oxide acts as vapor
phase flame retardant and molybdenum trioxide acts as a solid state
flame retardant. A higher oxygen index is found for the mixture of

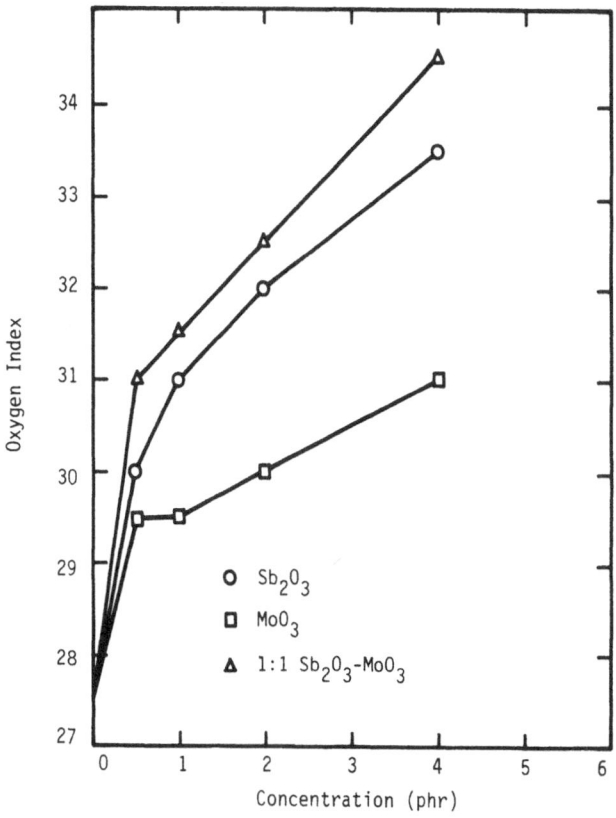

Figure 2. Oxygen Index of semi-rigid formulation.

the oxides than for an equivalent amount of antimony oxide indicating
an additive effect between the oxides, but the nitrous oxide index
data indicate that in nitrous oxide the two metal oxides act independently.[7]

In an attempt to further establish how molybdenum functions as
an additive, the effect of molybdenum trioxide on char formation of
an unfilled PVC formulation was examined.[7] Samples were burned in
the Arapahoe apparatus using the normal smoke measurement method and
the sample was weighed before and after decharring. As shown in Table
I, it can be seen that the addition of antimony oxide, a vapor phase
flame retardant, has little effect on char formation and increases
smoke formation at the 2 phr level. The use of MoO_3, alone or with
Sb_2O_3, increases the amount of char and decreases the amount of smoke.
These results confirm that molybdenum trioxide acts as a solid state
char former to reduce smoke.

Further evidence for the condensed-phase mechanism for molybdenum
has been obtained by us employing ignition studies of PVC and poly-

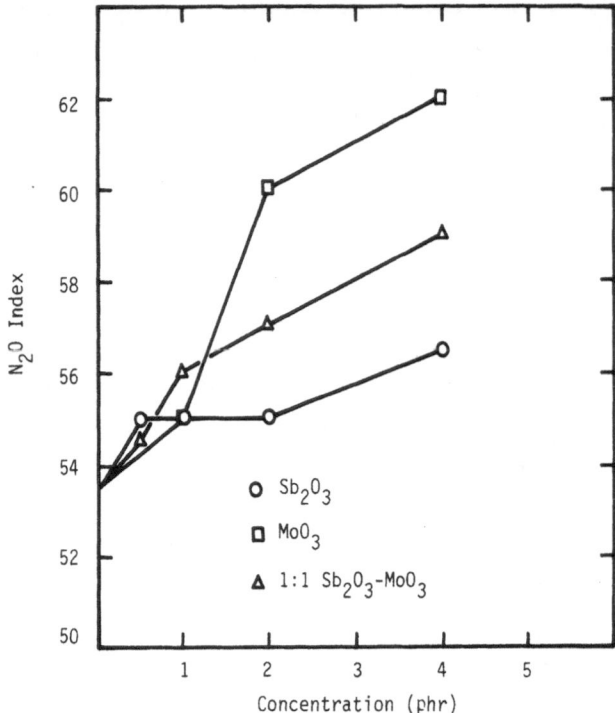

Figure 3. Nitrous oxide index of semi-rigid formulation.

esters. The data obtained are summarized in Tables II and III. On ignition, the Sb_2O_3-containing PVC and polyester samples lost over 90% of the original antimony confirming that antimony must function in the vapor phase, as also shown by others.[15-17] The chars from ignited MoO_3-containing samples, on the other hand, contained over

Table I. Char Formation Studies[a]

| | Arapahoe Chamber Data | |
Additives	Percent Char	Percent Smoke
No Additives	9.9	13.2
1 phr Sb_2O_3	11.6	13.9
2 phr Sb_2O_3	11.6	16.3
4 phr Sb_2O_3	10.6	13.4
2 phr MoO_3	23.5	7.6
4 phr MoO_3	23.7	7.5
2 phr Sb_2O_3 + 2 phr MoO_3	20.7	8.9

[a]Formulation: 100 parts PVC resin, 30 parts DIDP, 7 parts lead stabilizer, 0.8 part lubricants and additives as shown.

Table II. Ignition of PVC Samples

Sample	Ignition Technique[a]	Percent Ash	Percent of Original Metal in Ash
Sb$_2$O$_3$ Containing Formulations			
18.9% Sb$_2$O$_3$ - PVC Mixture[b]	1	12.0	4.7
5 phr Sb$_2$O$_3$ - 45 phr DOP - 35 phr CaCO$_3$	2	8.0	5.8
6 phr Sb$_2$O$_3$ - 20 phr DOP - 26 phr CaCO$_3$	2	29.8	5.4
3 phr Sb$_2$O$_3$ - 50 phr DOP	2	10.4	4.9
3 phr Sb$_2$O$_3$ - 50 phr DOP	1	10.7	6.5
MoO$_3$ Containing Formulations			
11% MoO$_3$ - Tetrabromophthalic Anhydride Mixture[b]	1	9.2	92.5
18.9% MoO$_3$ - PVC Mixture[b]	1	20.9	100.0
5 phr MoO$_3$ - 45 phr DOP - 35 phr CaCO$_3$	2	35.0	100.0
6 phr MoO$_3$ - 20 phr DOP - 26 phr CaCO$_3$	2	42.7	98.5
3 phr MoO$_3$ - 50 phr DOP	2	14.2	96.6
3 phr MoO$_3$ - 50 phr DOP	1	8.6	98.8

[a] Method 1 - Sample ignited in crucible with Meker burner with just enough heat to maintain burning.

Method 2 - Sample burned in air in a 500° furnace until flaming ceased.

[b] Mixtures contain 10 moles of halogen per mole of metal.

Table III.   Ignition of Samples Containing Mo and Sb

Sample	Percent of Original Metal·in Ash	
PVC Samples	Mo	Sb
50 phr DOP - 1.5 phr MoO_3 - 1.5 phr Sb_2O_3	90.4	5.1
20 phr DOP - 26 phr $CaCO_3$ - 3 phr MoO_3 - 3 phr Sb_2O_3	91.1	8.0
Polyester Samples		
Atlac 711-05A - 1% MoO_3 - 1% Sb_2O_3	92.2	2.3
Hetron 92C - 2.5% MoO_3 + 2.5% Sb_2O_3	92.5	9.1

90% of the starting molybdenum in agreement with the nitrous oxide index results that molybdenum acts as a flame/smoke suppressant in the solid state. It should be noted (see Table II) that this behavior is general as MoO_3 also remains behind during ignition of MoO_3-tetra-bromophthalic anhydride mixtures. The lack of 100% accountability for molybdenum in the chars obtained can be attributed to analyses. It has been shown by mass spectrometric studies that no MoO_3 was detected in the vapor phase of pyrolyzed plasticized PVC.[26] The oxide was observed in the vapor phase during pyrolysis only when large amounts were incorporated into the polymer.[26]

Since these studies were carried out, several investigations have appeared which concerned themselves with the mechanism of the action of MoO_3 as a flame/smoke suppressant for PVC and polyesters. These studies have shown that MoO_3 functions in the condensed phase by retarding the formation of volatile aromatics.[26-39] The nature of the chemical reactions that cause such aromatics suppression has been explained by a Lewis acid mechanism[26-30] and also by a "reductive coupling" reaction which causes the polymer to crosslink extensively in the early stages of thermal degradation.[31,32] Evidence to date indicates that MoO_3 acts as a flame suppressant in halogenated thermoset polyesters by accelerating the rate of loss of halogen from these polyesters while its smoke suppression properties depend on its ability to increase char levels.[39]

Polyvinyl Chloride Smoke Suppression

The data obtained to date at this laboratory on the smoke suppression of PVC are presented below.

Rigid Conduit Formulations. Data presented in Table IV show the effectiveness of molybdenum-containing additives as smoke suppressants in a rigid conduit PVC formulation. In the Arapahoe Smoke Chamber

Table IV. Smoke Formation of Rigid Conduit Formulations[a,b]

Formulation	Arapahoe Smoke Data, % Smoke	NBS Chamber Dmc Flaming Mode	OSRR Test Total Smoke Release (SMOKE/m^2)		Eternit Fumenometer mg[c]
			3 W/cm^2	5 W/cm^2	
Control	7.8	314	703	965	57[d]
2.5 phr MoO$_3$	5.3(32)	174(45)	108(85)	218(77)	105(46)
5 phr MoO$_3$	5.0(36)	135(57)	46(93)	158(84)	110(48)
5 phr Additive A	6.0(23)	269(14)	265(62)	478(50)	83(31)
10 phr Additive A	6.1(22)	214(32)	230(67)	289(70)	86(34)
5 phr Additive B	6.7(14)	180(43)	301(57)	550(43)	89(33)
5 phr Additive C	6.1(22)	186(41)	177(75)	343(65)	87(35)

[a]Formulation: 100 parts PVC resin, 25 parts whiting, 3 parts processing aid, plus stabilizer and lubricants.

[b]Percent reduction in smoke relative to control is shown in parentheses.

[c]Represents milligrams of sample to produce the same amount of smoke. Data taken from Ref. 4.

[d]Result estimated for this sample only as material tested was not exactly same as other controls.

the additives provide smoke reductions between 14 and 36%. However, greater reductions of up to 57% are seen in the NBS Smoke Chamber and reductions of 31-48% are seen in the Fumenometer test.[4] The OSRR test shows the greatest reductions, 43-93%. This test, which has been correlated by others[40-41] with results obtained in the larger scale Steiner tunnel (ASTM E-84-81a) and crib corner tests, also shows that MoO$_3$ further reduces the flammability of this already low flammability formulation. Maximum heat release rates for the 5 phr MoO$_3$-containing sample are 36 and 35 kW/m^2 at 3 and 5 W/cm^2, respectively, compared with 60 and 76 kW/m^2 for the control. Cumulative smoke release data obtained by the OSRR at 3 W/cm^2 are also shown in Figure 4. It should be noted that all the Mo-containing formulations provide a delayed smoke evolution as compared to the control. This fact is very pertinent to actual fire conditions where delayed effects on smoke evolution are necessary for adequate visibility to evacuate a building.

Semi-Rigid Formulations. The smoke formation data for a semi-rigid PVC formulation containing 2 phr Sb$_2$O$_3$ are presented in Table V. It can be seen that all of the molybdenum-containing additives provide good smoke suppression in this formulation as determined by all four tests. Additive A provides smoke reduction in this formulation which is as good as that provided by MoO$_3$, even though it has a much lower molybdenum content.

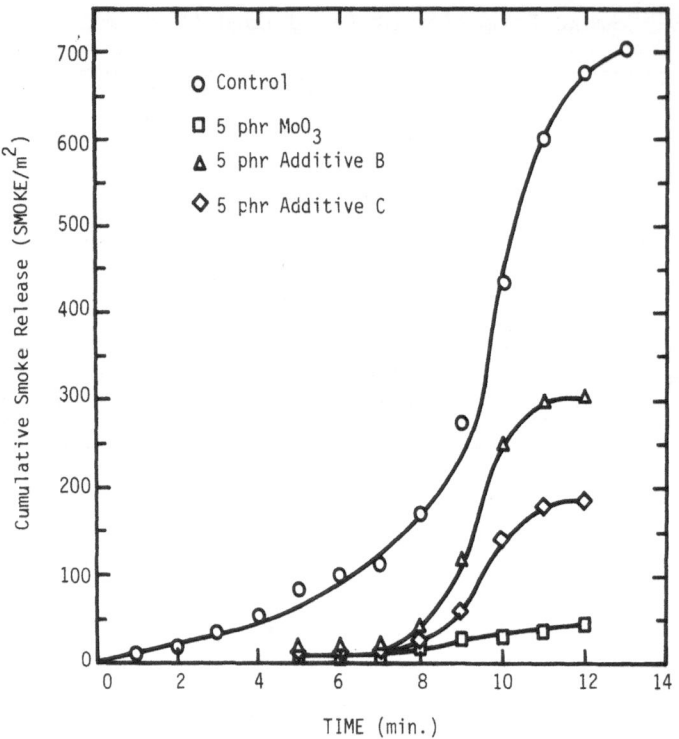

Figure 4.   Smoke release of rigid conduit at 3 W/cm^2.

In order to circumvent spalling problems which had been encountered in the OSRR test with semi-rigid plaques, samples were evaluated
using one-inch wire mesh over the face of the plaque. The data obtained are presented in Table VI. These data show that in samples
containing 2 phr Sb_2O_3, 4 phr of Additive A reduces smoke by 70%
and 42% at 2 and 3 W/cm^2, respectively, and 4 phr of Additive D reduces smoke by 73% and 53% at 2 and 3 W/cm^2, respectively. Also, in
addition to the smoke reduction, the use of Additives A and D reduces
the maximum rate of heat release. Thus, these two additives were
evaluated as a partial replacement for the Sb_2O_3. The data for samples containing 1 phr Sb_2O_3 plus either 1 phr Additive A or D show
that the maximum heat release rates are lower than the 2 phr Sb_2O_3.
Thus, these two materials are effective as partial replacements for
the Sb_2O_3 as flame retardants while also providing smoke reduction.
Cumulative smoke release data for this formulation, obtained by the
OSRR at 2 and at 3 W/cm^2, are presented in Figures 5 and 6, respectively.

The rate of heat release at these two heat fluxes is lower than
that of the Sb_2O_3-only containing formulation. Thus, in this formulation Mo provides for better flame/smoke suppression when used in
combination with Sb_2O_3.

Table V. Smoke Formation of Semi-Rigid
Formulations Containing 2 phr Sb_2O_3 [a,b]

Formulation	Arapahoe Smoke Data, % Smoke	NBS Chamber Dmc Flaming Mode	OSRR Total Smoke Release ($SMOKE/m^2$)		Eternit Fumenometer mg [c]
			2 W/cm^2 (10 min)	3 W/cm^2 (5 min)	
Control	16.3	502	458	625	31
2 phr MoO_3	9.4(42)	288(43)	173(62)	273(56)	49(37)
4 phr MoO_3	9.0(45)	242(52)	113(75)	268(57)	56(45)
4 phr Additive A	8.3(49)	237(53)	115(75)	227(64)	62(50)
4 phr Additive B	10.1(38)	338(33)	231(50)	223(64)	49(37)
4 phr Additive C	9.5(42)	279(44)	266(42)	298(52)	65(53)

[a] Formulation: 100 parts PVC resin, 30 parts DIDP, 7 parts lead stabilizer, 2 parts Sb_2O_3, 0.8 part lubricants.

[b] Percent reduction in smoke relative to control is shown in parentheses.

[c] Represents milligrams of sample to produce the same amount of smoke. Data taken from Ref. 4.

Table VI. Release Rate Data for Semi-Rigid Formulations[a,b]

Formulation	Maximum Rate of Heat Release (kW/m^2)		Total Smoke Release [c] ($SMOKE/m^2$)	
	2 W/cm^2	3 W/cm^2	2 W/cm^2 (10 min)	3 W/cm^2 (10 min)
2 phr Sb_2O_3 - Control	62	97	558	1028
2 phr Sb_2O_3 + 4 phr Additive A	45	61	169(70)	599(42)
2 phr Sb_2O_3 + 4 phr Additive D	49	63	148(73)	484(53)
1 phr Sb_2O_3 + 1 phr Additive A	56	82	315(44)	776(25)
1 phr Sb_2O_3 + 1 phr Additive D	55	83	246(56)	727(29)

[a] Formulation: 100 parts PVC resin, 30 parts DIDP, 7 parts lead stabilizer, 0.8 part lubricants.

[b] Samples tested with one inch mesh screen.

[c] Percent reduction in smoke relative to control is shown in parentheses.

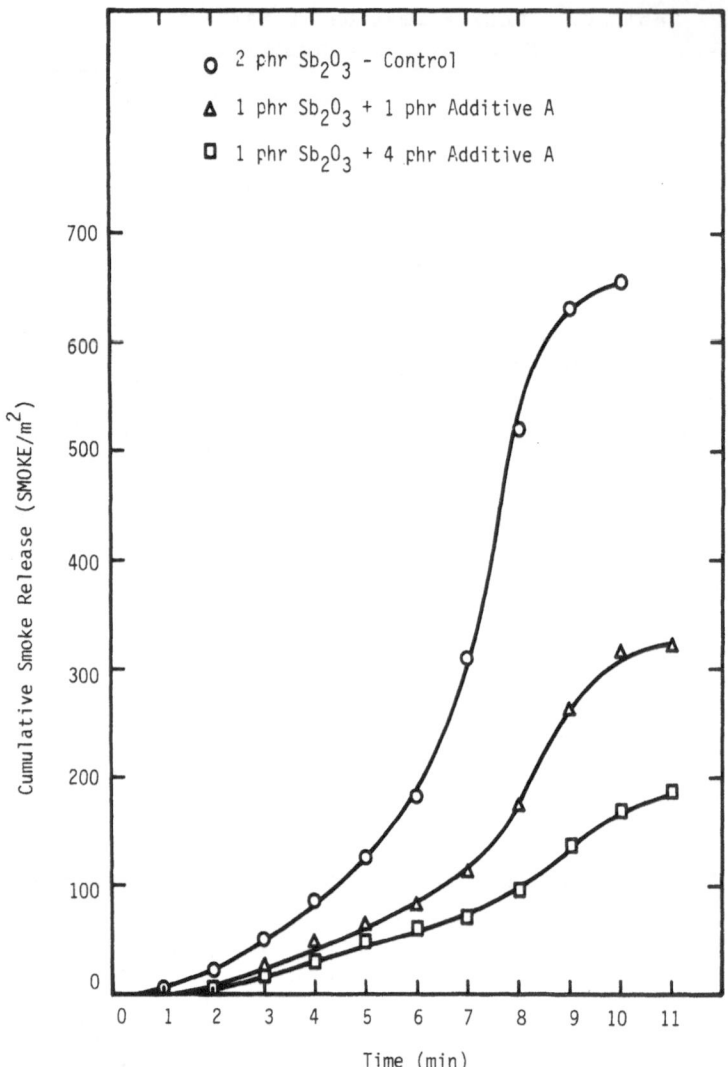

Figure 5. Smoke release of semi-rigid formulation at 2 W/cm^2.

Alumina Trihydrate-Filled Formulations. Alumina trihydrate is
an additive which is employed to improve the flame retardancy and
reduce smoke formation of PVC formulations. Smoke formation data
for an alumina trihydrate-filled PVC formulation containing 3 phr
Sb_2O_3 are presented in Table VII. These data show that the molybdenum-containing additives are effective smoke suppressants in this
formulation. Reductions of over 50% are obtained in the NBS Chamber
flaming mode for all additives except Additive B. In the OSRR test,
MoO_3 is the most effective additive; however, Additives A, B and C
are nearly as effective at 2 W/cm^2 and Additives A and D are quite
effective at 3 W/cm^2.

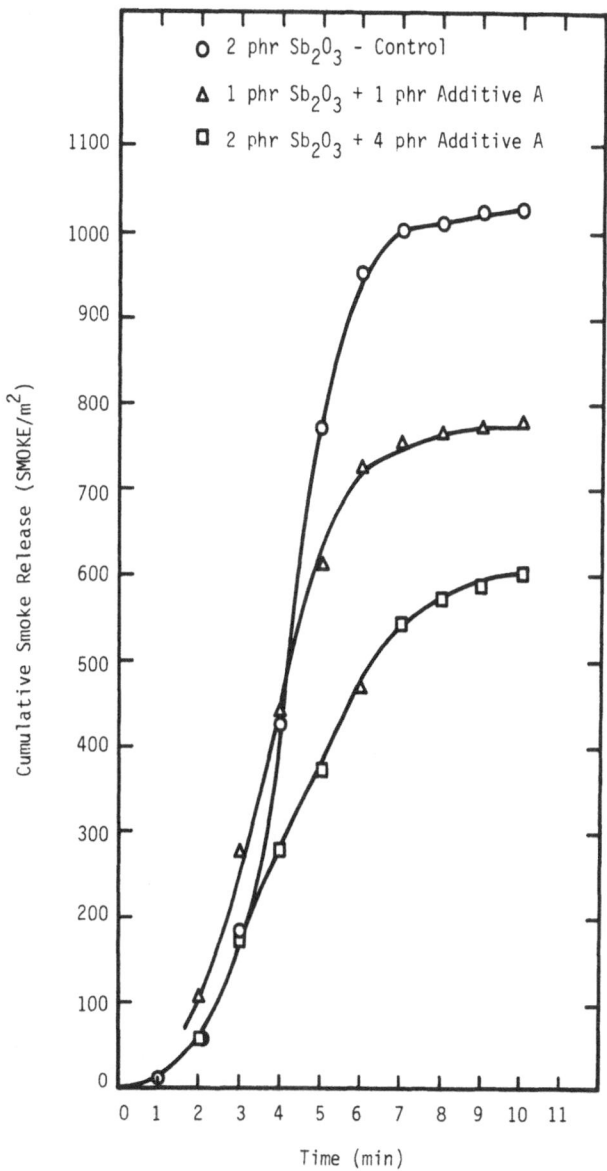

Figure 6. Smoke release of semi-rigid formulation at 3 W/cm^2.

    Figures 7 and 8 represent comparative smoke reductions obtained
in the alumina trihydrate-filled PVC formulation using a commercially
available ZnO/MgO smoke suppressant and Additives A, C and D. The
formulations contained 3 phr Sb_2O_3, the latter serving as the control.
The OSRR data obtained at 2 W/cm^2 (Figure 7) and at 3 W/cm^2 (Figure
8) demonstrate that all Mo-containing additives provide increasing
smoke reduction with increasing concentration whereas the ZnO/MgO

Table VII.   Smoke Formation of Alumina Trihydrate-
Filled Formulations with 3 phr Sb_2O_3[a,b]

Formulation	Arapahoe Smoke Data, % Smoke	NBS Chamber Dmc Flaming Mode	OSRR Total Smoke Release ($SMOKE/m^2$)	
			$2\ W/cm^2$ (10 min)	$3\ W/cm^2$ (5 min)
3 phr Sb_2O_3 - Control	11.3	536	392	487
6 phr MoO_3	7.5(35)	237(56)	66(83)	213(56)
6 phr Additive A	7.8(31)	264(51)	112(71)	329(32)
6 phr Additive B	8.1(28)	344(36)	104(73)	373(23)
6 phr Additive C	7.4(35)	256(52)	104(73)	350(28)
6 phr Additive D	6.9(39)	---	147(63)	295(39)

[a]Formulation: 100 parts PVC resin, 40 parts linear alkyl phthalate,
15 parts alumina trihydrate, 5 parts lead stabilizer, 3 parts Sb_2O_3,
plus lubricants.

[b]Percent reduction in smoke relative to control is shown in parentheses.

material yields about the same or less smoke reduction with increasing
concentration. At $3\ W/cm^2$ an increase in smoke evolution is obtained
with the ZnO/MgO additive beginning with the 6 phr level. This effect
can be attributed to the presence of ZnO which is known to act as a
smoke suppressant at low levels, but it contributes to smoke increase
at higher levels. The effectiveness of Mo smoke suppressants shows
even greater promise as requirements for higher smoke reductions in
plastics are instituted. Apparently, this can be achieved by higher
concentrations of Mo extended materials.

Molybdenum-containing additives were also found to be effective
smoke suppressants for this formulation when the Sb_2O_3 level was increased to 5 or 7 phr to raise the flame retardancy level (Table
VIII). The O.I. of this formulation was found to plateau at 33.5 at
7 phr of Sb_2O_3 for Sb_2O_3 levels up to 11 phr. Addition of 10 phr of
Additive B to the 5 or 7 phr Sb_2O_3 formulations raises the O.I. by
2.5-3.0 units and decreases the Arapahoe smoke values by 31 to 33%.
Similar results are obtained when 10 or 15 phr of Additive C is added
to the 5 phr Sb_2O_3 formulation. Both additives reduce the total
amount of heat formed at $2\ W/cm^2$ and yield about the same or lower
amounts of heat at $3\ W/cm^2$. The 5 phr Sb_2O_3 + 15 phr Additive C formulation yields the lowest amount of smoke with a 79% reduction at 2
W/cm^2 and a 75% reduction at $3\ W/cm^2$.

Formulations with Secondary Plasticizers. The effectiveness of
molybdenum-containing additives was examined in a PVC formulation
containing a chlorinated paraffin (50% Cl) as a secondary plasticizer

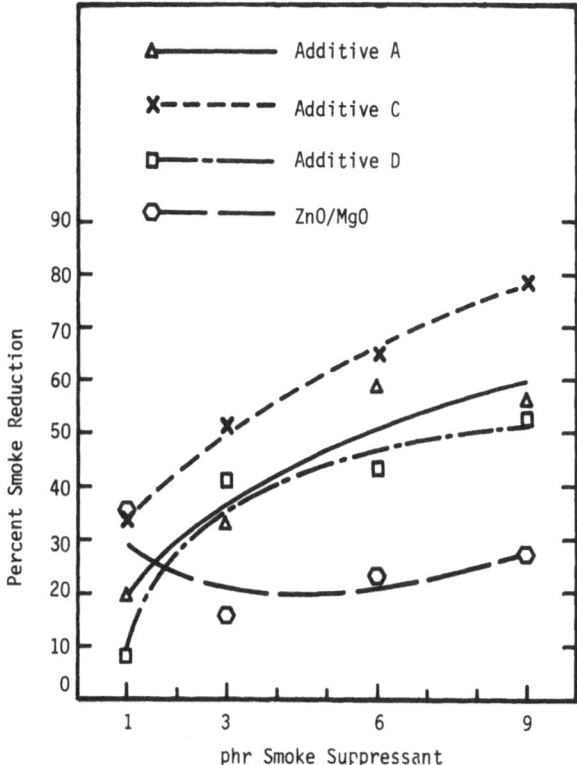

Figure 7. Reduction in smoke of alumina trihydrate-filled formulation at 2 W/cm^2 irradiation.

(Table IX). At the 3 phr level, MoO_3 is not as effective as Sb_2O_3 in raising the oxygen index, but the 1:1 MoO_3-Sb_2O_3 mixture is more effective than a comparable amount of Sb_2O_3 alone (O.I. 31.5). Ammonium octamolybdate is also very effective in this formulation. The addition of 3 or 5 phr of Additive A raises the oxygen index by 3 or 3.5 points, respectively. All of the additives increase the UL-94 rating from V-1 for the untreated formulation to the highest rating of V-0. The molybdenum-containing additives decrease the amount of smoke formed by 45 to 52% when used by themselves. Molybdenum trioxide is not as effective in reducing smoke when used in combination with Sb_2O_3, but ammonium octamolybdate is quite effective in the presence of Sb_2O_3.

The effectiveness of molybdenum with tricresyl phosphate used as a partial replacement for the phthalate ester was examined (Table X), since it has been reported[42] that the addition of Sb_2O_3 to phosphate ester-plasticized PVC does not substantially improve the flame retardancy. Molybdenum trioxide and ammonium octamolybdate alone or

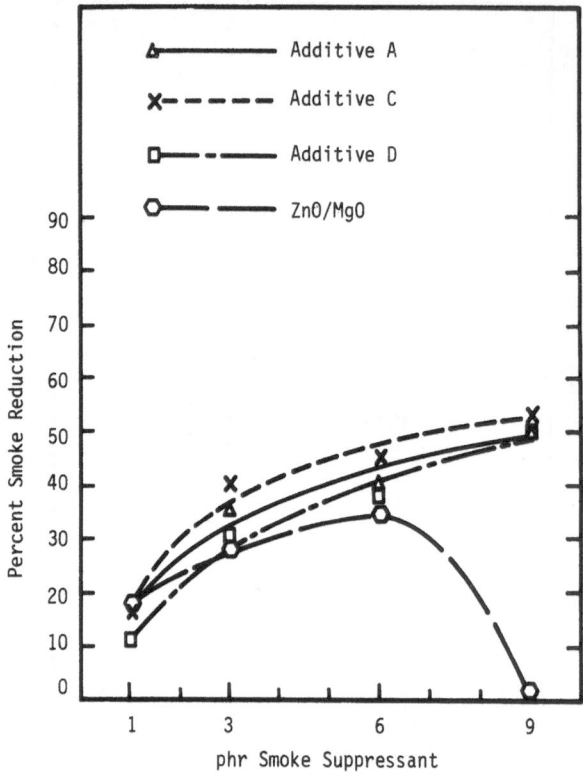

Figure 8. Reduction in smoke of alumina trihydrate-filled formulation at 3 W/cm^2 irradiation.

mixed 1:1 with Sb_2O_3 are as effective or more effective than Sb_2O_3
alone in raising the oxygen index of this formulation. Additive A
and Additive B are also effective in increasing the flame retardancy
as shown by increases in the oxygen index. All of the formulations,
including the control, have a UL-94 rating of V-0. Based on this
data it can be concluded that molybdenum compounds do not have an
adverse effect on the flame retarding ability of phosphorus-contain-
ing additives. The addition of the molybdenum compounds also decreases the amount of smoke formed by 25 to 39%.

CONCLUSIONS

    Molybdenum is a good smoke suppressant in rigid and plasticized
PVC as shown by the Arapahoe, NBS, Eternit Fumenometer and the inter-
mediate scale Ohio State Release Rate apparatus. As shown by the
OSRR apparatus, the use of MoO_3 alone in a rigid PVC formulation or

Table VIII. Properties of Highly Flame Retarded Alumina Trihydrate-Filled Formulations[a]

Formulation	Oxygen Index	Arapahoe Data % Smoke	OSRR Data[b]					
			Maximum Heat Release Rate (kW/m^2)		Cumulative Heat Release (MJ/m^2)		Cumulative Smoke Release (SMOKE/m^2)	
			2 W/cm^2	3 W/cm^2	2 W/cm^2	3 W/cm^2	2 W/cm^2	3 W/cm^2
5 phr Sb$_2$O$_3$ - Control	33.0	9.6	56	72	21	21	248	740
7 phr Sb$_2$O$_3$	33.5	8.9	55	77	20	22	194	718
5 phr Sb$_2$O$_3$ + 10 phr Additive B	35.5	6.4(33)	63	84	14	23	81(67)	397(46)
5 phr Sb$_2$O$_3$ + 10 phr Additive C	34.0	6.3(34)	47	73	12	23	40(84)	307(59)
5 phr Sb$_2$O$_3$ + 15 phr Additive C	35.5	6.3(34)	56	58	11	18	50(79)	185(75)
7 phr Sb$_2$O$_3$ + 10 phr Additive B	36.5	6.6(31)	68	84	16	23	113(54)	329(56)

[a]Formulation: 100 parts PVC resin, 40 parts linear alkyl phthalate, 5 parts lead stabilizer, 15 parts alumina trihydrate, 0.9 part lubricants, plus additives as shown.

[b]Sample covered with 1-inch wire mesh and tested for 10 min.

Table IX.   Flammability and Smoke Formation of Chlorinated
Paraffin Plasticized PVC[a],[b]

Additive	Oxygen Index	UL-94 Vertical Rating	Arapahoe Smoke Data	
			% Smoke	% Reduction[c]
None	25.0	V-1	11.2	--
3 phr Sb_2O_3	29.0	V-0	14.5	--
3 phr MoO_3	27.5	V-0	7.6	48
1.5 phr Sb_2O_3 + 1.5 phr MoO_3	31.5	V-0	11.8	19
3 phr Ammonium Octamolybdate	29.0	V-0	6.9	52
1.5 phr Sb_2O_3 + 1.5 phr Ammonium Octamolybdate	31.0	V-0	8.6	45
3 phr Additive A	28.0	V-0	12.1	17
5 phr Additive A	28.5	V-0	7.6	48
5 phr Additive B	28.0	V-0	7.4	49

[a]Formulation: 100 parts PVC, 38 parts DOP, 14 parts chlorinated paraffin (50% Cl), 5 parts epoxy soya, 1 part stabilizer, 35 parts $CaCO_3$, plus additives as shown.

[b]Samples were 75 mils thick.

[c]Reductions compared with 3 phr Sb_2O_3 sample.

Table X.   Flammability and Smoke Formulation of Tricresyl Phosphate Plasticized PVC[a],[b]

Additive	Oxygen Index	UL-94 Vertical Rating	Arapahoe Smoke Data	
			% Smoke	% Reduction[c]
None, Control	27.0	V-0	16.6	--
3 phr Sb_2O_3	29.5	V-0	16.0	4
3 phr MoO_3	29.5	V-0	11.9	28
1.5 phr Sb_2O_3 + 1.5 phr MoO_3	31.0	V-0	13.5	19
2 phr Ammonium Octamolybdate	30.5	V-0	10.5	37
3 phr Ammonium Octamolybdate	31.5	V-0	10.1	39
1.5 phr Sb_2O_3 + 1.5 phr Ammonium Octamolybdate	30.5	V-0	13.0	22
3 phr Additive A	28.0	V-0	12.1	27
5 phr Additive A	29.0	V-0	12.5	25
5 phr Additive B	30.0	V-0	11.2	33

[a]Formulation: 100 parts PVC, 30 parts DOP, 35 parts $CaCO_3$, 5 parts epoxy soya, 1 part stabilizer, 15 parts tricresyl phosphate, plus additives as shown.

[b]Samples were 75 mils thick.

[c]Reduction compared to control sample.

in a semi-rigid PVC formulation containing Sb_2O_3 reduces the maximum heat release rate and thus it improves the flame retardancy of these formulations.

Mechanistic studies show that molybdenum acts as a flame/smoke suppressant in the solid phase as ascertained by the amount of char formed, lack of volatilization of molybdenum, and by combustion studies using nitrous oxide.

FUTURE ASPECTS

Because of the ability of molybdenum to act as a smoke suppressant in the solid state via char formation, the technology developed for PVC can be extended into other polymer systems. Work is presently underway to smoke suppress thermoset polyesters. Other systems will include PVC alloys, polyolefins, epoxies and polyurethanes.

REFERENCES

1. America Burning, National Commission on Fire Prevention and Control, June, 1973.
2. Chem. Eng. News, July 16, 1973, p. 9.
3. P. C. Bowes, Ann. Occup. Hyg., 17, 143 (1974).
4. A. W. Armour and J. Le Foll, "Molybdenum Compounds as Smoke Suppressant Additives for Polymers," presented at the Symposium "Plastics and Fire," Paris, France, May, 1982.
5. F. W. Moore, "Molybdenum Compounds as Smoke Suppressants for Polyvinyl Chloride Formulations," Society of Plastics Engineers 35th ANTEC, Vol. 23, Montreal, Quebec, April, 1977.
6. F. W. Moore and G. A. Tsigdinos, "Advances in the Use of Molybdenum Additives as Smoke Suppressants and Flame Retardants for Polyvinyl Chloride," Proc. Int. Symp. Flammability Fire Retard., Technomic Press, Westport, New York, 160 (1978).
7. F. W. Moore, T. R. Weber and G. A. Tsigdinos, J. Vinyl Technology, 3, 139 (1981).
8. F. W. Moore and G. A. Tsigdinos, "The Role of Molybdenum in Flame Retardancy and Smoke Retardation," Proc. 2nd Int. Conf. on Chem. and Uses of Molybdenum, New College, Oxford, August 30-September 3, 1976.
9. F. W. Moore, T. R. Weber, G. A. Tsigdinos and J. G. Bilek, "Molybdenum Smoke Suppressants in Polyvinyl Chloride Formulations," preprints of papers presented by the Division of Organic Coatings and Plastics Chemistry at the American Chemical Society 183rd National Meeting, Las Vegas, Nevada, March 25-April 2, 1982, pp. 551-555.
10. G. A. Tsigdinos, T. R. Weber and F. W. Moore, U. S. Pat. 4,328,152 to AMAX, Inc., May 4, 1982.

11. J. Kracklauer, C. Sparkes and R. Legg, Plast. Tech., March, 1976, p. 46.
12. American Society for Testing and Materials, ASTM Method E662-79, American Society for Testing Materials, Philadelphia, PA, 1979.
13. National Fire Protection Association, NFPA No. 258, "Smoke Generated by Solid Materials," National Fire Protection Association, Boston, MA.
14. E. E. Smith, Fire Technology, 8, 237 (1972).
15. J. W. Hastie, J. Res. NBS, 77A, 733 (1973).
16. S. K. Brauman and A. S. Brolly, J. Fire Retardant Chem., 3, 66 (1976).
17. S. K. Brauman, J. Fire Retardant Chem., 3, 117, 138 (1976).
18. J. W. Hastie and C. L. McBee, "Mechanistic Studies of Triphenylphosphine Oxide-Poly(Ethylene Terphthalate) and Related Flame Retardant Systems," NBSIR 75-741, National Bureau of Standards, Washington, D. C. (1975).
19. S. S. Chernikov and B. M. Tarakanov, Russ. J. Inorg. Chem., 18, 22 (1973).
20. F. A. Schroeder and A. N. Christensen, Z. Anorg. Allgem. Chem., 392, 107 (1972).
21. D. E. Jensen and G. A. Jones, J. Chem. Soc., Faraday Trans., 1, 71, 149 (1975).
22. D. E. Jensen and B. C. Webb, AIAA J., 14, 947 (1976).
23. E. M. Bulewicz and P. J. Padley, Trans. Faraday Soc., 67, 2337 (1971).
24. G. R. Granzow, Accts. Chem. Res., 11, 177 (1978).
25. C. P. Fenimore and G. W. Jones, Combust. Flame, 10, 295 (1966).
26. R. M. Lum, J. Appl. Polym. Sci., 23, 1247 (1979).
27. W. H. Starnes and D. Edelson, Macromolecules, 12, 797 (1979).
28. D. Edelson, V. J. Kuck, R. M. Lum, E. Scalco, W. H. Starnes and S. Kaufman, Combust. Flame, 38, 271 (1980).
29. R. M. Lum, L. Seibles, D. Edelson and W. H. Starnes, Org. Coat. Plast. Chem., 176 (1980).
30. W. H. Starnes, L. D. Wescott, W. D. Reeuts, R. E. Cais, G. M. Villacosta, I. M. Plitz and L. J. Anthony, "Mechanism of Polyvinyl Chloride Fire Retardant by Molybdenum (VI) Oxide - Further Evidence in Favor of the Lewis Acid Theory," preprints of papers presented by the Division of Organic Coatings and Plastics Chemistry at the American Chemical Society 183rd National Meetings, Las Vegas, Nevada, March 25-April 2, 1982, pp. 556-561.
31. R. P. Lattimer and W. J. Kroenke, J. Appl. Polym. Sci., 26, 1191 (1981).
32. W. J. Kroenke, J. Appl. Polym. Sci., 26, 1167 (1981).
33. S. K. Brauman, J. Appl. Polym. Sci., 26, 353 (1981).
34. S. K. Brauman, J. Fire Retard. Chem., 7, 119 (1980).
35. A. Ballistreri, G. Montando, C. Puglisi, E. Scamporrino and D. Vitalini, J. Polym. Sci., Polym. Chem. Ed., 18, 3101 (1980).

36. A. Ballistreri, G. Montando, C. Puglisi, E. Scamporrino and D. Vitalini, J. Polym. Sci., Polym. Chem. Ed., 19, 1397 (1981).

37. T. C. Rees, "Some Aspects of the Mechanism of Smoke Suppression of Molybdenum and Zinc Compounds in Polyvinyl Chloride," presented at the 7th International Symposium on Flammability and Fire Retardants, New Orleans, LA, May, 1980.

38. F. W. Moore and G. A. Tsigdinos, J. Less-Common Met., 54, 297 (1977).

39. M. Das, P. J. Haines, T. J. Lever and G. A. Skinner, "The Role of Molybdenum Trioxide as a Flame Retardant and Smoke Suppressant in Halogenated Polyester Thermosets," Proc. 4th Int. Conf. on Chem. and Uses of Molybdenum, Colorado School of Mines, Golden, Colorado, August 9-13, 1982.

40. E. E. Smith, J. Fire Flamm., 8, 309 (1977).

41. G. R. Woolerton, J. Fire Flamm., 9, 317 (1978).

42. E. D. Weil in, "Flame Retardancy of Polymeric Materials," Vol. 3, ed. by W. C. Kuryla and A. J. Papa, Marcel Dekker, Inc., New York, 1975, p. 203.

MECHANISM OF POLY(VINYL CHLORIDE) FIRE RETARDANCE BY MOLYBDENUM(VI)

OXIDE. FURTHER EVIDENCE IN FAVOR OF THE LEWIS ACID THEORY

W. H. Starnes, Jr., L. D. Wescott, Jr., W. D. Reents,
Jr., R. E. Cais, G. M. Villacorta, I. M. Plitz and
L. J. Anthony

Bell Laboratories
Murray Hill, NJ 07974

INTRODUCTION

Many recent investigations have been concerned with the mechanism of action of MoO_3 as a smoke-suppressant and fire-retardant additive for poly(vinyl chloride) (PVC).[1-14] These studies have shown that MoO_3 functions within the polymer matrix[1-14] and retards the production of the volatile aromatics (especially benzene) that are the principal fuel and source of smoke under low-enthalpy-input conditions.[1-4,6-8,13,14] However, no consensus of opinion has been reached with regard to the nature of the chemical reactions that cause the aromatics suppression.

Two conflicting proposals[1-5] are now receiving attention in this respect. One, advanced by Starnes and Edelson,[1] is based on the simple scheme depicted below. In this mechanism, PVC initially undergoes

$$
\begin{array}{c}
\text{Benzene} \longrightarrow \text{Smoke} \\
\uparrow (2) \\
\text{PVC} \xrightarrow{(1)} \text{cis,trans Polyene} \xrightarrow{(3)} \text{Cross-linked Polymer} \\
(4) \searrow \quad \downarrow (5) \qquad \nearrow (6) \qquad \downarrow \\
\text{All-trans Polyene} \qquad\qquad \text{Char}
\end{array}
$$

dehydrochlorination (Equation 1) to generate linear polyene sequences having a low proportion of cis-substituted alkene units. Benzene is then formed from the polyene segments in which these units occur (Equation 2), via the stepwise thermolysis of cyclohexadiene intermediates. In MoO_3-containing systems, benzene formation is suppressed by the occurrence of one or more competing reactions that are subject

237

to Lewis acid catalysis by MoO_3 and/or its derivative species. These
reactions are: (a) the cross-linking of polyene segments containing
cis-substituted double bonds (Equation 3; cyclohexadienes derived
from these segments could experience cross-linking, as well); (b)
the direct formation of all-trans polyene segments from the starting
polymer (Equation 4; this result might be achieved in either of two
ways[1]); (c) the isomerization of cis-alkene moieties into the more
thermodynamically stable trans arrangement (Equation 5).

In the paper containing the first complete description of the
Lewis acid mechanism,[1] it was noted that good correlations of char
yield with smoke emission or Lewis acid content were not available
at the time, and that there was thus a lack of direct evidence to
indicate that metal-catalyzed cross-linking was the principal cause
of smoke suppression by MoO_3 and other metallic species. Better
smoke-char correlations have now been found for several metal-con-
taining systems,[4-6,10,13] and on this basis some workers have assumed
that a converse form of the argument just given[1] applies; viz., that
these correlations require metal-catalyzed cross-linking to be the
major mechanism of smoke inhibition.[4-6] However, an inspection of the
reaction diagram shown above reveals that this reasoning is specious.
Since the all-trans polyene would certainly cross-link (Equation 6)
in the same manner as the cis,trans polyene structure (though perhaps
at a different rate), good smoke-char correlations might yet ensue,
even if metal catalysis of reaction 3 (or metal catalysis of other
cross-linking reactions[4,5]) did not occur at all. Thus the relative
importance of reactions 3-5 must still be regarded as an open question, insofar as their ability to inhibit smoke is concerned.

The other mechanistic theory is that of Lattimer and Kroenke,[4]
who have suggested that aromatics formation is suppressed by a "reductive coupling" reaction which causes the polymer to cross-link extensively in a very early stage of its thermal destruction. This
reaction can be described by Equation 7, where R = alkyl or allyl,
Mo^n is a "low-valent" form of molybdenum, and the metal ligands
remain unspecified. In order for the coupling to achieve its req-
uisite degree of catalytic effectiveness, Mo^{n+1}

$$2RCl \;+\; 2Mo^n \longrightarrow R\text{-}R \;+\; 2Mo^{n+1}Cl \qquad\qquad (7)$$

$$2Mo^{n+1}Cl \;+\; \text{-CH=CH-} \longrightarrow \text{-CH=CCl-} \;+\; HCl \;+\; 2Mo^n \qquad (8)$$

must rapidly revert to Mo^n. The way in which this reduction occurs
has not been stated clearly, although it has been noted that the
overall redox cycle must somehow lead to an accelerated evolution
of HCl.[4] No metal-catalyzed reaction yielding HCl was proposed specifically for the molybdenum system,[4] although such a reaction was
suggested for systems containing copper. This reaction[4] is a polyene

chlorination process which, in the case of molybdenum, can be represented by Equation 8. However, reaction 8 reintroduces chlorine into
the polymer (one Cl is reincorporated for every cross-link formed),
and it is therefore inconsistent with the failure of MoO_3 to cause
significant reductions in the yield of HCl during the thermal degradation of PVC.[14],[15] (For example, under conditions where MoO_3 reduces
the amount of benzene by 60%, the total yield of HCl is not reduced
at all.[15]) Furthermore, if appreciable amounts of chlorinated polyenes were formed in the presence of molybdenum, then the thermolysis
of these structures should yield significant amounts of volatile aryl
chlorides (by analogy with the behavior of similar systems[16],[17]), in
contrast to the experimental observations of Lattimer and Kroenke.[4]
Finally, since reductive coupling would inhibit the "zipper" dehydrochlorination of the polymer,[4] it seems incapable of accounting for
the dehydrochlorination rate enhancements that occur in MoO_3-contain-
ing PVC.[1-3],[7],[13],[14] The "zipper" process yields many molecules of
HCl; whereas the redox cycle of Equations 7 and 8 produces only one.
Thus accelerated dehydrochlorination requires this cycle to be much
faster than the ordinary thermal loss of HCl. Under these circumstances, cross-linking of the polymer would have to be very extensive
indeed, and the HCl loss caused by rechlorination (Equation 8) should
certainly be large enough to be detectable.

Hence the reductive coupling mechanism seems to be at variance
with a number of well-established facts. We now wish to describe,
in a preliminary way, the results of further experiments that have
provided additional evidence against this mechanism and have reaf-
firmed the validity of the mechanism involving Lewis acid catalysis.

RESULTS AND DISCUSSION

Model Compound Studies

Compounds(I)-(V) were selected as convenient models for a cis
alkene linkage (I), an ordinary sec-C-Cl bond (II and III), and an
allylic chloride structure (IV and V) in PVC. These

(I)   (II)   (III)

(IV)   (V)

substances were allowed to react with catalytic amounts of MoO_3 and
other molybdenum compounds that were chosen for the reasons described
below. The reactions were carried out in sealed tubes using thor-

oughly degassed mixtures or, if convenient, in open flasks under nitrogen. Products were identified by GC and/or GC/MS methods, using
authentic samples for comparison whenever possible.

Infrared analysis showed that (I) experienced no conversion into
its trans isomer upon exposure to MoO_3 (7 wt. %) for 1 hr. at 250°C.
However, this observation does not exclude the possibility of cis-
alkene isomerization by a more acidic species created in situ during
the degradation of the polymer. Such a species is MoO_2Cl_2, which is
formed from MoO_3 and HCl at elevated temperatures[13,14,18] (e.g., at
207[18] or 250°C[13]) and has recently been identified among the volatile
pyrolysis products of MoO_3-PVC mixtures containing large amounts of
MoO_3.[15]

In fact, when (I) was stirred with a catalytic quantity of
MoO_2Cl_2 for 1 hr. at a temperature of only 100°C, the formation of
a considerable amount of trans alkene was revealed by the appearance
of a medium-intensity IR band at 965 cm^{-1}. Also, after a mixture
containing comparable molar amounts of (I) and (III) had been allowed
to react with MoO_3 (4 wt. %) at 200°C for 15 min., GC analysis showed
that trans-(I) and pentenes had been produced in yields of ca. 10
and 20%, respectively. This result is consistent with the isomerization of (I) by MoO_2Cl_2 that was generated in situ, although it can
be argued that the isomerization was actually caused by the liberated
HCl. However, control experiments (see below) indicated that the
dehydrochlorination of (III) is catalyzed by MoO_3 at 200°C. Thus,
when an alkyl chloride is present, the cis-to-trans conversion must
be facilitated by the introduction of MoO_3. This observation and
the other findings described above provide strong support for the
previously postulated[1,2] acceleration of reaction 5 in MoO_3-con-
taining PVC.

Reaction of (III) with MoO_3 (6 wt. %) at 160°C for 15 min. gave
a mixture of the three isomeric pentenes (yield, ca. 6%) but none
of the reductive coupling product, 3,4-diethylhexane (as determined
by GC analysis using an authentic reference specimen). In another
run performed at 190°C for 18 min. with 5 wt. % of MoO_3, very extensive reaction occurred to produce a great many species eluting
both before and after (III). No one species was prominent, and if
any 3,4-diethylhexane were present, it could only have been represented by a very small shoulder on one of the more than 50 GC peaks.
Control runs performed without MoO_3 showed that (III) was stable at
160°C and that it only experienced a minor amount of dehydrochlorination after 11 min. at 200°C (to produce pentenes in a yield of ca.
11%). On the other hand, when (III) was allowed to react with
MoO_2Cl_2 (19 wt. %) at 160°C for 15 min., considerable decomposition
again occurred; no 3,4-diethylhexane could be detected; and the GC
trace was essentially identical to the one obtained in the 190-°C
experiment with (III) and MoO_3.

More information was obtained with compound (II). After 15 min.
at 200°C in the presence of MoO_3 (1 wt. %) and an internal GC standard ($\underline{n}$-$C_{16}H_{34}$, 40 wt. %), (II) had been largely converted into several isomeric tridecenes (yield, ca. 80%), together with a few minor
products eluting near the starting chloride. Under GC conditions
ensuring the elution of an external standard, $\underline{n}$-$C_{28}H_{58}$, no peaks were
observed with retention times longer than that of the $\underline{n}$-$C_{16}H_{34}$. Thus
no reductive coupling product ($C_{26}H_{54}$) could have been formed in this
reaction. In a control experiment, (II) underwent essentially no dehydrochlorination after 1 hr. at 200°C. However, when (II) was
treated with MoO_2Cl_2 (17 wt. %) at 300°C for 15 min., a very complex
mixture resulted; and when 6-tridecene was treated similarly, an essentially identical GC trace was found, although 6-tridecene was quite
stable at 300°C in the absence of molybdenum compounds. Both of the
complex mixtures contained considerable amounts of heavy products
(e.g., $C_{26}H_{52}$ and $C_{39}H_{78}$), as well as a great many species eluting
before the starting substrates. By means of GC/(high-resolution MS)
analysis, the light substances were shown to be mostly alkanes and
alkenes with carbon numbers ranging down to at least C_6, together
with minor amounts of alkylaromatics and traces of oxygenated compounds. A similar complex mixture resulted when (II) was treated
with a catalytic amount of MoO_3 at 260°C for 30 min.

The nature of the light products and their formation in the
presence of metallic acids are certainly consistent with the operation of a cracking process involving carbocation intermediates.[19,20]
Nevertheless, a detailed study of the cracking mechanism would clearly
be worthwhile.

Experiments performed with (IV) and (V) showed that the dehydrochlorination of these compounds was accelerated by MoO_3 and MoO_2Cl_2
and that the latter substance was by far the most active catalyst
(even at room temperature, it was very effective). Results of particular significance were obtained from a reaction of (V) with MoO_3
(3 wt. %) at 100°C for 2 min. The resultant mixture was much too
complex for complete characterization, but GC and GC/MS analyses
showed that it contained (V) (about 50%), 1,3-pentadiene (1%), a
compound (or isomer mixture) having $\underline{m}/\underline{e}$ = 136 ($C_{10}H_{16}$), at least
three substances (the principal products) with molecular formulas of
$C_{10}H_{17}Cl$, and at least one substance with a molecular formula of
$C_{15}H_{25}Cl$. Only a trace (if any) of the reductive coupling product
could have been present, since no significant peak eluting in the
C_{10} region had $\underline{m}/\underline{e}$ = 138. Moreover, when (V) was treated with MoO_2Cl_2
(2 wt. %) at ambient temperature, a detectable amount of reaction
occurred after only 6 min. to yield a complex mixture whose GC trace
was very similar to the one obtained in the preceding experiment.

Under the conditions of the smoke-char test used by Lattimer
and Kroenke, molybdenum was stated to occur in the char residue as
MoO_2 and Mo_2C.[4] The data that led to this conclusion were not re-

ported, and we are unaware of any conclusive evidence for the formation of these substances in PVC at 200-350°C, which is the temperature
range where most of the benzene is evolved during programmed temperature pyrolysis in the absence of MoO_3.[2,3,7,14] Nevertheless, the
present study has included an examination of the reactions of MoO_2
and Mo_2C with compounds (II)-(V).

Small amounts (6-7 wt. %) of MoO_2 and Mo_2C were found to be
reasonably effective catalysts for the dehydrochlorination and isomerization of (II) at 200°C. After times ranging up to 1 hr., no GC
evidence was obtained for the occurrence of other reactions; and since
no peaks eluted after (II), no reductive coupling could have taken
place. Analogous results were forthcoming from similar experiments
performed with (III) at 160°C for 15 min. Again there was no GC evidence for the formation of heavy materials or cracking products, and
the compound that would have resulted from reductive coupling (3,4-
diethylhexane) was conclusively shown to be absent. In experiments
with (IV) that were carried out at 300°C for 15 min., MoO_2 (8 wt. %)
and Mo_2C (7 wt. %) were found to catalyze dehydrochlorination and
the formation of a complex array of heavy products. However, as
before, no fragmentation was detected, presumably because the Lewis
acidities of these molybdenum compounds (and their chlorinated derivatives?) are relatively low. Lastly, when (V) was allowed to react
with MoO_2 (4 wt. %) or Mo_2C (5 wt. %) at 100°C for 15 and 5 min.,
respectively, only catalyzed dehydrochlorination and the formation
of heavy materials occurred. On the basis of their GC retention
times, the principal products appeared to be monohalogenated dimers,
although the material with $\underline{m/e}$ = 136 was formed also in low yield
(cf. the results, already mentioned, that were obtained with (V) and
MoO_3).

Although many details are lacking, we believe that a clear general picture emerges from the experiments described above. Acting as
a Lewis acid (and, perhaps, even as a Brønsted acid, owing to the
presence of a small amount of surface hydroxyl), MoO_3 can catalyze
the dehydrochlorination of PVC. The resulting HCl converts MoO_3 into
MoO_2Cl_2, a more potent acidic catalyst, which accelerates reactions
3 and 5 and thus inhibits smoke production. Reaction 6 is indoubtably
accelerated also, and our model compound studies suggest that the
acid-catalyzed cross-linking processes are polyene dimerization (or
oligomerization) and polyene haloalkylation. (Metal catalysis of
reaction 4 remains as a reasonable possibility,[1,2] but the present
work has provided no evidence for or against its occurrence.) Simultaneously and/or subsequently, as the temperature of the system
increases, MoO_2Cl_2 (and, possibly, other metallic acids formed in
situ) can catalyze a cracking process which converts the cross-linked
polymer into hydrocarbon fragments that comprise an excellent fuel.
Other investigators[4,8,13,21] have obtained product evidence for the
occurrence of catalytic cracking in MoO_3-containing PVC, but the

likely involvement of carbocations in this reaction does not seem to
have been considered heretofore. Some of the best evidence for catalytic cracking has been provided by Lattimer and Kroenke,[4] who found
that MoO_3 increased the yield of volatile aliphatic hydrocarbons
"rather dramatically" during the flash pyrolysis of PVC. The formation of products resulting from catalytic cracking is almost certain
to be responsible, in part at least, for the accelerated flame spread
observed at very high temperatures with PVC containing MoO_3.[2]

Mechanism for the Formation of Volatile Aromatics from PVC

Using a GC/MS procedure, Lattimer and Kroenke[4,22] have carefully
measured the amounts of H/D scrambling in several of the volatile
products resulting from the flash pyrolysis of PVC-h_3/PVC-d_3 mixtures
at 550°C. The pyrolyses were conducted both in the presence[4] and in
the absence[22] of MoO_3, and from the results obtained, it was concluded that when MoO_3 is absent, "pure" aromatics (such as benzene
and naphthalene) are formed primarily in an intramolecular manner;
whereas "mixed" aliphatic-aromatic products (such as toluene) are
formed via intermolecular as well as intramolecular routes.[22] When
MoO_3 was present, the extents of H/D scrambling were enhanced.[4] This
finding was taken as evidence for the formation of larger proportions
of the volatile products from cross-linked segments that were created
by reductive coupling,[4] even though it was realized that the scrambling
could have resulted simply from intermolecular hydrogen transfers.[4,22]
Furthermore, the enhanced scrambling caused by MoO_3 was said to be
inconsistent with the "cis-trans" part of the Lewis acid theory of
additive action, a contention which was based on the belief that
this theory "does not predict or require cross-linking or intermolecular hydrogen exchange."[4]

The latter statement is extremely misleading. In none of our
earlier papers dealing with the Lewis acid theory[1-3] is there any as-
sertion, either direct or implied, to indicate that the cross-linking
of all-trans polyenes cannot occur. The scheme represented by Equations 1-6 is, in fact, precisely the one that was envisaged when the
Lewis acid theory was first proposed; although metal catalysis of
reaction 6 was not discussed explicitly at that time, because this
reaction does not necessarily contribute to benzene suppression (it
might contribute, of course, if reaction 5 were a mobile equilibrium,
a situation that is not unlikely). Moreover, the Lewis acid mechanism
certainly does not exclude the possible occurrence of side reactions
involving intermolecular hydrogen migrations. Thus the enhanced H/D
scrambling caused by MoO_3 cannot be used as evidence against the "cis-
trans" part of the Lewis acid hypothesis.

Nevertheless, the mechanism for the formation of volatile aromatics is still of intrinsic interest. Lattimer and Kroenke[4,22] have
noted that experiments with ^{13}C-enriched PVC should enable one to

determine whether the H/D scrambling in the volatile products is caused by cross-linking reactions or intermolecular hydrogen exchange. This possibility had occurred to us independently, and we now report the results of a brief study along this line.

Table I shows the parent-ion isotopic distributions for three products resulting from the flash pyrolysis of coprecipitated 1:1 mixtures of PVC-$\underline{h}_3$ with PVC-$\underline{d}_3$ or PVC-$^{13}C_2$. The experimental procedures used in this work were very similar to those of Lattimer and Kroenke,[4,22] except for our use of low-energy electrons (~14 eV) in order to obtain the mass spectra. The relative intensities (RI's) in Table I have been corrected for incomplete isotopic enrichment, molecular-ion fragmentation, and the natural abundance of ^{13}C; and the last column of the table contains the percentages of intermolecular product formation that the RI data imply.

The isotopic distributions obtained with the PVC-$\underline{h}_3$/PVC-$\underline{d}_3$ mixture are in reasonable agreement with those reported by Lattimer and Kroenke.[4,22] These results suggest that the toluene is formed primarily in an intermolecular manner and that most of the benzene and naphthalene are produced via intramolecular paths, although a significant amount of the latter product, in particular, seems to have been formed by an alternate route.

A very different picture is revealed by the results obtained with the mixture containing ^{13}C. For every product, the $^{12}C/^{13}C$ scrambling is much less than the H/D scrambling observed with the other mixture, and for toluene the discrepancy is especially severe. Instrumental noise may have contributed significantly to the measured intensities of the weaker ionic species, particularly with toluene and naphthalene, whose pyrolysis yields were relatively low. Thus the tabulated extents of $^{12}C/^{13}C$ scrambling must be regarded only as the maximum possible values, and we can conclude that, within the error limits of our measurements, all three products could have been formed exclusively by routes that are intramolecular with respect to the carbon skeletons. Most (or all) of the H/D scrambling must have been caused by intermolecular hydrogen exchange, and it is now clear that the observation of such scrambling is virtually useless as a means for establishing the formation of pyrolysis products from cross-linked polymer segments. Indeed, new pyrolysis experiments with PVC-$\underline{h}_3$/PVC-$\underline{d}_3$ mixtures[23] have provided strong support for this conclusion and have shown that the amounts of H/D scrambling observed with such mixtures are strongly dependent upon sample size.[23] Moreover, subsequent to our initial description of the work reported here,[24] Lattimer et al.[25] have revealed some pyrolysis results for PVC-$\underline{h}_3$/PVC-$^{13}C_2$ mixtures that are in good accord with ours. Thus there now seems to be general agreement that the volatile aromatics derived from PVC are formed primarily, or even entirely, by mechanisms that are intramolecular insofar as the carbon frameworks are concerned.

Table I. Isotopic Distributions in Some PVC Pyrolysis
Products Formed at 550°C

Benzene

	m/e:	78	79	80	81	82	83	84					Max. % Inter.[a]
RI[b] (PVC-$\underline{h}_3$/-$\underline{d}_3$):		100	4	3	2	2	5	83					8
RI[b] (PVC-$\underline{h}_3$/-$^{13}C_2$):		100	0	2	1	1	1	89					3

Toluene

	m/e:	92	93	94	95	96	97	98	99	100			Max. % Inter.[a]
RI[b] (PVC-$\underline{h}_3$/-$\underline{d}_3$):		73	81	55	28	28	39	76	100	52			77
RI[b] (PVC-$\underline{h}_3$/-$^{13}C_2$):		100	1	8	4	5	3	7	66	-			14

Naphthalene

	m/e:	128	129	130	131	132	133	134	135	136	137	138	Max. % Inter.[a]
RI[b] (PVC-$\underline{h}_3$/-$\underline{d}_3$):		100	11	7	4	5	2	3	10	48	-	-	22
RI[b] (PVC-$\underline{h}_3$/-$^{13}C_2$):		84	2	5	2	2	2	2	3	5	0	100	11

[a] Maximum possible percentage of intermolecular reaction, calculated
from the RI values. See text for discussion.

[b] Parent-ion relative intensity, corrected as described in the text.

Naphthalene might arise from an o-divinylbenzene-containing segment, via a mechanism analogous to that suggested for benzene production.[1] Toluene could be derived from a chain-end benzyl
moiety,[23,26] whose presence is, in fact, suggested by the results of
a recent ^{1}H NMR study.[27] The benzyl group could be produced by the
removal of a carbon-containing substituent and a tertiary hydrogen
from an o-disubstituted cyclohexadiene[1] which is a potential benzene
precursor, as well. Various workers[2,3,7,14,28] have shown that toluene and the other "mixed" aromatics are formed, to a major extent,
at pyrolysis temperatures which are much higher than those where most
of the benzene evolution occurs. This observation can account, at
least partially, for the relatively extensive H/D scrambling found
for the "mixed" aromatics, in general, under some experimental conditions.

Concluding Remarks

In addition to the points addressed in this paper, other objec-
tions to the Lewis acid theory of additive action have been raised
by Lattimer and Kroenke.[4] All of these arguments can be answered,
and we intend to respond to them fully at a later time. In view of
the results reported above and in earlier publications,[1-3] we believe
that the Lewis acid mechanism has now been verified abundantly in the
case of MoO_3. The scope of this mechanism will be discussed in subsequent communications from this laboratory.

EXPERIMENTAL SECTION

Materials

The MoO_3 was obtained from Climax Molybdenum, and the other
molybdenum compounds were supplied by Alfa Products. The PVC-h_3 was
also a commercial material (Geon 103EP from B.F. Goodrich). Samples
of PVC-d_3 and PVC-$^{13}C_2$ were prepared from the corresponding labeled
monomers (Merck) using standard polymerization procedures, and these
polymers were found to contain 99.7 atom % D and 99.0 atom % ^{13}C,
respectively, from the isotopic distributions of their pyrolysis products. The other chemicals were either highly purified commercial substances or specimens prepared in these laboratories using adaptations
of published methods. Structures and purities were established with
the usual variety of spectroscopic techniques.

Analyses

Gas chromatograms were obtained on a Varian Model 3700 instrument
that was equipped with a flame ionization detector, a CDS-111 data
collection system, and (usually) a 6-ft. x 0.125-in. (i.d.) column
containing 10% of OV-101 on 80/100-mesh Gas Chrom Z. Separation of
(I) from its trans isomer required the use of a 6-ft. x 0.125-in.
(i.d.) column containing 10% of β,β'-oxydipropionitrile on 80/100-
mesh Gas Chrom Z. Some of the reaction mixtures obtained with (II)
were also subjected to GC analysis on a Hewlett-Packard instrument
(Model 5880A) that was equipped with a flame ionization detector, a
capillary injection system, and a data processing system. These
analyses were performed with wall-coated open tubular fused-silica
columns containing OV-101 or OV-1 and having dimensions of 12 m x
0.2 mm (i.d.) and 25 m x 0.2 mm (i.d.), respectively. Most of the
GC/MS data were obtained with a Varian MAT-112 spectrometer that was
interfaced with a Varian Aerograph GC instrument (Model 1400) containing a 6-ft. x 0.125-in. (i.d.) column of SE-30 (10%) on Chromo-
sorb W. Polymer pyrolyses were carried out with a CDS Pyroprobe

(Model 100), using a ribbon probe and a heating rate of ca. 75°C/msec (i.e., the fastest rate available). The pyrolysis time was 20 sec., and the final temperature was 550°C. The detailed analysis of the mixture obtained from (V) and MoO_3 at 100°C was performed by M. Brownawell of the Rutgers University Department of Chemistry, using a Hewlett-Packard capillary-GC/MS instrument (Model 5985) that was equipped with a data processing system and a 0.2 mm (i.d.) fused-silica column coated with OV-1. The GC/(high-resolution MS) analyses were done by Shrader Analytical and Consulting Laboratories, Inc., Detroit, Michigan.

ACKNOWLEDGEMENTS

We are greatly indebted to Drs. D. Edelson and R. M. Lum for valuable comments and unpublished information, and to Dr. R. P. Lattimer for a preprint of Ref. 25.

REFERENCES

1. W. H. Starnes, Jr., and D. Edelson, Macromolecules, 12, 797 (1979).
2. D. Edelson, V. J. Kuck, R. M. Lum, E. Scalco, W. H. Starnes, Jr., and S. Kaufman, Combust. Flame, 38, 271 (1980).
3. R. M. Lum, L. Seibles, D. Edelson and W. H. Starnes, Jr., Org. Coat. Plast. Chem., 43, 176 (1980).
4. R. P. Lattimer and W. J. Kroenke, J. Appl. Polym. Sci., 26, 1191 (1981).
5. W. J. Kroenke, J. Appl. Polym. Sci., 26, 1167 (1981).
6. A. Ballistreri, G. Montaudo, C. Puglisi, E. Scamporrino and D. Vitalini, J. Polym. Sci., Polym. Chem. Ed., 19, 1397 (1981).
7. A. Ballistreri, S. Foti, P. Maravigna, G. Montaudo, and E. Scamporrino, J. Polym. Sci., Polym. Chem. Ed., 18, 3101 (1980).
8. S. K. Brauman, J. Appl. Polym. Sci., 26, 353 (1981).
9. S. K. Brauman, J. Fire Retard. Chem., 7, 119 (1980).
10. F. W. Moore, T. R. Weber and G. A. Tsigdinos, J. Vinyl Technol., 3, 139 (1981).
11. F. W. Moore and G. A. Tsigdinos, Fire Retard., Proc. Int. Symp. Flammability Fire Retard. 1978, 160 (1978).
12. F. W. Moore and G. A. Tsigdinos, J. Less-Common Met., 54, 297 (1977).
13. T. C. Rees, presented at the 7th International Symposium on Flammability and Fire Retardants, New Orleans, LA, May, 1980.
14. R. M. Lum, J. Appl. Polym. Sci., 23, 1247 (1979).
15. R. M. Lum, private communication, 1981.
16. S. Tsuge, T. Okumoto and T. Takeuchi, Macromolecules, 2, 277 (1969).
17. S. A. Liebman, D. H. Ahlstrom, E. J. Quinn, A. G. Geigley and

J. T. Meluskey, J. Polym. Sci., Part A-1, 9, 1921 (1971).

18.   N. Hultgren and L. Brewer, J. Phys. Chem., 60, 947 (1956).

19.   B. C. Gates, J. R. Katzer and G. C. A. Schuit, Chemistry of Catalytic Processes, McGraw-Hill, New York, 1979, Chapters 1 and 3.

20.   C. D. Nenitzescu, Carbonium Ions, 2, 463 (1970).

21.   T. Iida and K. Gotō, J. Polym. Sci., Polym. Chem. Ed., 15, 2427 (1977).

22.   R. P. Lattimer and W. J. Kroenke, J. Appl. Polym. Sci., 25, 101 (1980).

23.   R. P. Lattimer and W. J. Kroenke, J. Appl. Polym. Sci., 27, 1355 (1982).

24.   W. H. Starnes, Jr., L. D. Wescott, Jr., W. D. Reents, Jr., R. E. Cais, G. M. Villacorta, I. M. Plitz and L. J. Anthony, Org. Coat. Plast. Chem., 46, 556 (1982).

25.   R. P. Lattimer, W. J. Kroenke and R. H. Backderf, J. Appl. Polym. Sci., in press.

26.   W. H. Starnes, Jr., Dev. Polym. Degradation, 3, 135 (1981).

27.   S. K. Brauman and I. J. Chen, J. Polym. Sci., Polym. Chem. Ed., 19, 495 (1981).

28.   P. Burille, M. Bert, A. Michel and A. Guyot, J. Polym. Sci., Polym. Lett. Ed., 16, 181 (1978).

DIFFUSION OF ADDITIVES AND PLASTICIZERS IN POLY(VINYL CHLORIDE)-III

DIFFUSION OF THREE PHTHALATE PLASTICIZERS IN POLY(VINYL CHLORIDE)

P. J. F. Griffiths, K. G. Krikor and G. S. Park

Applied Chemistry Department
UWIST
Cardiff CF1 3NU, Wales, UK

INTRODUCTION

Most PVC finds commercial use when modified by the action of plasticizers. In assessing the usefulness of the plasticizer, four characteristics have to be considered. These are: (1) cost; (2) compatibility; (3) efficiency; and (4) permanence. Of these characteristics, the third and fourth are both related, at least in part, to the mobility of the plasticizer molecules in the plasticized system. Thus, the effectiveness of the plasticizer in lowering the glass transition temperature will be expected to be related to the self-diffusion coefficient of the plasticizer molecules. The probability of loss of plasticizer from the surface of any artefact will depend, amongst other things, on the rate of diffusion to the surface. In order to assess plasticizer mobility, the self-diffusion coefficients of di-n-butylphthalate, di-n-hexylphthalate, and di-n-decylphthalate have been obtained in PVC over a range of concentrations and temperatures.

THEORY OF EXPERIMENTAL METHOD[1]

When a slab of polymer of thickness, h, containing a β-emitting radioactive diffusant, is joined face to face with a slab of thickness, m, containing non-radioactive diffusant at the same concentration, the β-emission of the initially non-radioactive surface increases with time and that of the initially radioactive surface decreases. It can be shown that the β-activity, A, from the initially non-radioactive face at time, t, is given by:

$$A_\ell = A_\infty [1 + \frac{2\ell[\exp(-\mu\ell) + 1]}{\pi h}] \sum_{n=1}^{n=\infty} \frac{(\sin n\pi h/\ell)\exp(-n^2\pi^2 Dt/\ell^2)}{n[1+n^2\pi^2/\ell^2\mu^2][(-1)^n+\exp(-\mu\ell)]} \quad (1)$$

Here, $\ell = h+m$, A_∞ is the activity at the surface after infinite time, μ is the β-particle absorption coefficient, and D is the self-diffusion coefficient. It is possible, from measurements of A_ℓ at various times, to obtain values of D from this expression using a computer but more simply it is easy to show that for large values of t equation (1) reduces to:

$$\ln(A_\infty - A_\ell) = P - \pi^2 Dt/\ell^2 \quad (2)$$

where P is a time invariant parameter. This enables values of D to be obtained from simple linear plots if a value for A_∞ can be found or estimated. This method suffers from two disadvantages. A reliable estimate of A_∞ can be difficult and in some circumstances linear plots found only at relatively large values of t. The 1:2 double disk method (1) in which h:m = 1:2 or 2:1 does not have these advantages. In this method if A_m is the activity from the initially radioactive surface, then to very considerable accuracy the relationship

$$\ln(A_m - A_\ell) = P' - \pi^2 Dt/\ell^2 \quad (3)$$

holds even at quite small values of t.

EXPERIMENTAL DETAILS

[14]C labelled dialkylphthalates were made by reacting equivalent amounts of ^{14}C carbonyl labelled phthalic anhydride (10 or 60 μCig^{-1}) and the required alcohol using 0.1% of toluene sulphonic acid as catalyst. 20% of toluene was added to form an azeotrope with the water formed in the reaction and this was collected in a Dean and Stark trap as the mixture was refluxed in a current of nitrogen at 100-140°C. When the reaction was completed, as judged from the volume of water formed, the product was purified by several redistillations at low pressure (~10^{-3} torr). The purity of the products was assessed by elemental analysis, measurement of refractive index, and infrared and nmr spectroscopy. Unlabelled phthalates were made by the same technique.

A single batch of mass polymerized PVC having $\overline{M}_n$ of 46,000 and $\overline{M}_w$ of 154,000 was kindly provided by B. P. Chemicals, Ltd. It was purified by solution in tetrahydrofuran and precipitation into a large excess of methanol followed by prolonged drying at 30°C in vacuum.

Plasticized PVC film (0.15 to 0.50 mm) was made from solution of the
appropriate amounts of PVC and dialkylphthalate in peroxide-free tetrahydrofuran by casting in a glass ring resting on a flat glass plate.
The last traces of solvent were removed using silica gel and vacuum.
20 mm diameter discs were cut from the films for the diffusion measurements which were started by joining, face to face, a radioactive
disc to a non-radioactive one of similar composition but of twice (or
half) the thickness in the ratio 1:2. The composite disc was then
mounted in a reversible holder which could be firmed and reproducibly
mounted a few millimetres below the window of an end window GM tube
in a lead castle. The whole assembly was placed in a thermostat and
the β-particle activity from the sample was obtained using a scaler
unit. The sample was inverted after each count had been made so that
the variation with time of the two activities, A_m and A_ℓ, could be
obtained. The plasticizer self-diffusion coefficients were obtained
from linear plots using Equation (3).

RESULTS AND DISCUSSION

Effect of Concentration and Temperature

     The variation of the self-diffusion coefficients of di-n-butyl,
di-n-hexyl and di-n-decylphthalates with concentration are shown in
Figures 1, 2 and 3. The considerable increase that occurs with increasing temperature and with increasing concentration of diffusant
is reminiscent of the data obtained previously for the diffusion coefficients of dibutyltin dilaurate in PVC plasticized with diethy-
hexylphthalate.[2] Free volume theories have proved very useful for
correlating the diffusion coefficients of vapors in polymers at low
concentrations with vapor concentration and temperature. In these
theories the probability of occurrence of a unit diffusion jump in
the polymer is proportional to the probability of the occurrence, by
random thermal fluctuation, of a region of sufficiently high specific
volume in the diffusion medium. Perhaps the simplest treatment is
that due to Fujita[3] and this results in the relationship

$$\ln(D/RT) \;=\; \ln A - B/f_v \qquad\qquad (4)$$

Here, A and B are parameters and f_v is the fractional free volume in
the diffusion medium. In order to apply this relationship to the
diffusion of dibutyltin dilaurate in plasticized PVC systems, it was
assumed that the only contribution to the free volume came from the
plasticizer.[2] Here, we give a somewhat more sophisticated treatment
which takes into account the free volume of the polymer and enables
the effects of both concentration and temperature to be considered.

     The well-known relationship of Williams, Landel and Ferry[4]

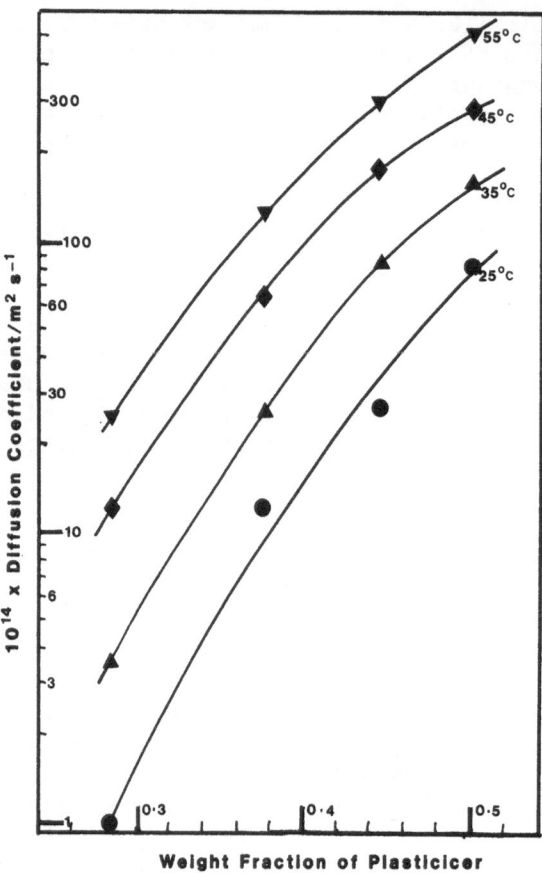

Figure 1. Self-diffusion coefficients of di-n-butylphthalate in PVC.

$$f_v = f_g + \alpha(T - T_g) \tag{5}$$

relates the free volume, f_v, of a polymer to the temperature, T. In
this relationship, f_g is the fractional free volume at the glass
transition temperature, and α is the thermal expansion coefficient
of the free volume. Both this and the glass temperature, T_g, will
depend on the plasticizer concentration. In order to enable free
volume, f_v, to be calculated from Equation (5), T_g values for all the
dibutylphthalate and didecylphthalate compositions were obtained
using a thermomechanical analyzer. The values are given in Table I.
Estimated values of α for polymer plasticizer systems have been made
from the simple additive relationship

$$\alpha = \alpha_2(1 - \phi_1) + \alpha_1 \phi_1 \tag{6}$$

Here, ϕ_1 is the volume fraction of plasticizer, while α_1 and α_2 are

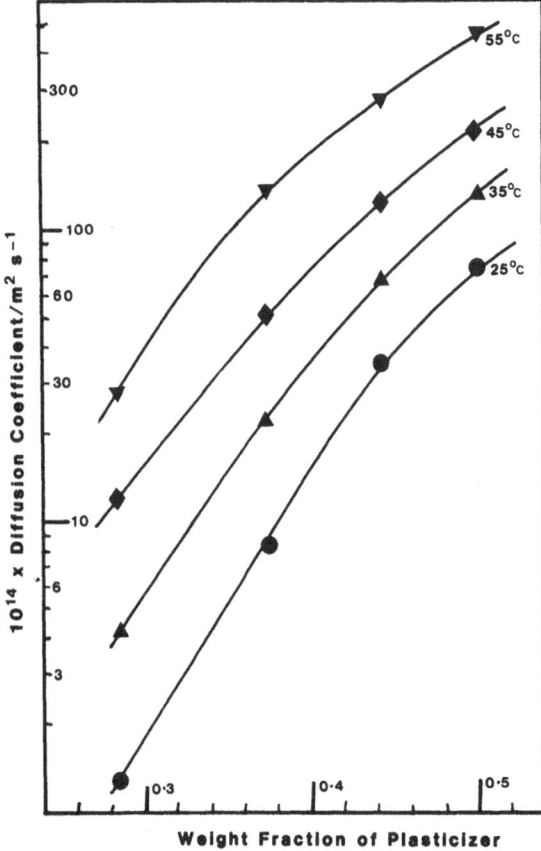

Figure 2. Self-diffusion coefficients of di-n-hexylphthalate in PVC.

the thermal expansion coefficients of free volume for the plasticizer
and for the polymer, respectively. Taking a value of $10^{-3}K^{-1}$ for
α_1[5] and assuming the universal value of $0.48 \times 10^{-3}K^{-1}$ for α_2, α
values were calculated for the polymer plasticizer compositions.
These were combined with the universal value of 0.025 for f_g and the
glass transition data of Table I to give the free volume figures in
that table.

     Plots of D/T against $1/f_v$ for the dibutylphthalate and the di-
decylphthalate are given in Figures 4 and 5. In both cases reasonably straight lines can be drawn through the points which cover four
plasticizer concentrations and four temperatures. From the slopes
of the regression lines, the β-values obtained were 0.41 for the
dibutylphthalate and 0.52 for the didecylphthalate. These plots show
that Fujita's free volume relationship fits reasonably well for the
self-diffusion of the plasticizers in plasticized PVC and it indicates that there should be a reasonable correlation between the self-

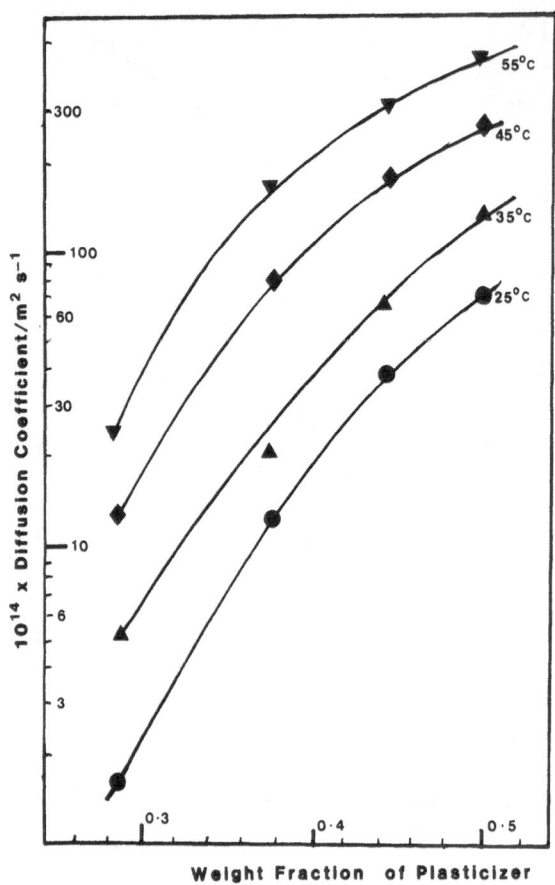

Figure 3. Self-diffusion coefficients of di-n-decylphthalate in PVC.

diffusion coefficients of plasticizers and their plasticizing efficiency as indicated by the extent of introduction of free volume and
the lowering of the glass transition temperature.

Effect of Size of Plasticizer Molecule

It is well established that diffusion coefficients in many polymeric systems decrease rapidly with increasing size of the diffusant.
It is clear from Table II, however, which gives all the present self-
diffusion coefficient data, that there is in general no clear dependence of D on size for these systems. This is presumably a result
of the dual role of the plasticizer which acts both as diffusant and,
to some extent, as diffusion medium. The free volume data in Table
I indicate an increase in free volume with increasing size of plasticizer at low plasticizer concentrations and a decrease in free
volume with increasing size of plasticizer at high plasticizer concentrations. These variations of free volume with plasticizer size

Table I. Glass Transition Temperatures, T_g, and Fractional Free Volume, f_v, of PVC Plasticized with the di-n-alkylphthalates $C_6H_5 (COO C_n H_{2n+1})_2$

PHTHALATE				$T_g/°C$	f_v x 100 at.			
n	PHR	Wt fraction	Vol fraction		25°C	35°C	45°C	55°C
4	40	0.286	0.230	2.5	3.85	4.45	5.05	5.65
4	60	0.375	0.309	-10.0	4.74	5.38	6.03	6.67
4	80	0.444	0.377	-27.5	6.05	6.73	7.40	8.08
4	100	0.500	0.431	-40.0	7.08	7.78	8.49	9.19
10	40	0.286	0.217	- 7.5	4.43	5.02	5.61	6.21
10	60	0.375	0.295	-13.5	4.94	5.57	6.21	6.84
10	80	0.444	0.359	-26.5	5.93	6.60	7.27	7.94
10	100	0.500	0.414	-35.0	6.67	7.37	8.06	8.76

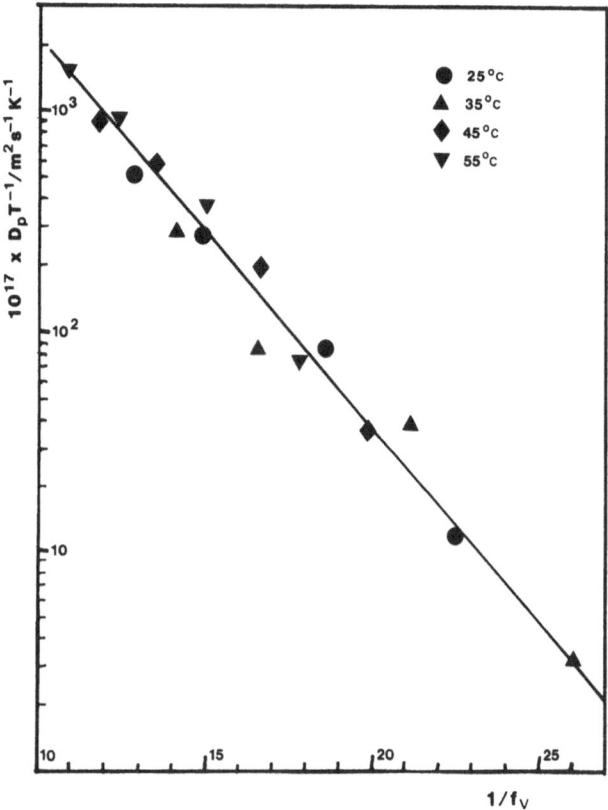

Figure 4. Relationship between the free volume, f_v, and the self-diffusion coefficient, D_p, of di-n-butylphthalate in PVC.

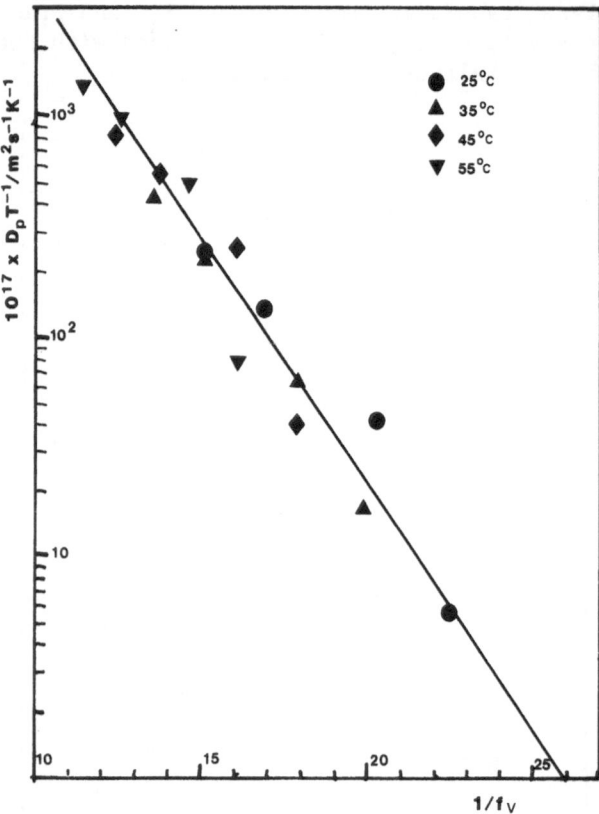

Figure 5.   Relationship between the free volume, f_v, and the self-
diffusion coefficient, D_p, of di-n-decylphthalate in PVC.

correlate quite well with the changes of the diffusion coefficient
with plasticizer size in Table II and could account for the different
variations of the diffusion coefficients at high and low concentrations.   In order to investigate this further, the effect of the diffusion medium on the diffusion coefficient was studied in isolation
by measuring diffusion coefficients of n-hexadecane in the different
plasticized systems using the 1:2 double disk method.   The data obtained are given in Table III.   Figure 6 shows clearly the increase
in the diffusion coefficient of the n-hexadecane brought about by
the increasing size of the plasticizer at low plasticizer concentrations showing that at 40 PHR of plasticizer the largest plasticizer
molecules introduce most free volume and can in these terms be regarded as the most efficient plasticizers.   At higher plasticizer
concentrations the variations with plasticizer size are less simple
but the D_h values in each series show the effect of plasticizer size
on the diffusion of a constant size diffusant.

A simple, if approximate, method of eliminating the effect of

Table II.   Effect of Molecular Size on Self-Diffusion Coefficient, D_p, C_6H_5 (COO $C_nH_{2n+1})_2$ Plasticizer

PHR	T/°C	10^{14} D_p/m^{2}s^{-1} for n =		
		4	6	10
40	25	0.97	1.30	1.65
40	35	3.65	4.25	5.0
40	45	11.6	12.0	13.3
40	55	24.7	28.5	25.7
60	25	11.8	8.55	12.9
60	35	26.1	21.7	19.4
60	45	65.2	52.2	83.3
60	55	124.9	134.5	165.7
80	25	25.5	34.7	41.2
80	35	84.8	67.8	69.6
80	45	185.0	123.2	180.5
80	55	303.2	277.5	319.9
100	25	86.8	73.9	74.2
100	35	160.4	133.7	142.3
100	45	284.1	227.7	269.9
100	55	511.8	477.5	455.2

Table III.   Effect of Plasticizer Size [n in C_6H_5 (COO $C_nH_{2n+1})_2$]
on Diffusion Coefficient D_h of n-Hexadecane and the
Self-Diffusion Coefficient D_p of Plasticizer Relative
to n-Hexadecane

PHR	T/°C	10^{14} D_h for n =			$10D_p/D_h$ for n =		
		4	6	10	4	6	10
40	25	7.7	12.0	23.6	1.24	1.09	0.68
40	35	19.4	39.6	77.8	1.89	1.14	0.64
40	45	59.6	84.3	157.3	2.06	1.43	0.84
40	55	149.1	204.6	239.3	1.66	1.39	1.07
100	25	461.8	503.7	520.7	1.88	1.47	1.43
100	35	868.9	827.4	1000	1.85	1.62	1.42
100	45	1423	1248	1598	1.98	1.82	1.69
100	55	2256	1561	1923	2.27	3.06	2.32

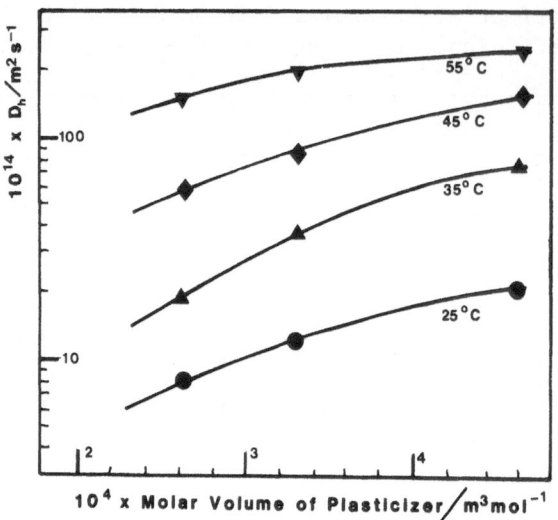

Figure 6. Diffusion coefficient, D_h, of n-hexadecane in plasticized
PVC containing 40 PHR of three phthalate plasticizers.

changes of the medium and isolating the effect of the diffusant's
size on the self-diffusion coefficient of the plasticizers, is obtained if the self-diffusion coefficient of the plasticizer is divided
by the diffusion coefficient of the standard probe molecule n-hexadecane. Values of the ratio D_p/D_h are included in Table III and there
is a clear fall of this ratio with increasing plasticizer size except
at 55°C (100 PHR).

Empirical Relationship Between Diffusion Coefficient and Activation Energy for Diffusion

For the limited range of temperatures covered in the present
study the diffusion coefficients can be well represented by the
Arrhenius relationship

$$D = D_0 \exp(-E_D/RT) \tag{7}$$

The E_D values for the self-diffusion coefficients of all three phthalates and for the diffusion coefficients of the n-hexadecane in all
the plasticized systems are given in Table IV. The empirical relationship

$$\log(D_0/m^2s^{-1}) = C_1 + C_2 E_D/RT \tag{8}$$

has been shown to apply to many diffusion phenomena and, in particular, Barrer and Chio[6] have collected together data for the diffusion
of small molecules in a number of elastomers and have shown that
values of $C_1 = -8.20$ and $C_2 = 0.272$ are obtained for Equation 8.

Table IV. Activation Energies, E_D, for Diffusion in Plasticized PVC

Phthalate		Self diffusion of plasticizer			Diffusion of n-hexadecane		
n	PHR	10^{14} D(45°C) /$m^2 s^{-1}$	E_D / kJ mol		10^{14}D(45°C) $m^2 s^{-1}$	E_D / kJ mol^{-1}	
			Measured	Calculated		Measured	Calculated
4	40	12.2	80	77	59.6	81	66
4	60	65.2	63	68	-	-	-
4	80	185.0	54	58	-	-	-
4	100	284.1	50	55	1423	42	51
6	40	12.0	85	77	84.3	77	63
6	60	52.4	75	67	-	-	-
6	80	123.2	57	61	-	-	-
6	100	227.7	49	56	1248	38	44
10	40	13.4	74	76	157.3	73	59
10	60	83.7	71	63	-	-	-
10	80	180.1	57	58	-	-	-
10	100	269.8	49	55	1598	36	42

Equations 7 and 8 can be combined together to give

$$\log(D/m^2 s^{-1}) = C_1 + (C_2 - 0.4343)E_D/RT \qquad (9)$$

Values of D at 45°C are included in Table IV and using these in Equation 9 together with the C_1 and C_2 values given by Barrer and Chio[6] the calculated E_D values also given in that table are obtained. There is considerable agreement between these calculated values from the self-diffusion coefficients of the plasticizers and the measured ones showing that even in these plasticized systems which are very different from the systems investigated by Barrer and Chio, their C_1 and C_2 values are satisfactory and enable estimates of the activation energies from single diffusion coefficient measurements. The agreement for the n-hexadecane diffusion is not as good but the calculated values still give a rough approximation to the true activation energy figures.

SUMMARY

The self-diffusion coefficients, D_p, of the plasticizers di-n-butylphthalate, di-n-hexylphthalate and di-n-decylphthalate, have been obtained at 25, 35, 45 and 55°C in PVC at concentrations of 40, 60, 80 and 100 PHR. Increases in D_p of about 50-fold occur over this concentration range and about 20-fold for a 30°C increase of temperature. These increases have been correlated with fractional free volumes calculated from glass transition temperatures measured by thermomechanical analysis of the di-n-butyl- and di-n-decylphthalate PVC mixtures. The effect of plasticizer size on the diffusion medium has been investigated by obtaining the diffusion coefficients, D_h, of n-hexadecane in the plasticized systems. It is found (with one exception) that the ratio, D_p/D_h, decreases with increasing plasticizer size. The activation energies, E_D, and preexponential terms, D_0, for the self-diffusion, can be correlated using an expression of the form

$$\log D_0 = C_1 + C_2 E_D/RT$$

ACKNOWLEDGEMENTS

We wish to thank Dr. Marianne Gilbert (Loughborough University of Technology) for the arrangements to use the thermomechanical analyzer.

REFERENCES

1. G. S. Park and V. H. Tran, European Polymer Journal, 15, 817 (1979).
2. G. S. Park and V. H. Tran, European Polymer Journal, 16, 779 (1980).
3. H. Fujita, Fortschr. Hochpolymer. Forsch., 3, 1 (1961).
4. M. L. Williams, R. S. Landel and J. D. Ferry, J. Am. Chem. Soc., 77, 3701 (1955).
5. L. J. Garfield and S. E. Petrei, J. Phys. Chem., 68, 1750 (1964).
6. R. M. Barrer and H. T. Chio, J. Polym. Sci., C10, 111 (1965).

EFFECT OF ADDITIVES ON PROPERTIES AND COATING PROCESSES

OF POLYPHENYLQUINOXALINES

Lu Fengcai, Wang Beiging and Chang Jinbiao

Institute of Chemistry
Academia Sinica
Beijing, China

INTRODUCTION

Studies have shown that polyphenylquinoxalines not only possess good thermal oxidative stability and good hydrolytic stability, but also good chemical and oil resistant properties.[1] They were soluble in tetrachloroethane, chloroform and cresol, but the solution exhibited high viscosity even though the concentration of solution was as low as 8-10% by weight. We studied methods of lowering the viscosity while increasing the concentration, and discuss the results in this paper.

Polyphenylquinoxalines were synthesized from the reaction of aromatic tetraamines with aromatic dibenzils according to Equation 1:

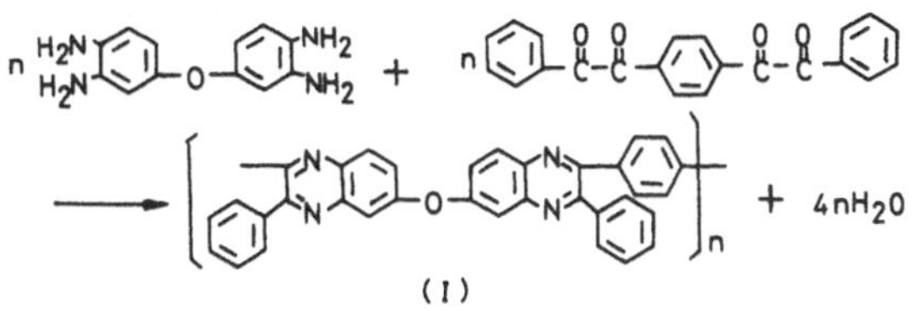

(I)

The general procedure for the preparation of these polymers involved mixing equal molar ratios of the two monomers as fine powders in cresol at ambient temperature. Due to the high purity of the monomers, a 10% (by weight) solution of stoichiometric quantities of the monomers became very viscous or extremely viscous during polymerization (η_{inh} 3.4 dl/g, 0.5% in cresol solution at 30°C). Therefore it is very difficult to use these polymers to form film or to coat wire. In order to improve the processability of these polymers,

studies were carried out of the effects of molar ratios of monomers;
temperature; solvent-diluent; and the additive on the viscosity and
solid content leading to a satisfactory processing method while
retaining the good properties of polymers. In addition, application
of 4,4'-bis(4"-cyanophenoxy)benzil as end-capped group in the synthesis of polymer resulted in large improvements of thermal stability,
mechanical properties as well as hydrolytic stability as compared
with those uncapped ones. Tg was also increased by more than 50°C
due to partial crosslinking of the cyano group to reduce thermoplasticity.

EXPERIMENTAL

Monomers

    3,3',4,4'-Tetraaminodiphenyl ether and 1,4-bis(phenylglyoxalyl)
benzene were synthesized by known methods.[1,2,3]

    4,4'-Bis(4"-cyanophenoxy)benzil was prepared from diphenyl ether
by bromination, followed by cyanation and Friedel-Crafts reaction.
The synthetic route is shown in Equation 2:

$$ (2) $$

    Bromination of Diphenyl Ether. A solution of bromine (87.9 g,
1.1 mole) in carbon tetrachloride was added dropwise to a stirred
solution of diphenyl ether (170.2 g, 1 mole) in the same solvent
(100 ml). The rate of addition was controlled so as to maintain the
reaction temperature at 0-2°C. After complete addition, the reaction
mixture was stirred for one and one-half hours, followed by succes-
sive washing with water, 5% sodium hydroxide solution, and water
again until it was neutralized. The resulting mixture was dried and
distilled under reduced pressure to yield 4-bromodiphenyl ether
(176.9 g, 71% yield,[4] 49%); B.P. 188-193% C/16mm, n_D^{20} 1.6088,[4] n_D^{20}
1.6088).

    Cyanation of 4-Bromodiphenyl Ether. 4-Cyanodiphenyl ether was
prepared according to the known procedure of Friedman and Schechter.[5]

    4,4'-Bis(4"-cyanophenoxy)Benzil. A solution of 4-cyanodiphenyl
ether (97.6 g, 0.5 mole), oxalyl chloride (25.39 g, 0.2 mole) in
carbon disulfide (50 ml) was added dropwise to a stirred suspension
of anhydrous aluminum chloride (66.68 g, 0.5 mole) in carbon disul-

fide (150 ml) at room temperature. After complete addition, the dark
red mixture was stirred at ambient temperature for 1 hour and then
refluxed for 2 hours. It was cooled and hydrolyzed by a mixture of
concentrated hydrochloric acid and ice water. The mixture was fil-
tered, and the residue was collected. The produce was recrystallized from ethanol-acetone (2:1 w/w) mixture with charcoal. 4,4'-
Bis(4"-cyanophenoxy)benzil was obtained as light yellowish crystal-
lites (377.8 g, 85% yield); Mp. 157.5-159°C. Analysis: $C_{28}H_{16}N_2O_4$.
Calcd: C, 75.69%; H, 3.60%; N, 6.30%. Found: C, 75.63%; H, 3.65%;
N, 6.15%.

Polymers

O-phenylene Diamine Terminated Polyphenyl Quinoxaline (II). This
was prepared by polymerizing 3.4235 g (0.0100 mole) of 1,4-bis(phenylglyoxalyl) benzene with 2.3260 g (0.0101 mole) of 3,3',4,4'-tetra-
aminodiphenyl ether in 60 ml of cresol at 10% solid content (w/w).
The mixture was stirred vigorously at ambient temperature for 2 hours
and was allowed to stand overnight. η_{inh} was equal to 1.07 dl/g.

(II)

4-Cyanodiphenyl Ether Terminated Polyphenylquinoxaline (III).
A cresol solution of polymer (II) was allowed to react under N_2 with
0.1067 g (0.00024 mole) of 4,4'-bis(4"-cyanophenoxy)benzil at 150°C
for half an hour. After cooling, the mixture was precipitated in
ethanol by using a Waring Blender. The polymer was then washed thoroughly with hot ethanol, and then acetone to remove the excess end-
capped reagent.

(III)

Benzil Terminated Polyphenylquinoxaline (IV). This was prepared
by polymerizing 3.4577 g (0.0101 mole) of 1,4-bis(phenylglyoxalyl)
benzene with 2.3029 g (0.0100 mole) of 3,3',4,4'-tetraaminodiphenyl
ether in 60 ml of cresol. The mixture was then treated in exactly
the same way as polymer (II), giving η_{inh} = 0.94 dl/g.

(IV)

RESULTS AND DISCUSSION

The solid content of polyphenylquinoxaline solution can have a significant influence on the viscosity and subsequently on the coating process. η_{inh} of the product resulting from the polycondensation of aromatic tetraamine and aromatic dibenzils in the molar ratio of 1.00:1.01 was 1.07 dl/g. Polymer solutions with different solid content were characterized at 20°C by GHB* (η_g) viscosity commonly used in coating factories. The results are shown in Table I. When the solid content was doubled, the GHB viscosity increased 35 times. This experimental phenomenon showed that a new way had to be found to reduce viscosity while keeping a high solid content.

Regulation of Molar Ratio of Monomers

Polyphenylquinoxaline (PPQ) was prepared by a one-step polycondensation of the two monomers. As indicated in Table II and in Figures 1 and 2, η_{inh} and GHB viscosity varies with molar ratio. It reached a maximum at a 1:1 molar ratio. However, the polymer solution became extremely viscous at this ratio. Viscosity of polymer can be decreased by upsetting the stoichiometry of either aromatic tetra-amine or aromatic dibenzil. The stoichiometry should be upset prefer-ably in favor of dibenzil, since the polymer with an amine end group displayed poor thermooxidative stability. If the molar ratio was upset by more than 2%, the molecular weight of polyphenylquinoxaline would decrease significantly. In order to investigate how the polymer properties varied with viscosity, the effect of different molar ratios of monomers was studied.

Table I. Effect of Solid Content on PPQ GHB Viscosity

Solid content (%)	8	10	12	14
GHB viscosity (min., sec.)[1]	35.6"	1'48"	4'31"	21'22"

[1] 1 second is equal to 90 centipoise.

*GHB viscosity : Gardner-Holdt Bubble viscosity.

Table II. Effect of Molar Ratio on PPQ Properties

Stoichiometry NH_2/CO		1.00:1.00	1.00:1.01	1.00:1.02	1.00:1.04
η_{inh}, dl/g[1]		2.9	1.0	0.7	0.4
$\overline{Mn}$[2]		6.06×10^4	2.98×10^4	-	1.34×10^4
Tg, °C[3]		310	305	304	296
Isothermal aging in air wt. loss, %	350°C/10 h.	0.73	0.79	0.93	0.92
	350°C/80 h.	8.38	9.44	9.72	14.55
	400°C/9 h.	4.29	4.33	4.98	5.15
	400°C/13 h.	7.50	8.97	9.32	10.21
Mechanical properties of film	Tensile tear strength, kg/cm^2	1253	1176	1109	-
	Elongation at break, %	25	12	10	8
	No. of repeated folding, (180°, 1 kg)	494	461	363	-
Properties of coating wire	Toughness after 300°C/6 h.	1 d[4]	3 d		
	Break down voltage, kV	15	8.5		
	Electric insulation in boiling water, ohm/meter in oil, ohm/meter	10^{10-12} 10^{10-12}	10^{10-11} 10^{10-11}		

[1] Inherent viscosity (0.5% in cresol at 30°C).

[2] $\overline{Mn}$: Number average molecular weight was determined by osmotic pressure method.

[3] Tg was determined by Torsional Braid Analysis (TBA), rate of heating: 1°C/min.

[4] d was the diameter of coated wire.

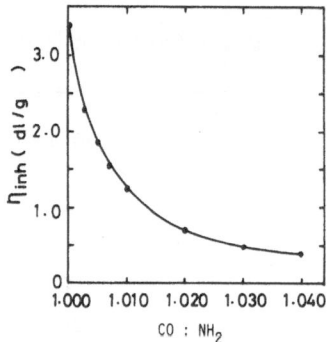

Figure 1.   Effect of molar ratio of monomers on inherent viscosity
            (0.5% in cresol at 30°C) of PPQ.

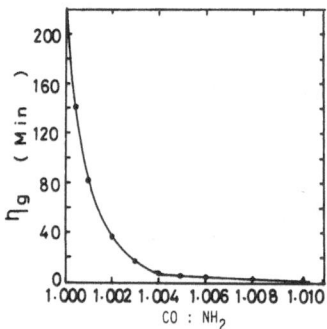

Figure 2.   Effect of molar ratio of monomers on GHB viscosity (10%
            in cresol solution at 30°C) of PPQ.

Variation of Environmental Temperature

    The relation between GHB viscosity and environmental temperature
of the coating process was investigated. The effects of environmental
temperature on the viscosity of the polymer with different molar
ratios but the same solid content and the same molar ratio with different solid content are shown in Figure 3 and 4, respectively. In
both cases the viscosity decreased as the environmental temperature
increased. When the environmental temperature changed from 5°C to
40°C, the GHB viscosity of polymer with 10% solid content would decrease 15 times while that of 16% solid content 25 times. Raising
the environmental temperature permitted higher solid content. But
the environmental temperature must not be too high, for otherwise
the solution would have evaporated too quickly.

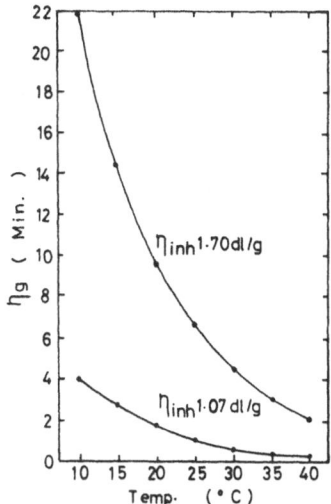

Figure 3. Effect of environmental temperature on GHB viscosity of
           PPQ with different inherent viscosity. (Solid content
           in cresol = 10% by weight.)

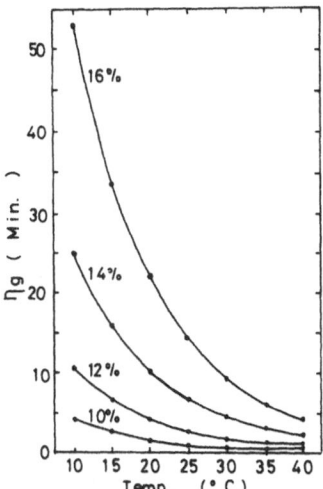

Figure 4. Effect of environmental temperature on GHB viscosity of
           PPQ with different solid contents. (η_{inh} = 1.07 dl/g
           in cresol at 30°C.)

Effect of Solvent and Diluent

The polycondensation of the two monomers took place in cresol
or in a mixture of cresol and xylene (6:4 by weight). The η_{inh} and
GHB viscosity of the polymer solution with 10% solid content are
shown in Table III and Figures 5 and 6. The effect of the diluent
was favorable in decreasing viscosity and increasing solid content.
For example, the GHB viscosity decreased 8 times at 25°C or 12 times
at 15°C by adding diluent during the synthesis of the polymer. From
the η_{inh} point of view, the molecular weight of the polymer was the
same, irrespective of the polymerization in cresol or in cresol-
xylene systems.

Table III. Effect of Solvent-Diluent on PPQ Viscosity

NH$_2$/CO	η_{inh}, dl/g	Solution System	GHB Viscosity, min. sec.					
			10°	15°	20°	25°	30°	35°C
1:1.01	1.07	Cresol	4'4"	2'40"	1'46"	1'13"	51"	35"
	1.07	Cresol-xylene	14.9"	11.4"	9.6"	7.5"	6.9"	5.0"
1:1	3.4	Cresol	1268'	855'	608'	403'15"	304'4"	214'4"
	3.4	Cresol-xylene	90'30"	76'21"	66'7"	54'48"	41'57"	41'30"

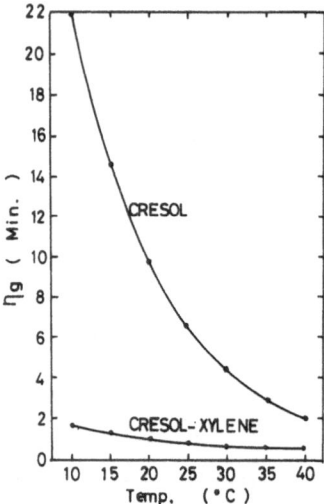

Figure 5. Effect of environmental temperature on GHB viscosity of
PPQ in different solvents. (η_{inh} = 1.70 dl/g; solid
content = 10% by weight.)

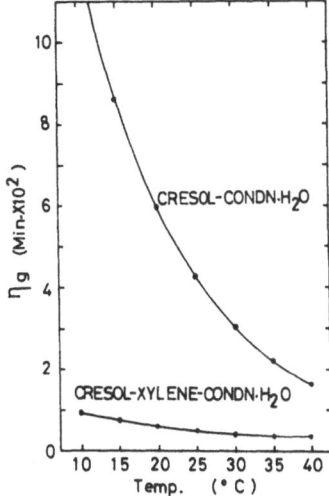

Figure 6.   Effect of environmental temperature on GHB viscosity of
            PPQ in different solutions.   (Solid content = 10% by
            weight.)

Effect of Additive

Additive was added to limit the rate of reaction to a certain
extent in order to obtain a polymer of sufficiently low viscosity
so that it could be used as a coating solution. During coating this
metastable polymer solution could react further to become a high
molecular weight polymer possessing good properties. Polyphenylquinoxaline was prepared by such a method. As shown in Table IV,
η_{inh} and GHB viscosity obtained by controlled polymerization was quite
different from that of uncontrolled ones. It was found that GHB viscosity measurement was more convenient than η_{inh} measurement in determining viscosity of the polymer solution. For example, the GHB
viscosity would decrease 25 times or even more than 50 times.

The properties of all polyphenylquinoxalines with different end-
capped groups are shown as follows:

Thermooxidative stability: The thermal properties of various
end-capped polymers are summarized in Tables V-VI. The thermooxidative stability of polymer (III) was not only better than that of
the aromatic diamine-terminated polymer (II) with the same molecular
weight (2.98×10^4), but also better than that of uncapped ones with
high molecular weight (6.02×10^4) polymer (I). Tg was determined
by Torsional Braid Analysis (TBA). In comparison, Tg of polymer
(III) was 50°C higher than that of the uncapped one. Its elongation
at break was 50% after the film had been treated in air for 1 hour

Table IV.   Effect of Additive on PPQ Properties

Polymerization		Controlled Polymn.			Uncontrolled Polymn.
		I	II	III	
NH_2/CO		1:1	1:1	1:1	1:1
η_{inh}, dl/g[1]		1.56	1.78	2.0	3.4
GHB viscosity, min., sec.[2]		2'26"	3'51"	8'40"	214'4"
	Toughness at:				
	room temp.	1d	1d	1d	1d
	300°C/1 h.	1d	1d	1d	1d
	300°C/6 h.	1d	1d	1d	1d
Properties of coated wire	Electric insulation in water:				
	25°C	10^{11-12}	10^{11-12}	10^{11-12}	10^{11-12}
	25°C/8000 h.	10^{10-12}	10^{10-12}	10^{10-12}	10^{10-12}
	100°C	10^{10-11}	10^{10-11}	10^{10-11}	10^{10-11}
	100°C/6000 h.	10^{10-11}	10^{10-11}	10^{10-11}	10^{10-11}

[1] η_{inh} (0.5% in cresol at 30°C).
[2] η_g (10% in cresol at 30°C).

Table V.   Thermal and Mechanical Properties of PPQ

Polymer Tg, °C	Isothermal Aging in Air at 350°C/20 h., Wt. Loss, %	300°C/6 h.	
		Tensile Tear Strength, Kg/cm^2	Elongation at Break, %
I	1.5	–	–
II	2.0	1253	15.3
III	1.5	1330	34.0
IV	1.7	1238	17.3

Table VI. Thermooxidative Stability of PPQ Coating Wire

Polymer	Toughness at Room Temp.	300°C/1 h.	300°C/6 h.
I	1 d	1 d	1 d
II	3 d	2 d	2 d
III	1 d	1 d	1 d
IV	2 d	2 d	2 d

at 300°C, and was 34% when treated for 6 hours at the same temperature. But that of the other polymers decreased significantly. Polymer (III) also exhibited better flexibility that other polymers because no cracks developed when the coated wire was wound into a
coil of diameter less than or equal to 1 d.

Hydrolytic stability, oil and chemical resistant properties:
In order to test these properties, a coating wire was made and was
treated in water to observe the change of electric insulation. As
shown in Tables VII and VIII, experimental results indicated that
the aromatic cyano-terminated polyphenylquinoxaline also exhibited
good hydrolytic stability, oil and chemical resistant properties.
There was little or no change in insulation after the wire was sub-
merged in boiling water, lubricating oil, crude oil, gasoline, 5%
hydrochloric acid, 5% sodium carbonate, 50% sodium hydroxide, 3% and
saturated sodium chloride solution, benzene or ethanol for one year
at room temperature.

Table VII. Chemical Stability of Polymer (III)
            (Insulating Resistance, ohm/meter)

	0 h.	One Month	Six Months	One Year
5% HCl	10^{10}	10^{10}	10^{10}	10^{10}
50% NaOH	10^{11}	10^{11}	10^{11}	10^{11}
5% Na_2CO_3	10^{11}	10^{11}	10^{11}	10^{11}
Saturated Na_2CO_3	10^{11}	10^{11}	10^{11}	10^{11}
3% NaCl	10^{11}	10^{11}	10^{11}	10^{11}
Saturated NaCl	10^{11}	10^{11}	10^{11}	10^{11}
C_6H_6	10^{11}	10^{11}	10^{11}	10^{11}
C_2H_5OH	10^{11}	10^{11}	10^{11}	10^{11}

Table VIII.   Oil Resistant Properties of Polymer (III)
(Insulation Resistance, ohm/meter)

	Lubricating Oil	Gasoline	Crude Oil
0 h.	10^{11}	10^{11}	10^{11}
One month	10^{11}	10^{11}	10^{11}
Six months	10^{11}	10^{11}	10^{11}
One Year	10^{11}	10^{11}	10^{11}

The thermoplasticity could be decreased to some extent owing to the existence of aromatic cyano groups which reacted with each other at elevated temperature to provide partial crosslinking. This effect could be seen when this polymer was analyzed by IR spectroscopy in which the absorption spectrum of the amino group at 2800-3200 cm^{-1} disappeared whereas the cyano group at 2240 cm^{-1} appeared. After treating this polymer at 360°C for 2 hours, the absorption spectrum of the cyano group gradually disappeared giving way to particular symmetric triazine ring at the absorption spectrum: 1375, 1445 and 1520 cm^{-1}. This polymer had lower molecular weight, so that it offered an advantage in processability.

In conclusion, it can be said that by controlling the molar ratio of monomers, environmental temperature, solvent-diluent, the additive and by adding a proper end-capping group, various properties such as thermal stability, mechanical properties and hydrolytic stability have been improved.

REFERENCES

1. F. C. Lu, et al, Gaofenzi Tongxun, 1980, (3), 146.
2. R. J. Foster and C. S. Marvel, J. Polym. Sci., 1965, A3, 417.
3. M. A. Ogliaruso and E. I. Becker, J. Org. Chem., 1965, 30, 3354.
4. C. M. Suter, J. Amer. Chem. Soc., 1929, 51, 2581.
5. L. Friedman and H. Schechter, J. Org. Chem., 1961, 26, 2522.

EFFECT OF PIGMENTS ON THE AGING CHARACTERISTICS OF POLYOLEFINS

H. M. Gilroy and M. G. Chan

Bell Laboratories

Murray Hill, NJ 07974

INTRODUCTION

Pigments are used extensively in polymers for a variety of reasons such as appearance, opacification, protection and identification. In many cases[1,2] the pigment additive can affect the thermal oxidative stability of a polymer and alter the long term aging characteristics of the material. The effect pigments have on polymers are often small and unimportant because the design lifetime of the product is short. However, the polyolefin resins used in telecommunication wire and cable are designed for long useful lifetimes, in many cases 40 years, and pigments can reduce these lifetimes. Field surveys of pigmented insulations and cable jackets show that white and red polyethylenes have twice the disintegration rate of other light colors,[3] while black polyethylenes rarely fail.[3,4]

Pigments can affect both the physical and chemical integrity of a polymer. Physical phenomena such as poor pigment dispersion or excessive loading of pigments can affect the electrical and mechanical properties of the resin.[5] Fortunately, these effects appear at the outset of fabrication and can be corrected early. More serious are the chemical effects of pigments which can develop with time and are difficult to detect. The pigments can oxidize and undergo degradation of their color properties. The oxidized pigments can subsequently catalyze polymer oxidation, and finally, deleterious interactions can occur between pigments and polymer stabilizers.[6]

Evaluation of the chemical effects of pigments can be difficult, particularly when limited time is available. The rapid determination of the effect of pigments on the oxidative behavior of hydrocarbon materials is often done at elevated temperatures (140-200°C).[7] Many

273

of these studies have shown that pigments can be both beneficial and
detrimental to polymers. However, low temperature, longer term
testing often gives quite different results and, in some cases,
actual reversals.[8] Polyolefin formulations containing carbon black
often show a drastic decrease in the efficiency of stabilizers when
their thermal oxidative stabilities are measured at elevated temperatures. Yet outdoor weathering studies at ambient temperatures prove
that similar resins can resist both thermal and photooxidation for
up to 50 years.[4]

We believe that pigments must be evaluated at as wide a range
of temperature as possible in order to give meaningful results. Pigment-polymer of pigment-stabilizer interactions can occur during compounding, processing and use and these must all be considered when
selecting a useful pigment. We have evaluated a number of commonly
used pigments at various temperatures representing the different
stages of compounding and processing in an effort to select pigment
additive systems that will maximize polymer longevity. We have also
compared our laboratory results obtained at temperatures as low as
90°C with field performance data to aid in the final evaluation of
the pigments.

Since carbon black has such unique properties we have considered
that separately. In the carbon black studies particular attention
has been paid to the furnace blacks which have replaced the channel
blacks used initially to protect polyolefins. The furnace black
effects have been studied in both polyethylene and polypropylene.

EXPERIMENTAL

Materials

Test formulations were prepared from commercial polyolefins
(Table I) and stabilizers (Table II), all of which were used as received from the manufacturers. The pigments were commercial grade
and are defined by Colour Index Number.[9] Three furnace blacks, supplied by Cabot Corporation, were used: Vulcan 9 (19 mμ, pH 6.2),
Regal 660 (24 mμ, pH 6.2) and Black Pearls L (24 mμ, pH 3.0). All
samples selected for long term testing, except the black samples,
contained titanium dioxide (Type IV, ASTM D476).

Sample Preparation

All pigments were added to the appropriate polymer as high concentrate masterbatches to insure the best dispersion and then diluted
to 2.5% pigment concentration for black samples and 1% pigment concentration for all other colors. LDPE samples containing varying
concentrations of carbon black were prepared in the same fashion.
Polyethylene compounding was carried out on a two roll laboratory

Table I. Base Resins[a]

Resin	Composition	ASTM Designation (D-1248)	Supplier
LDPE	branched homopolymer	Type I, A5	Union Carbide Corp.
MDPE	ethylene-hexene copolymer	Type II, A5	Phillips Co.
HDPE	ethylene-mono-olefin copolymer	Type III, A5	E. I. duPont de Nemours and Co.
PP	propylene-ethylene copolymer		Amoco Chemicals Corp.

aResins were unstabilized except PP which contained 0.1% 3,6-ditert butyl-p-cresol.

Table II. Stabilizers

Designation	Structure	Supplier
DNPD	N,N'-Di-ßnaphthyl-p-phenylenediamine	R. D. Vanderbilt Co., Inc.
THPM	Tetrakis[methylene(3,5-di-tert-butyl-4-hydroxyhydrocinnamate)] methane	Ciba-Geigy Corp.
TBHC	Thiodiethylene bis-(3,5-di-tert-butyl-4-hydroxy)hydrocinnamate	Ciba-Geigy Corp.
TMBP	4-4'-Thiobis(3-methyl-6-tert-butylphenol)	Monsanto Industrial Chemicals Co.
DBOD	N,N'-Dibenzal oxalyl dihydrazide	Eastman Chemicals Co.

mill at 140-160°C, for 10 min. for masterbatch preparation and 5 min. for final concentration milling. Polypropylene formulations, compounded under nitrogen on a Brabender Plasticorder, were mixed for a maximum of 7 min. Specimens were molded at 160°C for 2-3 min. and cooled under pressure. Pigmented HDPE samples were molded against aluminum sheets while all other samples were molded against Mylar films.

Wire samples were extruded onto 22 AWG copper wire using a 0.8 in Welding Engineers twin screw extruder with a flat temperature profile of 225°C. The insulation thickness was 6 mils.

Test Methods

Oxidative stabilities of the black samples were determined at 210 and 200°C on aluminum pans using a duPont 990 Thermal Analyzer equipped with a Differential Scanning Calorimetry (DSC) cell). Stabilities of these samples at 140, 120 and 100°C were determined by oxygen absorption experiments. Standard test methods were used for both the thermoanalytical[10] and oxygen-absorption studies.[11]

The stabilities of the samples other than black were all obtained on the thermal analyzer described above. Two measurements were made:

1. Oxidative Decomposition Temperature (ODT), where samples were heated in aluminum pans at 20°C/min. in oxygen until the onset of oxidation. The temperature at which oxidation starts, as shown by the first change in the thermogram base line, was taken as the ODT.

2. Oxidative Induction Time (OIT), where samples are heated in aluminum pans at constant temperature (200°C) and the time for oxidation to occur measured.[10]

The sample size in both tests was 6-8 Mg. Insulated wire samples were tested at 110, 100 and 90°C in static air ovens.[12] Samples were monitored periodically for loss of oxidative stability using DSC and the insulation stressed at each testing interval by wrapping around a steel mandrel the same diameter as the insulated wire.

RESULTS AND DISCUSSION

Black Pigment (Carbon Black)

Studies of the effect of furnace black on stabilized polyolefins were carried out with 50 mil films of resins containing 0.10% by weight of four typical antioxidants (Table II). They were: an amine, DNPD; a hindered phenol, THPM; a thiobisphenol, TMBP; and a second sulfur bridged phenol, TBHC. The antioxidants were studied in a medium density polyethylene and in polypropylene. Preliminary

results for medium density polyethylene have been reported elsewhere.[13]

In all these studies, the effect of carbon black on the stabilizing ability of the antioxidant is found to depend strongly on temperature. At 200°C and above the furnace black at times increases
the induction time for the onset of oxidation while at lower temperatures it decreases it. A typical temperature profile, that of
medium density polyethylene containing the thiobisphenol, THPM, is
shown in Figure 1. The oxidative induction times plotted in Figure
1 are obtained by extrapolating the rapid, steady state portion of
the oxidation curve back to the baseline. The crossover in the
effect of furnace black on the ability of the thiobisphenol to stabilize polyethylene is seen clearly.

However, as the temperature of oxidation is lowered, an interesting phenomenon is observed. At 120°C and below it becomes apparent that for most samples the furnace black not only affects the
induction period of the oxidation reaction but also influences the
rate at which oxidation progresses. This is especially striking
in the case of polyethylene (Figure 2). The non-black samples undergo a typical autocatalytic reaction while the black samples do not.
At 120°C black polyethylene samples containing the amine, DNPD, or
the phenol, THPM, undergo a short period of rapid oxidation at substantially lower induction times than similar non-black samples.
However, the oxidation is suppressed eventually and continues at an
inhibited rate. Rapid oxidation is never observed for the black
samples containing the sulfur-bridged phenols. The thiobisphenol
is particularly effective in suppressing oxidation of the black
sample and after 10,000 hours this sample has absorbed only 10 cm^3/gm
oxygen. At 100°C there is no discernible oxidation in any of the
black samples after 40,000 hours of testing while one of the non-

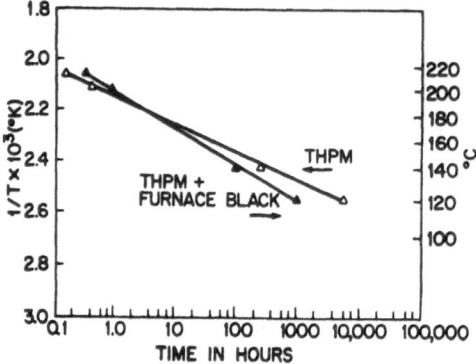

Figure 1.   Variation of oxidative induction time with temperature
            for medium density polyethylene containing 0.1% THPM
            (Δ) or 0.1% THPM + 2.5% furnace black (▲).

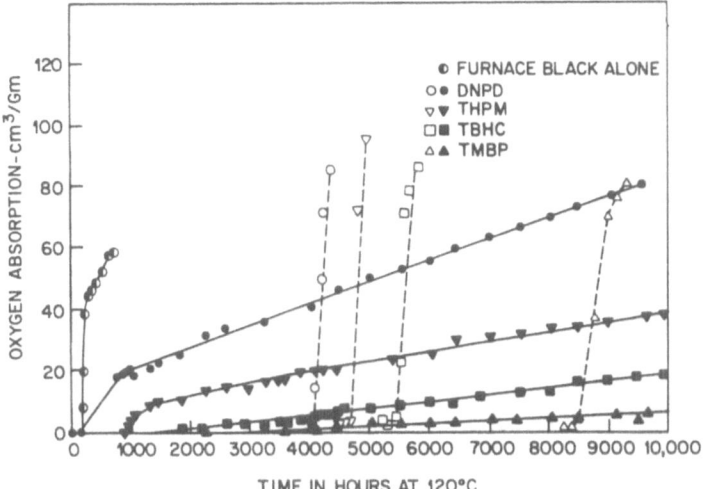

Figure 2.   Oxidative stability of medium density polyethylene con-
            taining 0.1% antioxidant (---) or 0.1% antioxidant +
            2.5% furnace black (——).

black samples, that containing TMBP, has oxidized extensively.

The lower temperature phenomena observed for the polyethylene
are not as evident in the polypropylene samples (Figure 3). At 120°C
the retardation in rate is distinctly observed for the thiobisphenol,
TMBP, and minimally apparent for the sulfur-bridged phenol, TBHC.

Carbon blacks can react with stabilized polyolefins in a variety
of ways. Part of this interaction is believed to be due to the
nature of the carbon black surface. The furnace black discussed
above was a fine particle, slightly acidic black which is reported[14]
to have a volatile content of 1.5% and to contain weak surface acids
in the form of phenols and hydroquinone derivatives.[15] Studies of
this black and of a more acidic black show that the nature of the
black surface has a marked effect on polyolefin stability. The
acidic nature of the black enhances the stability of unstabilized
polyethylene to a marked dgreee (Figure 4). It has a reverse effect
on a stabilized polyethylene (Figure 5).

The observed effect of the furnace black on oxidative induction
times is probably due to absorption and subsequent reaction of the
antioxidants with the furnace black. As the oxidation temperature
is lowered enhanced absorption and subsequent deactivation of the
antioxidants could be expected to occur. This would explain the
effect of the black on the 120 and 140°C oxidative induction times.

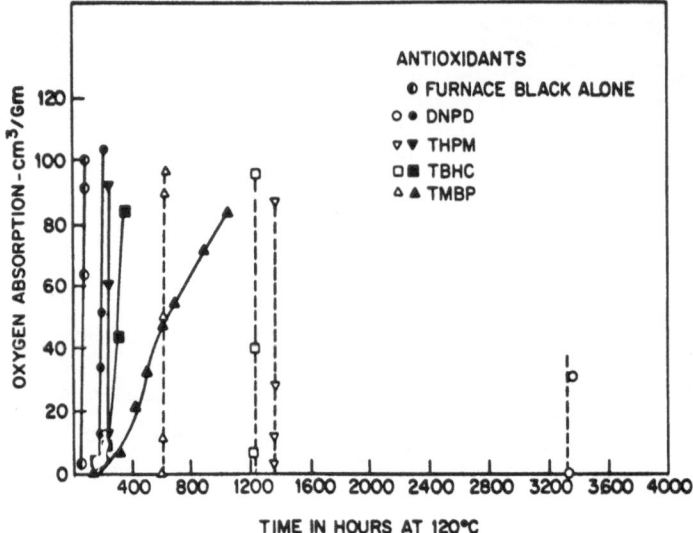

Figure 3. Oxidative stability of polypropylene containing 0.1%
          antioxidant (---) or 0.1% antioxidant + 2.5% furnace
          black (——).

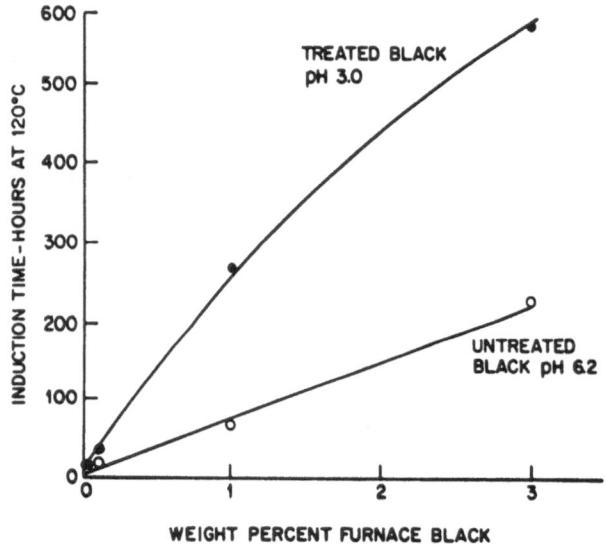

Figure 4. Oxidative stability of low density polyethylene con-
          taining furnace blacks.

On the other hand, the antioxidant activity of carbon black
itself is enhanced as the reaction temperature is lowered. We believe that this explains the suppression in the rate of oxidation

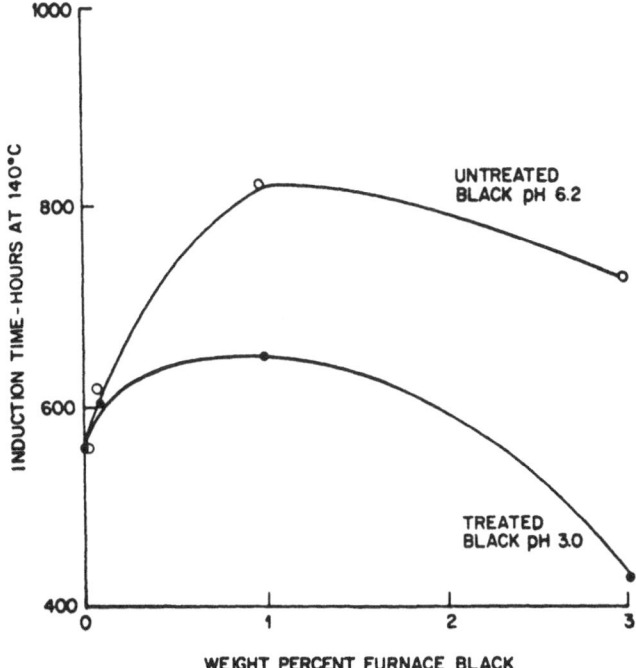

Figure 5. Oxidative stability of low density polyethylene con-
           taining 0.1% TMBP and furnace blacks.

at 120°C and below. While morphological effects may contribute to
this retardation, since the MDPE melts around 120°C, they do not
entirely explain it. The reduction in rate of oxidation is first
observed at 140°C well above the melting range of the polymer. It
is also logical that the antioxidant properties of the carbon black
are observed more readily in polyethylene than in polypropylene which
is more susceptible to oxidation.

Other Pigments

     Life testing of polyethylene wire insulating compounds is a
lengthy process, since all testing must be done below the melt
(110°C). Extrapolation of high-temperature (120 to 200°C) data to
use temperatures has been shown to be invalid for wire insulation.[16]
Because of the large number of pigments available, an accelerated
test was required to determine which pigments should be selected for
long-term aging.

     The method selected for screening pigments was based on the
antioxidant behavior of carbon black. At use temperatures (40°C)
carbon black is an excellent thermal antioxidant in the low concentration (0.4%) used in wire insulation. When this concentration

of carbon black is evaluated for thermal stability at high temperatures (180-200°C), essentially no effective stabilization effect is found. However, if a masterbatch concentrate (20-50%) is tested by thermal analysis at high temperatures, the mixture is extremely stable. In an effort to find pigments that might have similar beneficial properties, initial screening tests were carried out on pigment masterbatch concentrates using thermal analysis.

Evaluation of commercial masterbatch concentrates by thermal analysis showed that several colors improved the thermal stability of polyethylene. As shown in Table III, unstabilized polyethylene alone will remain unoxidized during programmed heating in oxygen until about 200°C, at which time rapid oxidation takes place. Several of the commercial masterbatch formulations retarded this oxi-

Table III. Oxidative Decomposition Temperature (ODT) and Oxidative
           Induction Time (OIT) of Commercial Masterbatches
           Pigment Concentration of 20 - 50%

Color	ODT °C[1]	OIT - min. @ 200°C[2]
Black	248	52
Brown	230	22
Violet	225	10
Red	220	13
Slate	220	13
Orange	205	6
Yellow	203	3
Unstabilized polyethylene	200	<1
Blue	190	<1
Green	180	<1
White	160	<1

[1]Heated at 10°C/min in oxygen.

[2]Heated isothermally in oxygen.

dation until higher temperatures were reached, indicating a stabilizing effect on the polyethylene. Several other commercial masterbatches had the reverse effect, causing oxidation to occur at temperatures lower than for that of polyethylene alone. When these commercial masterbatches were evaluated by measurement of OIT, the trends remained the same. As shown in Table III, the same pigment formulations that exhibited stabilizing effects in the ODT test had OIT values greater than several minutes at 200°C, again indicating a stability effect.

It is difficult to determine which pigment is affecting the oxidative stability of these commercial masterbatches in these tests, since the masterbatches contain more than one pigment.

In order to determine the individual pigment effect on thermal oxidative decomposition, new masterbatches were prepared, using only one pigment per masterbatch. These masterbatches were then evaluated by thermal analysis in the same manner as the commercial ones. The results (Table IV) show a wide range of values, similar to those found for the commercial masterbatches. However, in this case the pigment responsible is known. Several of the pigments, notable Red #38 and Orange #13, show outstanding stabilization effects, approaching the value found for carbon black.

Measurement of OIT values for the individual pigment masterbatches (Table IV) show similar trends to those found in ODT measurements.

Several pigments (Yellow 60, Blue 16 and 22) were eliminated from further consideration when severe migration and bleeding were observed from the masterbatch concentrates. Pigments Red 2 and Yellow 101 did not show noticeable bleeding from the masterbatch concentrates but after short term aging in ovens as wire insulation bleeding became apparent. These pigments were also removed from the test program.

Pigmented Wire Insulation

Pigments evaluated in masterbatches were used to prepare wire samples for long-term evaluation in ovens. All samples tested contained one percent pigment, one percent titanium dioxide, and 0.1% each antioxidant and metal deactivator.

The criteria for wire insulation performance is based on oven data for production cables. Older cables of low-density polyethylene with a low molecular weight phenolic stabilizer have been oven tested and failures produced over a temperature range of 110 to 40°C (Figure 6). The curve generated by this data has been shifted to reach a target of 40 years at 40°C. This 40 years satisfies the design life

Table IV.   Oxidative Decomposition Times (ODT) and Oxidative
            Induction Time (OIT) of Single Pigment Masterbatches
            Pigment Concentration of 20%

Color	Pigment Number	ODT in °C[1]	OIT Minutes @ 200°C[2]
Black	7	265	>60
Red	38	260	>60
Orange	13	255	23
Green	8	218	3
Yellow	101	216	6
Red	48	214	2
Brown	23	213	2
Violet	23	210	6
Blue	60	208	7
Orange	61	203	2
Orange	31	203	2
Red	177	201	1
Yellow	93	201	1
Unstabilized polyethylene		200	0
Red	220	198	<1
Yellow	17	197	4
Blue	16	193	<1
Yellow	14	193	<1
Green	7	185	<1
Green	36	183	<1
Blue	15	173	<1

[1]Heated at 10°C/min in oxygen.

[2]Heated isothermally in oxygen.

currently required for manufactured cable and any data that meets or
exceeds the times defined by this curve are considered satisfactory.

The test results for single pigment samples (Table V and Figure

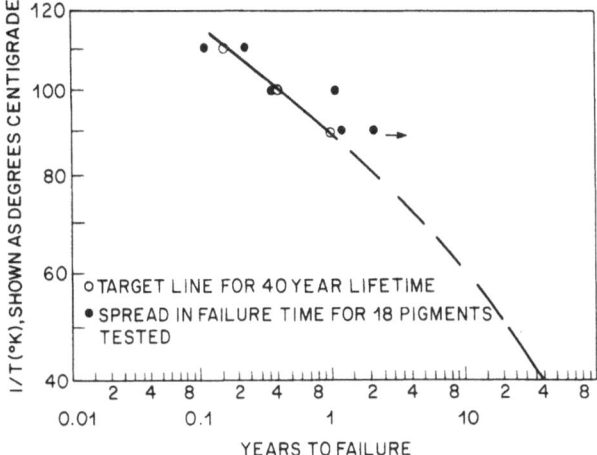

Figure 6. Range of pigmented wire insulation failures tested at
110°, 100°, 90°C.

6) show that most samples exceed the minimum time acceptable, as
defined above, at 110, 100 and 90°C.

Color Stability

All the pigments used in the preparation of insulated wire samples were color stable through the processing into masterbatch and
extrusion as insulation.

The wire insulation samples all retained their color during the
aging at 110°, 100° and 90°C until the polymer oxidized. Then two
pigments, Orange #31 and Green #8, also oxidized and change color.
After extensive oxidation of the polymer most pigments darkened and
extensive color changes occurred in several, with the orange turning
yellow, blues to green and red to brown. The color changes, coupled
with the darkening and fading of all colors made identification of
specific wires impossible.

CONCLUSIONS

Pigments added to polyolefins can affect the thermal oxidative
stability, both adversely and beneficially. Measurement of these
effects at high temperature, while rapid, does not always reflect
the conditions found at lower use temperature. Evaluation of a furnace black pigment in polyethylene shows that at temperatures near
or below the melt the rate of oxidation is suppressed and autocatalytic oxidation does not occur. This effect is also seen in polypropylene but to a much lesser degree. Other pigments, selected after

Table V. Oven Test Results - Single Pigment Wire Samples

Pigment Number		Time to Failure - Years		
		110°C	100°C	90°C
Red	48*	0.14	0.64	>2.0
	38	0.22	0.90	>2.0
	177	0.10	0.58	>2.0
	220	0.22	1.0	1.8
Orange	13*	0.22	1.0	>2.0
	31	0.22	1.0	>2.0
	61	0.22	1.0	>2.0
Yellow	14*	0.10	0.57	1.3
	17	0.20	1.0	>2.0
	93	0.14	0.65	1.5
Green	7*	0.21		1.5
	8	0.07	0.40	1.1
	36	0.20	0.60	1.5
Blue	15*	0.20	1.0	1.5
	60	0.08	0.57	1.5
Violet	23*	0.11	0.60	1.5
Brown	23	0.18	1.0	1.5
Control - no pigment		0.19	0.90	1.5
- TiO_2		0.18	0.90	1.5
Minimum time acceptable		0.16	0.42	1.0

*Currently specified for use in polyethylene wire insulation.

preliminary screening in concentrate form, show no detrimental pigment-polymer or pigment-stabilizer interactions, suggesting that long useful lifetimes will be obtained.

REFERENCES

1. C. W. Uzelmeier, SPE Journal, 26 (5), 69 (1970).
2. W. L. Hawkins, M. G. Chan and G. L. Lin , Polym. Eng. Sci., 11 (5), 376 (1971).

3. J. B. Howard, Proc. 21st Int. Wire and Cable Symp., 329 (1972).

4. H. R. Gilroy in: "Durability of Macromolecular Materials,"
    R. K. Eby, Ed., ACS Symp. Series 95, ACS, Washington, DC,
    (1979), p. 63.

5. E. J. G. Balley and R. W. Gould in: "Polythene," A. Renfrew
    and P. Morgan, Eds., Interscience, New York (1960), p. 419.

6. D. W. Holtzen, Proc. SPE ANTEC, 22, 488 (1976).

7. See for example: E. Kovacs and Z. Wolkober, J. Polym. Sci.,
    Symp. No. 57, 171 (1976).

8. F. H. Winslow and W. L. Hawkins in: "Crystalline Olefin Poly-
    mers Part II," R. A. V. Raff and D. W. Doak, Eds., Inter-
    science, New York (1964), p. 361.

9. "Colour Index," Third Ed., Society of Dyes and Colorist,
    London (1971).

10. J. B. Howard, Proc. SPE ANTEC, 19, 408 (1973).

11. W. L. Hawkins, R. H. Hansen, W. Matreyek and F. H. Winslow,
    J. Appl. Polym. Sci., 1, 37 (1959).

12. H. M. Gilroy, Proc. 23rd Int. Wire and Cable Symp., 42, 1974.

13. M. G. Chan and L. Johnson, Proc. SPE ANTEC, 26, 494 (1980).

14. F. R. Williams, M. E. Jordan and E. M. Dannenberg, J. Appl.
    Polym. Sci., 9, 861 (1965).

15. D. Rivin, Paper presented at the 4th Rubber Technology Confer-
    ence, London, 1962.

16. J. B. Howard and H. M. Gilroy, Polym. Eng. Sci., 15 (4), 296
    (1975).

DYNAMIC MECHANICAL INVESTIGATIONS OF HIGHLY FILLED POLYETHYLENE

R. Kosfeld, Th. Uhlenbroich and F. H. J. Maurer*

FB 6 - Physical Chemistry
Duisburg University
German Federal Republic

INTRODUCTION

The properties of filler polymers can be influenced by specific
surface interactions between fillers and polymers and are mainly determined by volume fractions, shape of filler particles and bulk properties of the components.[1,2,3,4,5] Though there is still considerable
uncertainty about the existence and detailed structure of the interphase for several filled polymers, the occurrence of "bound polymer"
is direct evidence of polymer-filler interaction.[1,6,7,8] Bound polymer is the portion of a polymer in a composite which cannot be extracted with solvent, although the polymer as such (without filler)
is quickly and completely dissolved by the solvent used. Recently
several authors have demonstrated the presence of bound polymer in
various polyethylene-filler composites.[9,10,11,12]

For kaolin-filled high-density polyethylene we were able to
show a correlation between the amount of bound polymer present in
the composites and fracture properties.[11] Furthermore we found the
heat of fusion of the bound semi-crystalline polymer to be considerably lower than the value of 225 J/g for the unfilled high-density
polyethylene. In Figure 1, the slope of the line representing the
heat of fusion of the total undissolved fraction (kaolin + HDPE) as
a function of the amount of HDPE in this fraction equals the heat of
fusion of the bound polymer ΔH_{BP} in kaolin - HDPE composites, which
is about 120 J/g. Similar results were obtained for the heat of

---

*Central Laboratories, DSM, Geleen, The Netherlands

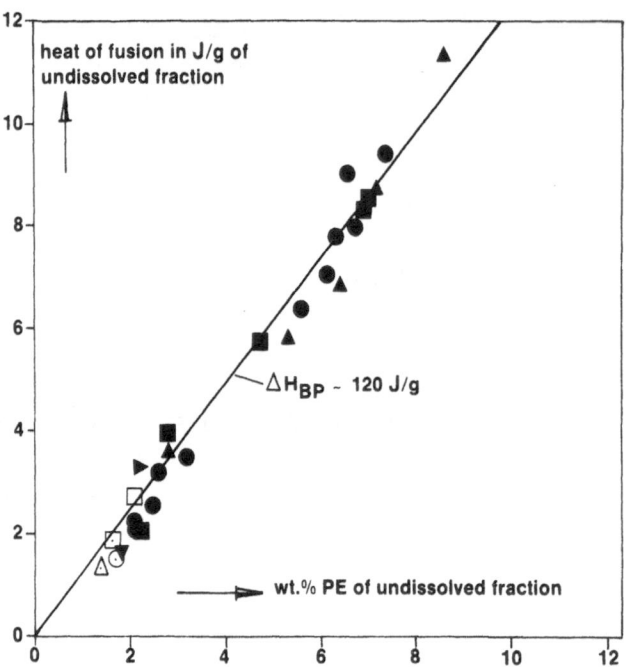

Figure 1. Heat of fusion of the undissolved kaolin and polyethylene
          after extraction with xylene as a function of the per-
          centage by weight of polymer in this fraction. Composites
          with ●34%, ▼36%, ▲38%, ▶40%, ■42% kaolin by volume;
          white symbols represent untreated kaolin, black symbols
          represent, vinylsilane-treated kaolin, after.[11]

fusion of the bound polymer in Aerosil-HDPE composites, while the
heat of fusion of the total polymer in the composites as a function
of the volume fraction of filler changes by the amount of bound polymer present, in agreement with these results.[12] These experimental
results gave strong evidence for the existence of a bound polymer
shell around the filler particles with properties different from the
properties of the matrix material in filler polyethylenes.

     The objective of the work reported here is to investigate wheth-
er and how the interphase in kaolin-polyethylene composites affects
the dynamic-mechanical properties as measured by torsional pendulum
technique with special attention to the influence of the filler on
the γ- and β-transition of polyethylene. The β-relaxation mechanism,
which has been interpreted as the glass transition in polyethylene,
is often not detectable in high density polyethylene. Illers,[13-16]
Pauwe[17] and Arai[18] have discussed in detail several relaxation mechanisms in unfilled polyethylene and made the β-mechanism visible by
swelling unfilled high-density polyethylene samples in a non-polar
swelling agent. Following their procedure we have investigated some

kaolin and other particulate-filled polyethylene composites.

MATERIALS AND EXPERIMENTAL

Linear polyethylene (Stamylan 9309) with a MFI of 8 g/10 min., ρ_{23} = 963 kg/m³, was filled with, respectively untreated and vinyl-silane-treated kaolin, in amounts ranging from 0 to 42% by volume. (Kaolin Icecap K and kaolin KE, Burgess Co., Sandersville, U.S.A.). The composites were compounded on a two-roll mill for 15 minutes, at 170°C. Afterwards, they were pressed to sheets of 1 mm thickness using a pressure of 300 kg/cm², with cooling from 170°C to room temperature at a rate of 40°C/min.

By the same procedure composites of high-density polyethylene were blended with barium sulphate (∅ < 1 µ) and glass beads (Potters Ballotini, 3000, without coupling agent, ∅ < 50 µ). The samples (1 mm x 6 mm x 50 mm) were investigated by the method of free torsional vibration using an automated torsional pendulum. (Systeem ATM 3, Myrenne, Roetgen, GFR). The storage modulus G', loss modulus G" and tan δ were determined over a temperature range from -180°C to 120°C at a frequency of 1 ± 0,1 Hz and an angle of torsion of 1 degree.

RESULTS AND DISCUSSION

The storage modulus G', the loss modulus G" and the loss factor tan δ = G"/G' for unfilled high-density polyethylene are shown in Figure 2. The well-known α- and γ-dispersions at +50°C and -120°C are represented by the maxima of the loss modulus G". The β-transition, which is believed to be the glass transition of the amorphous polyethylene phase at -70°C,[13,14,19,20] cannot be observed in this high-density polyethylene sample. The α-peak has been attributed mainly to intra crystalline relaxation mechanisms, the γ-peak to contributions from the crystalline as well as the amorphous phase.

Both α- and γ-transitions have been discussed in detail by several authors, leading to a more detailed description in terms of α_I, α_{II}, α_{III} and γ_I, γ_{II}, γ_{III} relaxation mechanisms.[13,18,21,22] The multiple relaxation of the γ-loss peak of polyethylene is extensively discussed by, for instance, Illers.[13] According to Illers there exists a microstructure of 3 relaxations for the γ-process, with the temperature positions of these relaxations at 1 Hz having the following values: T_{max} (γ_I) = -110°C; T_{max} (γ_{II}) = -135°C; T_{max} (γ_{III}) = -160°C ∸ -170°C. According to Illers[13] the γ_I-relaxation is caused by the local molecular motion of the amorphous regions; the γ_{II}-process is attributed to the motion of very short CH$_2$-sequences which is restrained; the γ_{III} relaxation will be caused by molecular motion within the lamellar crystals.

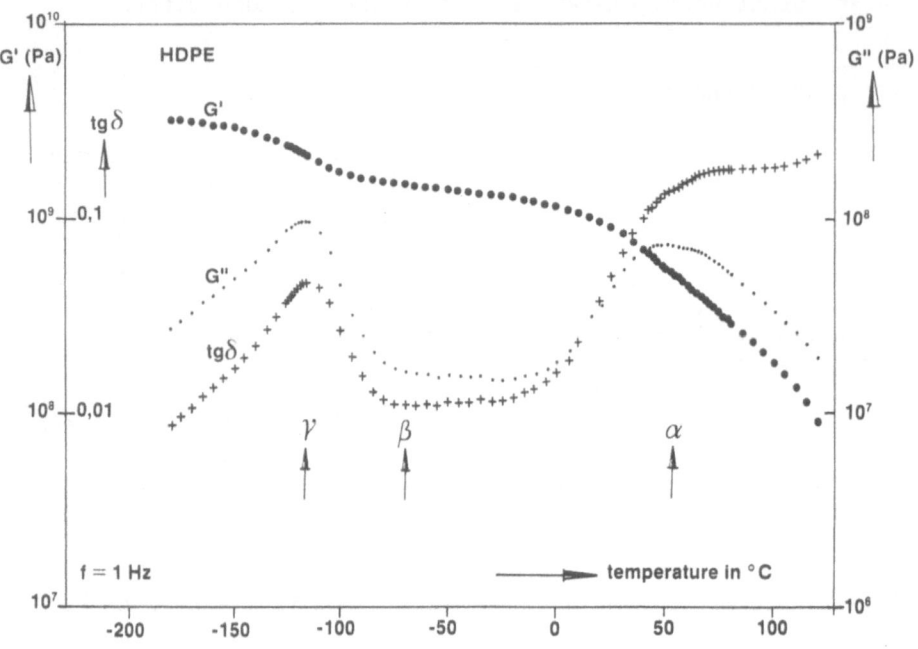

Figure 2.   The storage modulus G', loss modulus G" and loss factor
            tan δ of the unfilled HDPE.

If CCl_4-molecules are incorporated into the polymer, the γ_{II}- and
the γ_{III}-process will not be influenced by these molecules, whereas
the γ_I-relaxation is strongly diminshed.[13]

    For composites with different volume fractions of coated kaolin
the loss modulus G" as a function of the temperature is shown in
Figure 3. With increasing filler content the high temperature relaxation peak (α) shifts to higher temperatures whereas the maximum of
the γ-transition moves to slightly lower temperatures.

    The shape of these relaxation regions remains virtually the same.
Between the above-mentioned two significant relaxation processes no
other can be detected, but on the high-temperatures side of the γ-
relaxation an additional increase of the G"-values with increasing
filler concentration can be observed. Nevertheless it cannot yet be
decided from these measurements whether this increase is due to the
β-relaxation mechanism.

    As mentioned before, it is possible to make the β-maximum vis-
ible by swelling the samples in a low-molecular and non-polar solvent.
In our experiments we have used carbon tetrachloride as swelling
agent. The absorption of swelling agent for various polyethylene
kaolin-composites as a function of time is shown in Figure 4. Less

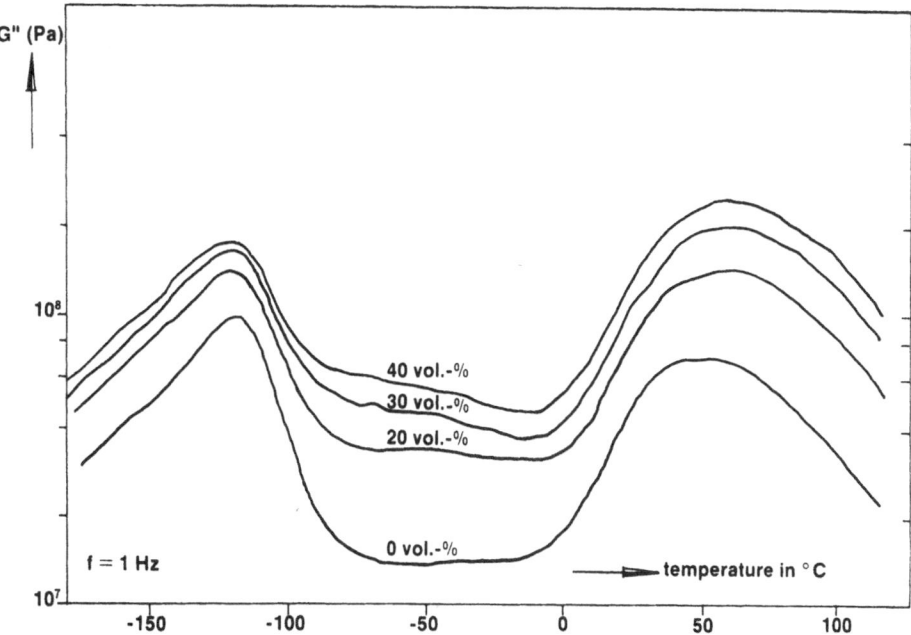

Figure 3. Loss modulus G" of various HDPE - kaolin composites as
            a function of the temperature.

CCl_4 is absorbed as the filler content increases. After approximately
one day of swelling at room temperature no significant change in
swelling agent absorption appears.

     Swelling of polyethylene-kaolin composites for a definite time
(in this case one hour) in CCl_4 reveals a temperature dependence of
the loss moduli G" as shown in Figure 5. The swollen unfilled system
exhibits an α- and γ-maximum with a lower intensity than those of the
corresponding unswollen systems. The pronounced maximum at -120°C
which is, the γ-peak or, according to Illers, γ_I-peak, finds its
origin in the amorphous polyethylene phase. Illers relates the reduced absolute value of this γ-maximum in comparison with the unswollen system to the immobility of the embedded foreign solvent molecules. In the temperature range around -65°C a pronounced β-maximum
can be seen. With increasing filler content the β-maximum shifts to
a higher temperature. This shift is combined with a strong broadening of the maximum. The effect of broadening and shifting of the
glass transition to higher temperatures by filler action has been explained by several authors[2,3,24,25,26] in terms of reduced mobility
of polymer molecules adsorbed at the filler surfaces.

     Figure 6 shows the G"-behavior of the filled systems as compared
with the behavior of the filled-and-swollen systems. The following
facts can be noticed. For unfilled samples the γ_I-process is strongly

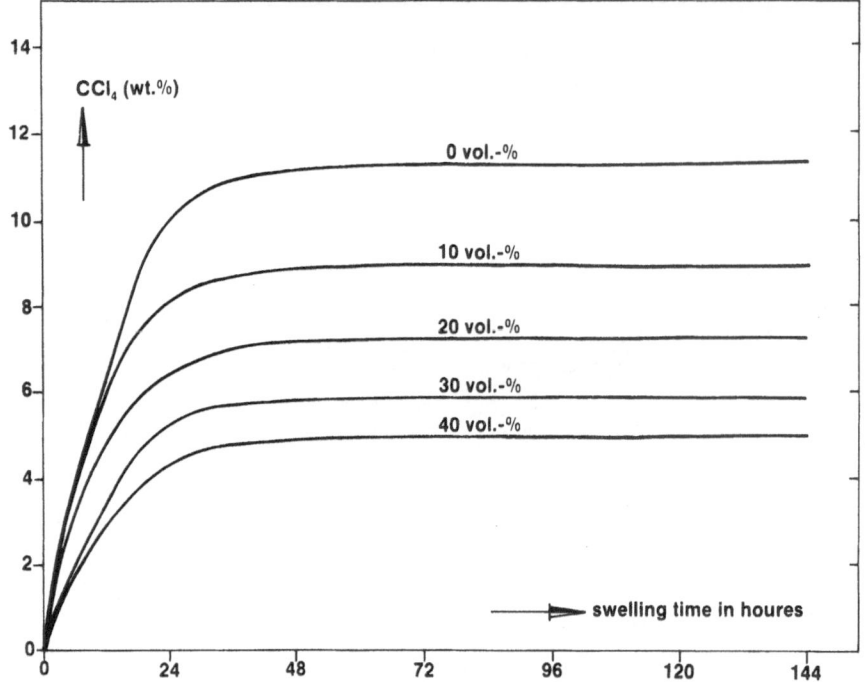

Figure 4.   The absorption of carbon tetrachloride in HDPE – kaolin
            composites as a function of time.

reduced by swelling whereas the influence on the γ_{II}-process at –135°C
is very small or nonexistent and the γ_{III}-process at –165°C is not
at all influenced. It is difficult to make a statement about the
γ_{II}-process because of the strong influence of the γ_I-process on this
peak. Filled systems exhibit a modified γ_{III}-relaxation, the modification being such that the γ_{III}-contribution of the filled and swollen system lies above the γ_{III}-relaxation of the filled but not swollen system.

These results indicate that swelling of the filled systems causes
a slight increase in the molecular mobility in the crystalline polyethylene regions or at least at crystal surfaces. Since unfilled
systems do not exhibit a change in the γ_{III}-relaxation after having
been swollen, it follows that a part of the crystalline phase in the
HDPE kaolin composite has been modified by the action of the filler,
as was to be expected from thermal analysis measurements.[12] This
conclusion is in agreement with the measurements of the heat of fusion
of the composites.[12] What effect does the silane coupling agent have
on the dynamic mechanical behavior of the composite?

The difference in swelling agent absorption between the coated
and uncoated system is shown in Figure 7. It is clear that, with

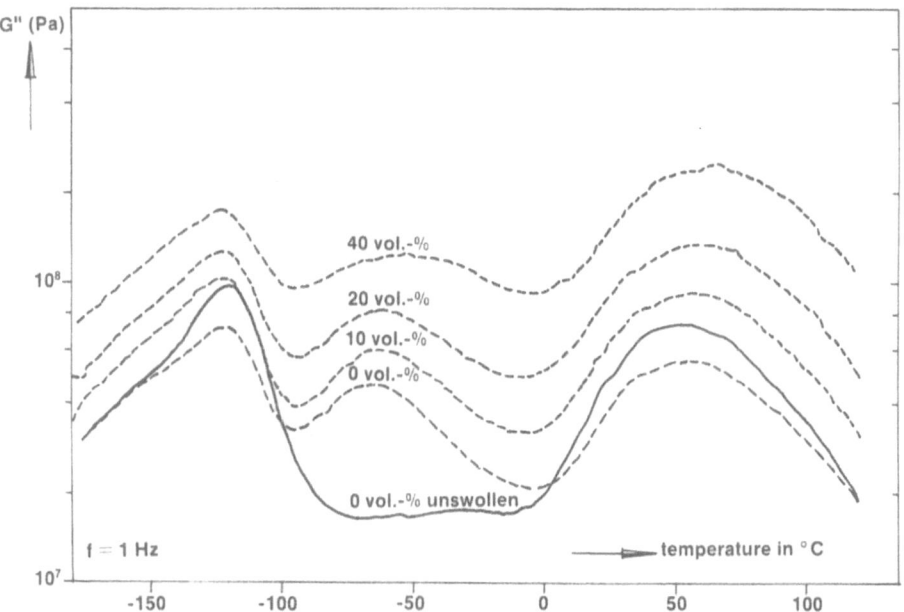

Figure 5.   The loss modulus G" of various HDPE - kaolin composites,
            which were swollen in CCl$_4$ for 1 hour before measurement.

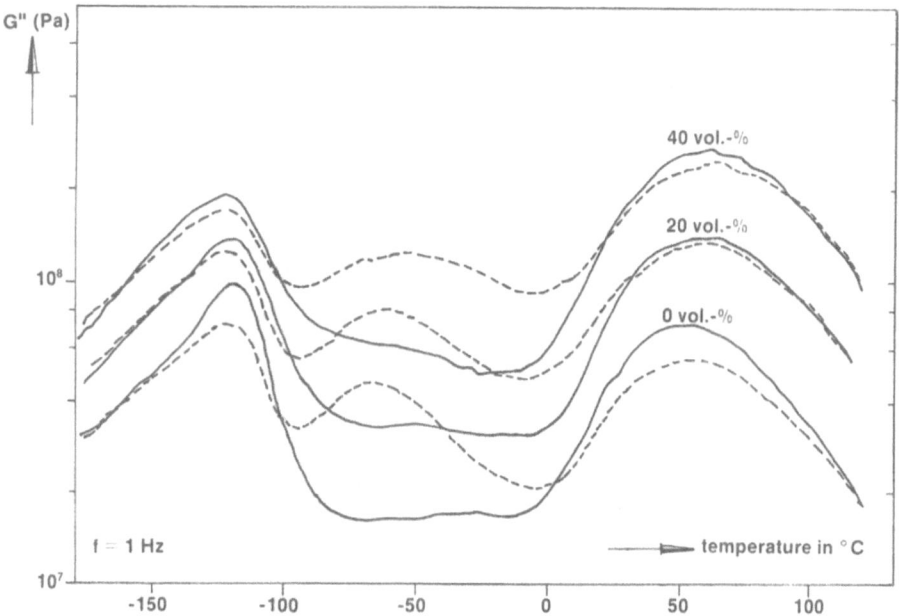

Figure 6.   The loss modulus G" of swollen (——) and non-swollen
            (---) HDPE - kaolin composites.

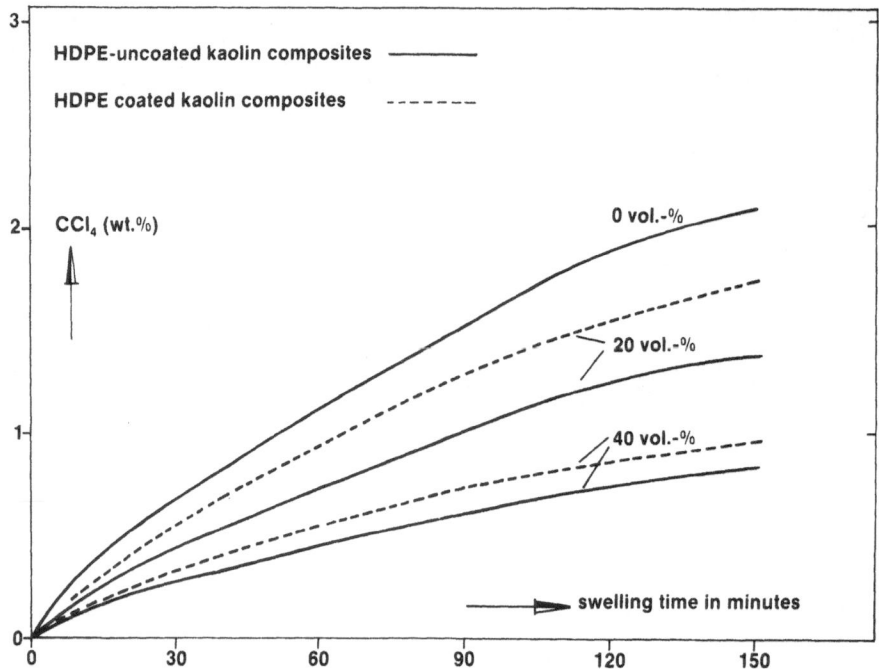

Figure 7.   The absorption of CCl$_4$ in HDPE with coated or uncoated
            kaolin as a function of time.

equal residence times, coated systems are impregnated by more solvent
than uncoated systems. Since the swelling agent is absorbed preferentially in amorphous regions, the coated system must possess a larger
amount of these regions. This is confirmed by the lower heat of
fusion we found for coated kaolin-filled polyethylene in comparison
with the uncoated filler (see Figure 1). In the region of the γ-
and α-relaxation (-120°C and +50°C, respectively) the temperature
dependences of G" for the coated swollen system (Series K) and the
uncoated system (Series I) are not significantly different (Figure
8). In the region of the β-relaxation, G" for the HDPE filled with
coated kaolin lies above the loss modulus of the uncoated system.

     Here, too, we can conclude that the coated system possesses an
increased amount of amorphous PE-phase. Since the coupling agent
is believed to influence only the immediate neighborhood of the filler
particle, this surplus of amorphous phase (as compared with the uncoated system) must be found on the filler particle or in its immediate neighborhood. The broadening of the β-maximum accompanied
by a shift to higher temperatures leads to the conclusion that this
surplus of amorphous phase must have an at least partially changed
mobility. This means that this "disturbed" amorphous phase is not
identical with the amorphous phase of unfilled polyethylene.

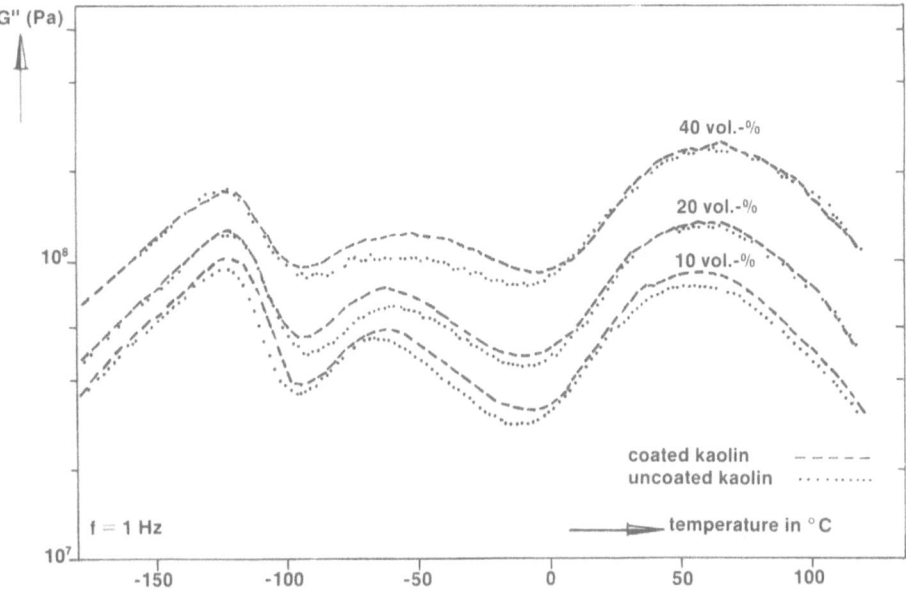

Figure 8. The loss modulus G" for the swollen HDPE-composite with coupling agent and without coupling agent.

As can be seen from Figure 1, the specific heat of fusion of the bound polymer is nearly the same for all investigated samples. Hence it can be concluded that the thickness of the layer of bound polymer is constant. In our opinion, the greater amount of bound polymer found for the coated system can be atttributed at least in part to better dispersion of the particles.

Comparing several other fillers, G', G" and tan δ are presented in Figure 9. The fillers were coated kaolin, $BaSO_4$ and micro-glass beads. The samples were kept in CCl_4 for 1 hour at room temperature before measurement. The β-relaxation of the polyethylene-kaolin system and of the polyethylene-$BaSO_4$ composite is strongly broadened as compared with the glass-beads-filled HDPE composite. The β-relaxation of the glass-beads-filled system is nearly identical with the β-relaxation for the unfilled but swollen PE. For the glass-beads-filled polymer no significant modifications in the relaxation behavior can be detected with this method. This could be due to the very small specific surface of the glass-beads, in comparison with the other fillers used.

CONCLUSIONS

The β-relaxation peak which is interpreted as the glass transition of the amorphous PE-phase, intensified by swelling in CCl_4,

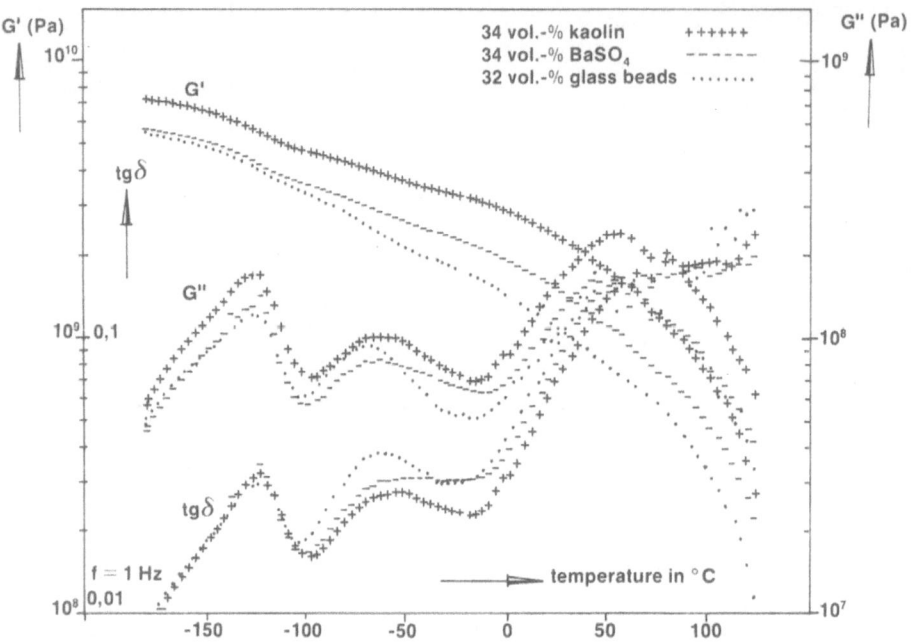

Figure 9.   Storage modulus G', loss modulus G" and loss factor tan
δ for HDPE composites filled with kaolin, barium sulphate
and micro-glass beads.

broadens with increasing kaolin filler content and is shifted to
higher temperatures. The temperature shift and broadening has been
explained in terms of a partially reduced mobility caused by the
filler particles. Fillers with a coupling agent give rise to more
bound polymer in the composites, which is in agreement with the results of solvent extraction reported elsewhere.[11] The amount of
bound polymer depends strongly on the specific surface of the filler
particles.

ACKNOWLEDGEMENTS

    The authors express their special thanks to the "Arbeitsgemein-
schaft Industrieller Forschungsvereinigungen (AIF), Köln, for gener-
ous financial support (Research No. 4227 and No. 5019).

REFERENCES

1. G. Kraus, Reinforcement of Elastomers, Wiley, London, 1965.
2. J. A. Manson and L. H. Sperling, Polymer Blends and Composites,
    Plenum Press, New York, 1976.

3. Yu. S. Lipatov, Advances in Polymer Sci. 22, Phys. Chem., 1, (1977).

4. Yu. S. Lipatov, Physical Chemistry of Filled Polymers, the British Library, 1979.

5. A. Galeski and P. Kaliniski, Polymer Blends, ed. by E. Martuscelli, R. Palumbo and M. Kryszewski, p. 431, Plenum Press, New York, 1980.

6. G. Kraus, Rubber Chem. Techn., 38, 1070 (1965).

7. C. M. Blow, Polymer, 14, 309 (1973).

8. B. B. Boonstra, Polymer, 20, 691 (1979).

9. V. Dolakova and F. Hudecek, J. Macromol. Sci. Phys., B15, .337 (1978).

10. K. Kendall and F. R. Sherliker, Brit. Polym. J., 12, 85, 111 (1980).

11. F. H. J. Maurer, R. Kosfeld, Th. Uhlenbroich and L. G. Bosveliev, Proceedings of the 27th Intern. Symp. of Macromolecules Strasbourg (F), 1251, (1981).

12. F. H. J. Maurer, H. M. Schoffeleers, R. Kosfeld, Th. Uhlenbroich, Proceedings of the Fourth International Conference on Composite Materials, ICCM-IV, Tokyo, (1982).

13. K. H. Illers, Kolloid - Z. u. Z. Polym., 231, 622 (1969).

14. K. H. Illers, Kolloid - Z. u. Z. Polym., 250, 426 (1972).

15. K. H. Illers, Coll. and Pol. Sci., 251, 394 (1973).

16. K. H. Illers, Coll. and Pol. Sci., 325, 1 (1974).

17. M. A. Pauwe and J. R. Knox, Pol. Eng. Sci., 16, 36 (1976).

18. R. Arai and I. Kuriyama, Coll. and Pol. Sci., 254, 967 (1976).

19. R. F. Boyer, J. Macromol. Sci. Phys., B9 (2), 187 (1974).

20. R. F. Boyer, Polymer, 17, 996 (1976).

21. Kajiyama, T. Okada, A. Sakoda, M. Takayanagi, J. Macromol. Sci. Phys., B7 (3), 583 (1973).

22. S. A. Paipetis, Coll. and Polym. Sci., 258, 42 (1980).

23. V. K. Bergmann and K. Nawothi, Kolloid - Z. u. Z. Polym., 219, 132 (1967).

24. M. Kahizahi and T. Hideshima, J. Macromol. Sci. Phys., B8 (3-4), 367 (1973).

25. L. C. E. Struik, Physical Aging in Amorphous Polymers and Other Materials, Elsevier, Amsterdam, 1978, p. 52.

26. G. J. Howard and B. A. Shanks, J. Macromol. Sci. Phys., B19 (2), 167 (1981).

THE USE OF TITANATE COUPLING AGENTS FOR IMPROVED PROPERTIES

AND AGING OF PLASTIC COMPOSITES AND COATINGS

Salvatore J. Monte and Gerald Sugerman

Kenrich Petrochemicals, Inc.
East 22nd Street
Bayonne, NJ, 07002

INTRODUCTION

Coupling Agents are molecular bridges at the interface between an inorganic filler and an organic polymer matrix. According to Gent[1]: "Coupling Agents or adhesion promoters are widely used to make adhesive joints between polymers and metal or glass that are able to withstand severe conditions of high temperature and humidity. They have a dual functionality so that they are capable of interlinking the two adherends by reacting with surface atoms of both substances." It is proposed that titanium derived coupling agents react with free protons at the inorganic interface resulting in the formation of organic monomolecular layers on the inorganic surface according to the following chemical mechanism:

$$(Y-R-X-O)_3Ti-OR' + MOH - (Y-R-X-O)_3Ti-OM + R'OH$$

Once the hydrolyzable portion [OR'] of the titanate molecule reacts with the proton on the inorganic bearing species or substrate [M], it is theorized that the titanium oxygen bond [Ti-(O)] may possibly provide blocked acid catalysis via a transesterification mechanism. The [X] portion of the titanate molecule effects composite performance as determined by the chemistry of alkylate, carboxyl, sulfonyl, phenolic, phosphate, pyrophosphate and phosphite groups. The [R] function provides polymer compatibilization and van der Waals' entanglement via aliphatic or aromatic long carbon chains. In addition, the titanate molecule may contain [Y] functionality typically provided by either an amino, methacrylic or acrylic group. A chemical description of the titanate coupling agents discussed in this paper is shown in Table I. A convenient letter-number code is used throughout the paper for convenience and brevity.

301

Table I.   Chemical Description of Titanate Coupling Agents

Titanate Coupling Agents	Chemical Description
KR TTS	isopropyl, triisostearoyl titanate
KR 39DS	isopropyl, triacryl titanate
KR 44	isopropyl, tri (N ethylamino-ethylamino) titanate
KR 26S	isopropyl, 4-aminobenzenesulfonyl di(dodecyl-benzenesulfonyl) titanate
KR 37BS	isopropyl, di(4-aminobenzoyl) isostearoyl titanate
KR 12	isopropyl, tri(dioctylphosphato) titanate
KR 212	di(dioctylphosphato) ethylene titanate
KR 34S	isopropyl, tricumylphenyl titanate
KR 134S	titanium di(cumylphenylate) oxyacetata
KR 38S	isopropyl, tri (dioctylpyrophosphato) titanate
KR 138S	titanium di(dioctylpyrophosphate) oxyacetate
KR 238S	di(dioctylpyrophosphato) ethylene titanate
KR 238M	KR 238S + Methacrylic functional amine
KR 262A	KR 262 + Acrylic functional amine
KR 55	tetra (2, 2 diallyloxymethyl-1 butoxy titanium di (di-tridecyl phosphite)
KR 138D	KR 138S + 2-Dimethylamino-2-methyl-1-proponol

It is proposed that dispersion of an inorganic in an organic phase, in the presence of titanate coupling agents, is enhanced by the replacement of the water of hydration at the inorganic surface with a monomolecular layer or organo functional titanate causing inorganic/organic phase compatibilization at the interface, thereby increasing the degree of displacement of air by the organic phase in the voids of the inorganic component. Figure 1 shows the proposed effect of coupling an agglomerated inorganic with a monoalkoxy titanate in an organic vehicle. Typically, dispersion tests show that 1/2% titanate coupling agent on calcium carbonate will reduce the viscosity of 50% calcium carbonate filled dioctyl phthalate from 177,000 cps to 2,600 cps. When successful coupling and dispersion is achieved through proper processing, new adhesion, property and rheology performance standards of fiber reinforced pigmented and/or extender filled polymer composites and coatings are possible.

RESULTS AND DISCUSSION

Processing and Rheology

Processing of composites with titanate coupling agents requires consideration of the following elements: titanate selection, titanate amount, titanate treatment of reinforcement or filler and good compounding practice.

Titanate selection is determined by the chemistry of the composite ingredients and end properties desired and references[2] are available to assist the compounder.

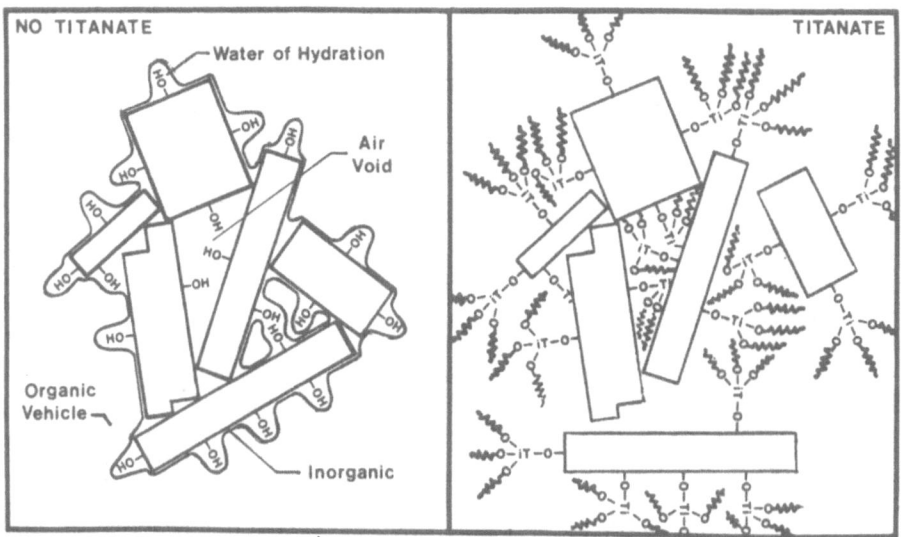

Figure 1. The proposed mechanism for deposition of a mono-layer of
triorganofunctional titanate to effect the elimination
of inorganic water of hydration and air voids resulting
in deagglomeration.

Titanate amount may vary from 0.2% by weight of glass fiber to
1.0% for fine particle size silica and carbon black and is determined
by, among others, surface area, specific gravity, particulate pore
velocity, and end composite properties desired.

Titanate treatment of reinforcement or filler requires a technique to provide uniform distribution of a small amount of liquid
titanate over a large filler surface area and can be accomplished by
fluidizing or agitating the filler bed as diluted (with solvent, plasticizer or polymer) titanate is dripped or undiluted titanate is
sprayed onto the fluidized bed. Mixers of the Henschel, Papenmeir,
V-shaped cone with intensifier bar, Littleford type are preferred.
Ribbon blenders, drum rollers, cement mixers and hand stirring often
give poor results. When batch mixers such as two roll mills or Ban-
burys are used for mixing, a powdered form of the titanate coupling
agent may be mixed in with the polymer before the addition of filler.
The powdered form is made typically by mixing 72% of liquid titanate
with 28% Johns Manville Micro Cel E in a Henschel mixer. Titanates
may be used to couple "in situ" by dispersing directly into a compatible solvent, plasticizer, water or liquid polymer followed by particulate to be coupled.

Good compounding practice will provide maximum dispersion for best possible, consistent end composite physical properties and is accomplished efficiently by producing the highest amount of mixing torque in the shortest time cycle. A power curve plot-torque (vertical axis) vs. time (horizontal axis) is useful for guidance. Since titanates often effect systems rheology and subsequent torque significantly, close attention to extrusion or mixing temperatures, cooling, rpm's and back pressure conditions is necessary to assure adequate internal shear development for proper dispersion. A discussion of fiber and filler reinforced polystyrene, polyphenylene sulfide, nylon and polyolefins is helpful.

G. R. Kritchevsky[3] subjected silane and titanate treated glass fiber reinforced polystyrene to 48 hour water boil and by SEM analysis noted that the points at which delamination, adhesive failure and corrosion begin are where islands of water appear on the silica (glass fiber) surface. Kritchevsky concluded, "....boiled polystyrene samples show evidence of the diffusion of water through the resin. Islands are evident on fiber fracture surfaces in all polystyrene samples[4] except for the titanate." S. J. Monte and G. Sugerman[5] reported a reduction of injection mold pressure at 450°F from 1500 psi to 800 psi in $CaCO_3$, carbon black filler polystyrene using KR 34S titanate. A patent[6] showed a reduction of Braebender Plasticorder torque values at 145°C from 250 meter-grams without titanate to 20 meter-grams with titanate. A patent[7] states:

"The use of organo-titanate treated Ryton R10 provided reduced viscosity, substantially lowered the required injection temperatures and pressure and also resulted in improved encapsulated component quality, such as much less and fewer surface blemishes.

Blends of 35% by weight of polyphenylene sulfide and 65% by weight of glass fiber, talc filler (5:3 Ryton R10) were tumble dry blended with various additions of KR 134S predissolved in methylene chloride. The blended mixtures were air dried for 2 hours under a vented hood to remove the methylene chloride and viscosities of the various blended mixtures are shown in Table I. [See Table II]. The viscosity reductions achieved of from 50 to 80% as compared to the control are significant.

A similar composite was prepared in which KR 134S was predissolved in methylene chloride as a 3% solution. The blended mixture was air dried for 2 hours under a vented hood to remove the methylene chloride. The resulting blend was used to conventionally encapsulate ceramic capacitors using a Newbury Eldorado Reciprocating Screw Plastic Injection Molding Machine Model VI-75 ARS (Newbury Industries Inc.). The capacitors,

before encapsulation, had dimensions of 0.180 x 0.130 x
0.030 in. The encapsulated capacitors had dimensions of
0.300 x 0.300 x 0.100 in.

The procedure was repeated except that organotitanate
was not used. The injection molding parameters achieved
for each procedure is shown. [See Table III].

The initial injection pressure of the reinforced
polyphenylene sulfide was reduced from 1300 to 700 psi
and the injection rate increased from 3 to 4 parts by use
of the organotitanate when compared to the untreated
control."

Han[8] reported,

"....More specifically, the titinate KR 44 increased the
viscosity of the nylon 6/$CaCO_3$ composite by the greatest
amount, whereas the increases caused by the three other
coupling agents (silane A-1100, silane Y-9187, and titanate
KR 138S) were about equal. However, when used in the nylon
6/Wollastonite (50% filled) composite, some coupling agents
decreased its melt viscosity significantly. It is seen
[in Figure 2 of Reference 8] that, at high shear stresses,
titanate KR 44 brings the melt viscosity of the nylon 6/
Wollastonite (Wollastokup[R] KR 44, 1.0 - NYCO) composite
below the viscosity of pure nylon 6."

The monolayer of nonextractable, titanium metal-bound isostearoyl
functionality as typified by the KR TTS titanate coupling agent provides a method of achieving improved process rheology and an overall
balance of good properties for filled polyolefins without the negative
side effects often experienced when using extractable conventional
physical wetting agents such as stearates. For example, when high
levels of stearic acid coated calcium carbonate are used in films,
blisters and pinholes are caused by the free stearic acid. Titanate
treated calcium carbonate at the same filler level eliminates the
problem entirely.[9] Han[10] suggests the possibility that the titanate
coupling agent has a direct effect on the morphology of the polypropylene polymer.

Table II. KR 134S Effect on Ryton R10 Melt Viscosity

Wt. % KR 134S Organotitanate	Melt Viscosity at 600°F., Kpoise
0	38.0
0.66	17.64
1.33	10.01
3.3	7.95

Table III. KR 134S Effect on Injection Molding of 65% Glass Fiber:
Talc (5:3) Filled Polyphenylene Sulfide (Ryton R10)

Formulation	Control Untreated Ryton R10	Organotitanate KR 134S Treated Ryton R10
Glass Fiber	40.6	40.6
Talc	24.4	24.4
KR 134S 1.3% (3% MeCl Sol.)	–	0.85
Polyphenylene Sulfide (Ryton R10)	35.0	35.0
Molding Conditions		
Hopper drying time @ 250°F	2	2
Rear Zone temperature	640°F	620°F
Front zone temperature	655°F	630°F
Nozzle temperature	665°F	640°F
Mold temperature	125°F	125°F
Initial injection pressure	1300 psi	700 psi
Overall injection time	5 sec.	5 sec.
Initial injection time	N.A.	N.A.
Injection rate	3 (press setting)	4 (press setting)
Knockout time	Max. (press setting)	Max. (press setting)
Screw speed	3 (press setting)	3 (press setting)
Screw back pressure	200 psi	200 psi
Cycle time	25 sec.	25 sec.

Monte and Bruins[11] reported superior reduction of torque during
compounding of calcium carbonate-filled LDPE in a Brabender operating
at 82 rpm, 100°F with a 5 kg. weight ram pressure. The torque reduction results comparing the titanate, the control, and nine other classical wetting agents are shown in Table IV. The titanated system
achieves lowest torque in 60 s while the second best performer required 150 s. Faster Banbury cycles are therefore made possible.
In addition, melt index is improved fifteenfold over the control.
Significant energy efficiency and productivity effects in alkyds,
acrylics, epoxy, epoxy ester, chlorinated rubber, and vinyl plastisols have been reported.[12]

Figures 4 and 5 are plots of rheology response under varying
shear conditions using a Han slit rheometer.[13] The HDPE plot shows
that the 40% calcium carbonate-1% KR TTS-filler system mimics virgin.
Also, the slope and character of the rheology curve changes when the
filler level with titanate coupling agent is increased from 40 to 55%
and 70%. It appears that at the 55% filler level for this system,
the CPVC (critical pigment volume concentration) has been exceeded.
Figure 2 suggests that if one is to mimic the rheology and properties
of virgin HDPE polymer, the filler level should stay below 55% and

TABLE IV. Effect of KR TTS and Conventional Wetting Agents on the Brabender Mixing Torque of CaCO$_3$-Filled LDPE[a,b]

Additive	Torque (g/m^2)						Melt index (g/10 min)	Particles per 144 in.^{2}
	30 s	60 s	90 s	120 s	150 s	180 s		
None	2,000	2,000	1,900	1,750	1,750	1,750	0.27	312
Emcol H31-A[c]	2,150	1,400	1,150	1,000	1,000	1,000	2.30	6
Emcol 117[c]	1,900	1,400	1,300	1,250	1,100	1,000	1.90	16
Glycerol monostearate	2,150	1,400	1,300	1,250	1,250	1,150	3.87	8
Emcol 116[c]	1,900	1,400	1,300	1,150	1,150	1,100	1.69	20
Aluminum oleate	1,650	1,600	1,400	1,300	1,250	1,250	1.80	44
Aluminum stearate	1,900	1,400	1,300	1,250	1,250	1,250	1.60	32
Tween G5[d]	2,500	1,500	1,250	1,000	900	900	1.25	7
Zinc stearate	1,900	1,650	1,400	1,150	1,150	1,050	2.30	44
Paricin 220[e]	1,900	1,400	1,250	1,150	1,100	1,100	0.95	16
Ken-React TTS	1,250	900	900	900	750	750	4.17	16

[a]Formulation: 9 parts of LDPE and 1 part of dry blend consisting of CaCO$_3$ (Camel-Tex) 70%, LD 501 28%, and additive 2%.

[b]LDPE 501, Exxon Chemical Co.

[c]Witco Chemical Co.

[d]Imperial Chemical Industries Ltd. (ICI).

[e]Baker Castor Oil Co.

in the neighborhood of between 40 and 50%. Maximum impact values of
$CaCO_3$ filled HDPE have been noted at or near the CPVC (approximately
50% loading with a 2.5 micron $CaCO_3$) with impact values lower at too
low or too high a filler loading. Also, higher molecular weight HDPE
polymers showed[14] a 5.5 fold higher impact than a lower molecular
weight HDPE at the CPVC of a 50% $CaCO_3$/HDPE composite.

Monte and Sugerman[15] reported positive dispersion effects on surfaces of carbon black, phthalo pigments, red lake pigments, yellow
chrome oxide, etc. Figure 3 for 70% calcium carbonate-filled polypropylene shows the same characteristic curve as in Figure 2 when the
calcium carbonate level is above the CPVC of the system. Also, the
rheology of the system is close to or lower than virgin across the
entire shear range using 1% KR TTS coupling agent. No additional
rheological benefits are gained above the 1% level. Two percent coupling agent would only be used for higher elongation; higher impact
values; or further low temperature performance.

Effects of titanate selection, amount and treatment on adhesion
and subsequent polyolefin, PVC, epoxy, acrylic and urethane composite
physical, conductive, aging and corrosion properties follows.

Adhesion and Composite Properties

Reinhart[16] investigated titanates as primers for adhesion of 1-2
polybutadiene modified epoxy to anodized aluminum. The results are
shown in Table V. The HME resin (high vinyl modified epoxy) was ap-

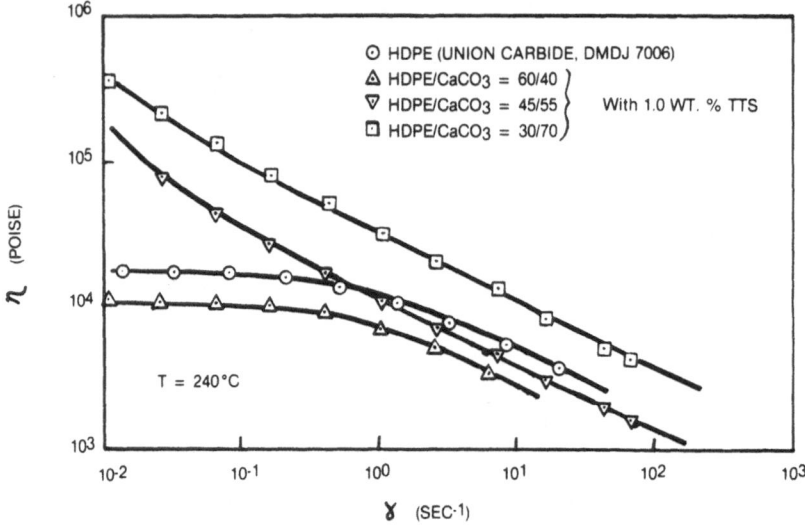

Figure 2.   Rheology of 40, 55 and 70% $CaCO_3$ (1% KR TTS) filled HDPE
            as compared to virgin using a Han slit rheometer.

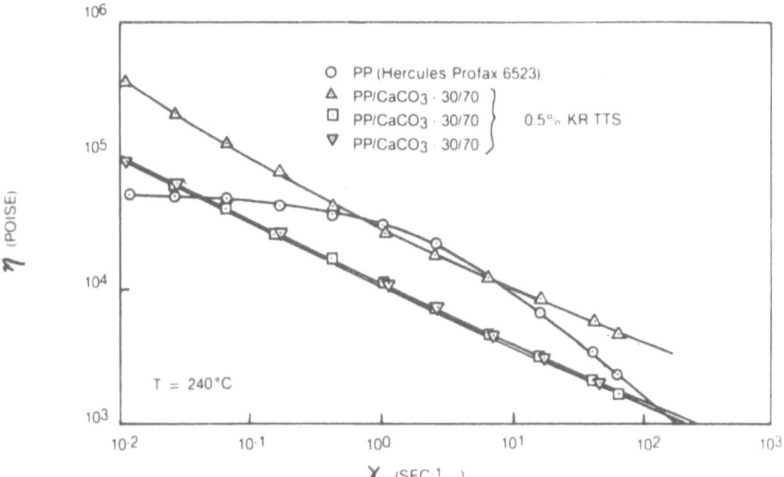

Figure 3.   Rheology of 70% CaCO₃ (0.5, 1.0 and 2.0% KR TTS) filled
            polypropylene as compared to virgin using a Han slit
            rheometer.

Figure 4.   Proprietary pigmented virgin HDPE copolymer (left) and
            50% CaCO₃ (Atomite) 0.5% KR TTS filled HDPE homopolymer
            (DMDJ-7006 Natural 7) (right) injection molded milk crates.

plied to 2024 bare aluminum preslotted finger panels surface prepared
by phosphoric acid anodization. Various primers were made using titanate and silane coupling agents in isopropyl alcohol, xylene and
deionized water solvents. A later report[17] compared silanes and titanates and the results are shown in Table VI. Reinhart stated:
"....As can be seen, the breaking strength of the specimens is af-
fected strongly by the amount of titanate as well as the predrying

Figure 5.   Injection molded Plating Rack made of 40% Wollastokup[R]
            KR TTS 0.5 (Nyco) filled PP homopolymer (2.0 gms/10 min.
            melt index) replaces virgin PPO (polyphenylene oxide)
            engineering thermoplastic prior art.

Table V.   KR 44 and KR 38S Titanates as Adhesion Promoters
           in Unfilled 1-2 Polybutadiene Modified Epoxy on
           Anodized Aluminum - Al/Al Lap Shear Specimens,
           1/2" Overlap

Titanate	Drying Conditions	Strength (MPa)
None	None	9.56
KR 44 1% (iso-prop.)	R.T.	7.58
KR 44 1% (iso-prop.)	30 min.- 66°C	12.06
KR 44 1% (iso-prop.)	30 min.- 122°C	8.54
KR 44 .2% (iso-prop.)	R.T.	17.56
KR 44 .2% (iso-prop.)	30 min.- 66°C	20.75
KR 44 .2% (iso-prop.)	30 min.- 122°C	13.53
KR 38S 1% (Xylene)	R.T.	13.40
KR 38S 1% (Xylene)	30 min.- 66°C	6.06
KR 38S 1% (Xylene)	30 min.- 122°C	11.27
KR 38S .4% (Xylene)	R.T.	11.56
KR 38S .4% (Xylene)	30 min.- 66°C	12.36
KR 38S .4% (Xylene)	30 min.- 122°C	16.96

Table VI. Titanates and Silanes as Adhesion Promoters in
           Unfilled 1-2 Polybutadiene Modified Epoxy on
           Anodized Aluminum-Al/Al Lap Shear Specimens,
           1/2" Overlap

Coupling Agent (0.2% Solution)	Drying Conditions	Ultimate Strength (MPa)
A-151[a], Vinyltri-ethoxysilane	R.T. Dry + 30 min.@66°C	14.06
KR 238M	R.T. Dry + 30 min.@66°C	17.21
XZ-8-5066, Lot 100[b]	R.T. Dry + 30 min.@66°C	18.03

a. Union Carbide
b. Dow Corning

conditions of the specimens prior to adhesive application and cure.
The KR 44 material has provided the highest lapshear results to date.
The 20.75 MPa value is considered excellent for an adhesive to polymer having no additives, fillers or toughening agents." Reduction
of KR 44 from a level of 1% to 0.2% in isopropanol, drying the titanate primer at 66°C as compared to room temperature results in
higher adhesive strength. At 66°C drying conditions for both KR 44
and KR 38S, the adhesive strength is higher at lower titanate levels.
A possible explanation is that once a monolayer is formed on the surface of the aluminum, excess non-metal bound titanate in the interphase creates an adhesive effect.

Plueddemann has often commented on the efficacy of simple razor
blade-glass slide models for adhesion determination. Hutchins and
Spell[18] reported results of a water permeation test for a filled
Kynar (PVDF) coating on glass at 90°C for 500 hours comparing an
acrylic titanate (KR 39DS) and gamma-Methacrylocypropyltrimethoxy
silane (A-174). After 500 hours, the coating containing silane could
be easily removed from the glass with a razor blade cut while the
coating containing titanate was unremovable, even with scratching.
Also, stability of methacrylic titanate towards oxidation was cited
for successful adhesion to steel. The titanate had better resistance
to hydrolysis when compared to the best silane control. Skidmore[19]
reported that in successful bonding of polyurethane to HDPE, concentrations of KR TTS or KR 55 larger than 0.2% by weight resulted in
their acting as lubricants. The HDPE surface was subjected to oxidation before encapsulation. Baker[20] reported significant adhesion
results with titanates in several epoxy, silicone and urethane systems. Work in ongoing and quantitative data will be available to
Monte for publication upon completion of the work for the Dept. of
the Navy, U.S.A. Briggs[21] reported 0.5% KR 238M by weight of total

epoxy formulation, successfully adheres coating to concrete under water. Also, KR 55 gave excellent results as a corrosion inhibitor for steel surfaces under water when used in an acetate solution to coat grit onto the steel. Research work has been forwarded to Ministry of Defense, U.K. Westerman[22] reported 0.25% KR 238M by weight of polyester resin, showed improved adhesion of carbon fiber to polyester to the extent that compression strength and modulus increased by 15% compared to the untreated control. N. Eickman[23] reported significant adhesion of fiberglass to a developmental polymer using KR 38S. J. Gabbert[24] reported a 25% improvement in wet strength, tensile elongation, modulus and izod impact evaluating KR 12, KR 26S and KR 37BS in amino silane sized fiberglass reinforced Nylon 6 block copolymer system. SEM's were to be available for the October, 1982 SPE NATEC. Geddes[25] reported increased bond strength, even when samples are immersed in water, when studying KR 44 as an adhesion promoter for priming stainless steel surface for epoxy bonding. Further comments were: "When the steel surface is etched first, the primer doesn't seem to have as much of a positive effect. The research is not complete yet, at this time." Chattopadhyay[26] reports best performance for adhesion to steel using KR 38S in a heavily filled polyurethane coating spray applied from 125 to 250 mil thickness for tank car insulation. The coating is charred with a bunsen burner at 1100°C and once pyrolized, is subjected to high pressure water hose spray under simulated fire fighting conditions. 4' x 8' steel panels coated in this manner have passed the U.S. Department of Transportation specifications for structural integrity after pyrolysis and water pressure exposure. M. Crystal[27] commented:

> "Uniplex produces a variety of profile intrusions primarily based on Mylar and/or vinyl laminated tapes to ABS or HIPS and occasionally vinyl. These extruded profiles are used in such applications as picture frames, decorative trim, label tapes, and point of sale advertising assemblies. Many of the profiles are post formed (often in a vacuum tank) after attachment of the decorative Mylar or vinyl top layer tape. Such post forming and/or exposure to humidity has frequently resulted in rejects as a consequence of delamination of the decorative tapes from the profile substrate. Addition of 0.2% by weight of ABS of Drimix 65% KR 134S/ HiSil to the lightly pigmented (5% TiO_2, 5% iron oxide) pipe grade ABS (Cycolac T-1000) increased the adhesion of the profile to the decorative tape to the degree that the failure mode changed from the profile tape interface to the interface of the laminations used to make the decorative tape itself. For example, whereas adhesion failure occurred in the past at the interface between the ABS profile and the aluminum plated foil coated with Mylar, the titanate containing ABS causes the Mylar to aluminum and not ABS to aluminum to be the first point of failure."

Polyolefins

N. H. Sung[28] has discussed adhesion promotion effects of coupling agents in aluminum oxide-polyethylene joints. He noted that heating the titanate above 70°C results in significant increase in peel strength.

A possible mechanistic explanation for the observation of the formation of carboxylate ester observed by Sung and Calvert for substrates that have been reacted with monoalkoxy tricarboxylate titanate subjected to vacuum, and perhaps of lesser importance temperature, is proposed by the authors:

a) Coupling agent applied to surface   b) Surface equilibrium:

$$(RC-O)_3 Ti-OR' + H-Surface-H \longrightarrow$$

with the carbonyl $\overset{\|}{O}$ on the RC group.

Right side (b):

R'O-H Surface-H

$$RC-O-Ti-O-CR$$ with carbonyls and O below leading to C=O, R.

c) Vacuum to remove ester:

Surface-H

$$R'OCR\uparrow + HO-Ti-O-CR$$

with O, C=O, R below; and

d) New Surface equilibrium:

Surface

$$HOH + RC-O-Ti-O-CR$$

Peel strength was also effected by the amount of titanate added at the interface. The amino functional silane developed higher peel strength when compared to the isostearoyl titanate. The data in Table VII invites a direct comparison of both amino functional titanate and silane in the aluminum oxide-polyethylene joint model. In a study of PP-CaCO$_3$ fibers, Han[10] reported that as the draw-down ratio V_L/V_0 is increased, the titanate coupling agent TTS gives rise to higher fiber tensile strengths than the silane coupling agents, Y-9187 and A-1100.

Several patents[29-32] have shown significant increases in elongation of mineral filled LDPE, HDPE, Polybutene and Ethylene Copolymer composites while maintaining tensile strengths near or above controls. Table VIII[33] takes advantage of coupling and process rheology benefits to allow use of a higher molecular weight PE to increase elongation 14.5 fold, thereby providing a tougher composite. Thermoplastic

Table VII. Adhesion Results Using Pencil Hardness
Test of Acryloid B 66 to Mild Steel

Coupling Agents

```
Isopropyl tri(N-ethylamino-              Definitely superior
  ethylamino titanate..............1.............to standard
Isopropyl tri methacryl                  Equal or slightly
              titanate..............2....better than standard
Isopropyl tri acryl titanate.......3.......Equal to standard
Control..........................3........................
Isopropyl di(4 amino benzoyl)
          isostearoyl titanate.....4..........:.Clearly worse
-Amino propyl tri ethoxy silane....5...........Clearly worse
Isopropyl isostearoyl diacryl
              titanate......5...........Clearly worse
Isopropyl tri isostearoyl
              titanate......6...........Clearly worse
Isopropyl tri (dioctyl                   Very much worse
  phosphato) titanate..............7...........than standard
Isopropyl tri methyl-
  ricinoloyl titanate..............8...........Total Failure
```

toughness may be defined as the area under the stress/strain curve obtained by plotting the stress (tensile strength-force per unit area) on the vertical axis versus strain (elongation per unit length) on the horizontal axis.

Milk crate prototypes (see Figure 4) consisting of a proprietary pigmented virgin high density polyethylene copolymer of typical manufacture and 50% $CaCO_3$ (Atomite) 0.5% KR TTS treated filled HDPE homopolymer DMDJ-7006 Natural 7 (6.0 grams/10 minutes melt index) were compounded on a 53 mm W&P twin screw at Adell Plastics. The crates were then injection molded on a 700 ton press at a major polymer producer location with no change required in the injection molding conditions when going from virgin to the 50% filled system. With the use of filler and titanate, the weight of the crate was increased from 1,380 grams to 1,959 grams. Nevertheless, there was a total pound volume savings in raw materials of $.215/crate before calculating cost of compounding. The $CaCO_3$, KR TTS filled HDPE homopolymer

Table VIII. 50% $Al(OH)_3$-Filled Polyethylene

	No Titanate	Titanate
Sholex 5008 (MI = 0.65 g/10 min)	100	-
Sholex 4002 (MI = 0.20 g/10 min)	-	100
$Al(OH)_3$	100	100
KR TTS, 3%	-	3
Properties (kneaded at 150°C, 1 mm sheets):		
Tensile strength, kg/cm^2	118	120
Elongation, %	35	508

crate withstood a falling impact test of a ten pound ball from a five foot height at room temperature, but failed the -40°F impact resistance test. Investigators were extremely encouraged since the impact resistance, ft.lb. at -40°F for the unfilled homopolymer is 30 ft.lb. and the 50% calcium carbonate filled crate tested at 29 ft.lb. The copolymer had a specification impact resistance at -40°F of 34 ft. lb. Subsequently, the use of the HDPE copolymer, in conjunction with titanate treated calcium carbonate, passed the -40°F freeze impact test requirements.[34]

Plating racks (see Figure 5) have been made until recently of PPO (polyphenylene oxide - Noryl[R]). Alternate lower cost thermoplastics such as ABS, Nylon, Polystyrene and Polyolefins invariably failed in one or more field impact, water immersion, plating bath, solvent resistance, vibration under load and dimensional stability at 190°F tests. A compound, produced by Suntech Polymers, Inc., Basking Ridge, NJ, consisting of 60 parts PP (polypropylene homopolymer - 2.0 gms/10 min. melt index) and 40 parts Wollastokup KR TTS 0.5 (NYCO) and injection molded by Fancourt Industries, Inc., Fairfield, NJ, appears satisfactory at substantially lower cost.

Thermoforming highly filled polypropylene sheet typically presents some production problems. Itap S.A., Sao Paulo, Brazil, thermoformed a polypropylene (PP) container for margarine from sheet made by its Cromex Div. from a formulation that consists of 60 parts PP homopolymer and 40 parts precipitated calcium carbonate ($CaCO_3$) of 0.9 micron size. (See Figure 6). This high filler load, without coupling agent, caused loss of flexibility and impact strength in the finished container, Itap reports,[35] leading to a high failure rate. To upgrade the sheet's physical properties and cut the reject rate, Cromex incorporated 0.2 parts titanate as a coupling agent. The $CaCO_3$ is a pure grade; the treatment generally is applied before intensive mixing of the filler with the PP resin.

In an attempt to overcome some of the problems inherent in thermoforming virgin PP, Itap undertook a study of seven different additives, including stearates and silanes.

According to Itap, the titanate treatment provides all the physical properties required, plus improved productivity and handling characteristics necessary to replace pigmented polystyrene prior art. An additional feature: Itap normally pigments containers of this type with 2 to 3% of titanium dioxide (TiO_2), but because of the good finish obtained with the filled PP, now is considering reducing or eliminating this costly pigment.

Titanate coupling agent efficacy in self-extinguishing polypropylene compounds has been previously reported by the authors based on information supplied by Japanese[4] and United States investigators[36]

Figure 6.   Vacuum thermoformed 40% pptd. $CaCO_3$ (0.9 micron) filled
            PP homopolymer margarine tub produced by Itap S.A., Sao
            Paulo, Brazil, using 0.5% KR TTS.

and a U. S. patent.[37] Efficacy in $BaSO_4$ and $CaCO_3$ filled polyolefin
composites has also been reported.[38-41] A paper[42] indicated titanate
coated $CaSO_4$ plus antimony oxide and 5,6-dibromomonoborane,2,3 di-
carboxanilide yields "...a flexible, UL-94 V-O rated polypropylene
compound with good dimensional stability, high heat distortion temperature, high tensile modulus, and the flow characteristics of the
natural resin- and the price is competitive."

Polyvinyl Chloride

      Monte and Sugerman have reported in several papers[43,44,14] on
the effects of titanates in PVC rigid, flexible and plastisols.
Typical applications include rigid PVC injection molding and sheet-
ing, shoe foxing, tablecloths, shower curtains, hose and wire and
cable. Figure 7 typifies usage wherein the $CaCO_3$ loading has increased threefold over prior art for construction expansion joints
while maintaining suitable properties. Maximum impact values are
achieved using titanates at a level of 1% by weight of filler.
Levels above 0.5% titanate in plastisols often produce heat instability problems. A study[45] in 30 to 100 phr 0.4 micron $CaCO_3$ filled
rigid PVC indicated, at the 100 phr loading, the Decomposition Time
(Plastograph - 170°C, 40 rpm) for 1% KR 238S was 85 min. as compared to 48 min. for 1% stearic acid coating and 45 min. for the untreated control. Also, Equilibrium Torque (plastograph - 170°C, 40
rpm) at the 100 phr loading was 440 kg-cm for 1% stearic acid coat-

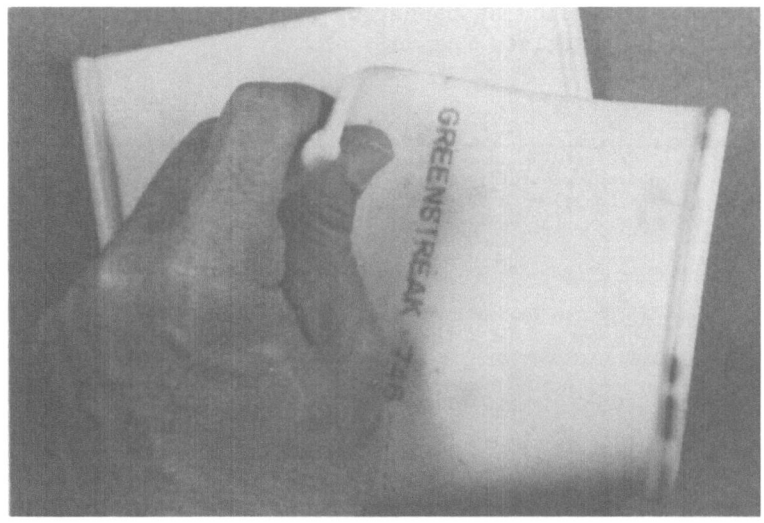

Figure 7. Flexible PVC construction expansion joint uses $CaCO_3$ 0.5%
           KR TTS to increase filler load significantly.

ing, 435 kg-cm for the untreated control and 393 kg-cm for 1% KR 238S.
Yelten[44] reported, "There was no noticeable overloading in the compounding extruder (Cincinnati Milacron twin screw extruder Model K-
2X120) and injection molding equipment during the production of the
parts, despite the high level (60 phr 5 micron $CaCO_3$ 1.0% KR 38S) of
filler employed."

Epoxy

J. J. Jakubowski and R. V. Subramanian[46] reported results of the
use of phosphato titanates in electrochemical coatings for prevention
of carbon fiber release from epoxy polymer composites as determined
by thermogravimetric analysis of organophosphorus coating on carbon
fibers, thereby providing a novel approach to protection of said composites exposed to fire.

A patent[47] (Table IX) reports the effect of KR TTS on the adhesion of copper to an epoxy impregnated fiberglass prepreg used to
make printed circuit boards. Peel strength is increased significantly. In another printed circuit board patent[48] (Table X), electrical
and water uptake property improvements were noted.

Tables XI and XII indicate significant benefits by use of titanate coupling agents in amide and anhydride cured Kevlar-epoxy composites. Reduced viscosity, increased elongation, flexural strength
and falling ball impact strength were observed. The titanate appears

Table IX. Adhesion of Copper by Epoxy Impregnated Fiberglass Prepreg

1. Blend:

	No Titanate	Titanate
DER 511 Epoxy	90.00	90.00
Epikote 828 Epoxy	10.00	10.00
Dicyandiamide Curative	3.20	3.20
Dimethylbenzylamine Hardner	0.16	0.16
KR TTS Titanate Coupling Agent	-	1.50

2. Dissolve Into: Methyl, Ethyl Ketone and Cellosolve

3. Varnish Coat: Epoxysilane-treated glass cloths, dried @ 150°C for 5 min. to give prepregs having:

RESIN VISCOSITY, POISE @ 130°C	130	82

4. Laminated (3 plies): Sandwiched between 2 etched Cu laminates, and hardened at 130-170°C and 5-40 KG/cm^2

PEEL STRENGTH, KG/cm	0.54	1.28

Table X. Properties of Copper to Epoxy Impregnated Synthetic Paper Prepreg

1. Impregnate Sheet with Resin:

   Sheet: 20% Cellulosic and 80% glass fiber
   Impregnate: to 10% resin content w. methylolphenol resin contg. 1% KR TTS and dried
   Impregnate: to 55% resin content w. epoxy resin

2. Prepare 1.6 mm thick laminate

   9 sheets of prepreg w. Cu foil on both sides

3. Properties

   Electrical Resistance (depending on treatment):
   No titanate $\geq 10^7$(C-90/20/65) or $2x10^2$ MΩ(D-2/100)
   Titanate $\geq 10^7$(C-90/20/65) or $5x10^3$ MΩ(D-2/100)

   Water Uptake:
   No Titanate - 0.30%
   Titanate   - 0.12%

to be reacting in the same way as it would on an inorganic filler, i.e., $CaCO_3$/DOP. Silanes are non-reactive in this instance.

The Datsun Motor Company, Johannesburg, S. Africa, uses a 79,000 liter tank filled with a 48 to 52% iron oxide solids/epoxy water based primer formulated for use as part of an anionic electrophoretic system for dip coating steel stampings. (See Figure 8). Settling of iron oxide solids had been a continuous problem partially overcome by constant stirring. A power failure would cause shutdown while the sludge was dug out by hand from tank, pipes and valves. Approxi-

Table XI. Properties of Amide Cured Aramid Fiber Reinforced Epoxy

Formulation:

Resin (Epicast 31D - Furane Plastics)	63
Hardner (Furane Hardner 927)	12
Reinforcement (Milled 1/16" Kevlar)	25
Titanate Coupling Agent	0.2 as shown

Coupling Agent added to resin-hardner blend prior to Kevlar incorporation.
Cure: 4 hrs. @ 160°C, after deaeration.

Formulation No.	1	2	3	4
Titanate Employed	None	KR 44	KR 238M	KR 55
Flexural Strength, MPa	152	269	200	165
Elongation, %	0.9	2.3	4.2	5.8
Initial Mix Viscosity @ 20 RPM (cps) x 10^3	105	72	86	29

Table XII. Properties of Anhydride Cured Aramid Fiber Reinforced Epoxy

Formulation:

Resin (Bakelite ERL-2774 UCC)	40
Hardener (Methylnadic anhydride)	20
Catalyst (Benzyldimethylamine)	0.5
Reinforcement (Milled 1/16" Kevlar)	40
Titanate Coupling Agents	0.3

Coupling agent added to resin-hardener-catalyst composite prior to Kevlar incorporation.
Cure: 2 hrs. @ 150°C followed by 4 hours at 200°C after deaeration.

Formulation No:	1	2	3	4
Titanate Employed	None	KR 44	KR 238M	KR 55
Flexural Strength, MPa	234	572	799	269
Elongation, %	0.3	0.5	1.7	2.5
Falling Ball Impact Strength Nm/m^2 x 10^3	193	571	826	890

mately R140,000 ($150,000.00) had been spent during a typical vacation shutdown period emptying the tank into drums and cleaning the lines. The use of 0.5% KR 262A (a water soluble salt formed by the reaction of an acrylic functional amine with a pyrophosphato organo titanate chelate) by weight of black iron oxide, eliminated the settling and caking problem. Datsun then combined KR 44 and KR 262A in a 1:3 ratio at a total amount of 0.15% titanate by weight of the paint and obtained the following primer performance advantages: (1) 25% greater cover of parts due to improved run off; (2) corrosion rate in the humidity chamber test was reduced 2.5 fold; (3) salt spray failure occurred at 1,000 - 1,300 hours as against 100 - 150 hours,

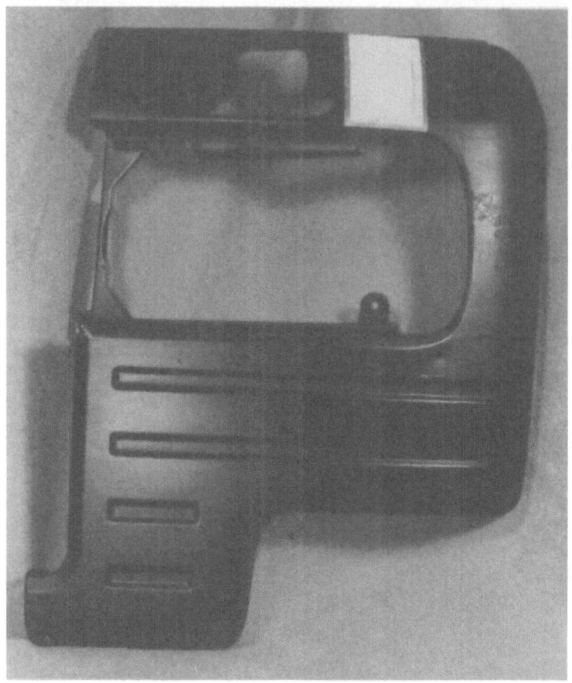

Figure 8.   Steel stamping of Datsun Magnis truck door coated with
            iron oxide/epoxy water based primer employing KR 44 and
            KR 262A in a 1:3 ratio.

which is typical for the untitanated control.

    KR 44 appears to speed the bake by catalyzing the amine cure and
KR 262A deagglomerates the iron oxide filler, thereby increasing the
available buffer surface area allowing a substantial reduction in the
amount of ammonia needed for pH control which in turn eliminates the
disadvantages of water sensitivity and reduced corrosion resistance
usually associated with the use of ammonia. Optimum performance for
the titanate films has been noted when the pH of the system is maintained in the range of 5 to 6 as opposed to 8 to 9 in the absence of
the titanates.

Acrylic

    An investigation[49] of the organic titanates on the adhesion of
two simple acrylic coatings to steel coupons indicated that substitution of the neat organic moiety in the coating produced the same
results as when combined with the titanium atom. Therefore, it was
concluded by the investigators that it was not necessary to use the

respective titanates for the effects reported as shown in Table VII. However, a direct comparison of the isopropyl tri-(N-ethylamino-ethyl-amino) titanate and the gamma-amino propyl tri-ethoxy silane in Table VII indicate the amino functional moiety claimed not to be connected to the titanium center is superior to the amino moiety connected to the silicon center. Also, although the ESCA results show that no titanium was found at the metal surface from which the coating containing the isopropyl tri-(N-ethylamino-ethylamino) titanate was removed, in contrast, both titanium and phosphorus were present under steel surface after removal of the coating containing the isopropyl tri-(dioctyl phosphato) titanate. The investigation further states: "In general, then it can be concluded that the titanates which were studied exhibited surface activity since they produce both positive and negative results, depending upon the degree of polarity of the substituent groups. The question arises is whether these substituent moieties are still attached to the titanium molecule or have been removed by hydrolysis with water at or near the interface."

Data from a patent[50] indicating efficacy of isostearoyl and pyrophosphato based titanates to promote adhesion of acrylic resin to slate is shown in Table XIII. A patent[51] demonstrates significant adhesion improvement and maintenance of aged properties of marble dust filled acrylic emulsion to plywood and further demonstrates titanate efficacy for adhesion to cellulosic fiber. The patent also states: "Binder systems developed for applications having high inorganic solids loadings such as in solid propellants are of particular utility. Systems developed recently having delayed cure onset, such as those of U.S. Patent Nos. 4,110,135[52] and 4,098,626[53] which permit long pot lives and mild initiation of cure to allow void free castings with little shrinkage, are of special interest." This statement confirms results reported[54] by the authors as to titanate efficacy as urethane catalysts and a detailed discussion as to effects in urethane composites follows.

Table XIII. Adhesion of Acrylic Resin to Exterior
Coating to Slate

Formulation	No Titanate	Titanate
Acrylic resin hydrosol (Aron HD-2)	200	200
TiO_2	50	50
White Marble	500	500
$CaCO_3$	100	100
Water	150	150
KR TTS (or KR 38S)	6.5	6.5
Methyl hydroxypropyl cellulose	2	2
Additives	15	15

Properties, Chalky Emulsion Sprayed Coating on a Slate Board		
Adhesion - Initial, kg/cm^2	9.6	13.2
12 mos. exposure, kg/cm^2	9.9	13.6

Urethanes

Titanates may act in an NCO/polyol urethane system as catalysts according to the following proposed mechanism:

a) Polyol Alcoholysis:

$$(Y-R-X-O)_3Ti-OR' + H-OR^2O-H \rightarrow (Y-R-X-O)_3Ti-OR^2O-H + R'OH$$

titanate catalyst polyol   polyol, catalyst   alcohol
                                      complex

b) NCO Alkylation:

$$(Y-R-X-O)_3Ti-OR^2OH + O=C=N-R^3-N=C=O \rightarrow (Y-R-X-O)_3Ti-\overset{\displaystyle O}{\overset{\|}{N-C}}-OR^2O-H$$
$$\underset{R^3-N=C=O}{\vert}$$

polyol, catalyst   isocyanate   urethane, catalyst
    complex   complex

$$(Y-R-X-O)_3Ti-\overset{\displaystyle O}{\overset{\|}{N-C}}-OR^2O-H + H-OR^2O-H \rightarrow HN-\overset{\displaystyle O}{\overset{\|}{C}}-OR^2O-H + (Y-R-X-O)_3Ti-OR^2OH$$
$$\underset{R^3-N=C=O}{\vert} \qquad\qquad \underset{R^3-N=C=O}{\vert}$$

urethane, catalyst   polyol   urethane   polyol, catalyst
    complex   complex

The [OR'] portion of titanate catalyst may react with the polyol via an alcoholysis mechanism while the titanium atom [Ti] acts as an electron sink. This polyol alcoholysis mechanism does not occur at significant rates with either organo siloxanes or organostannates (tin catalysts). The [X] portion of the titanate catalyst will control the rate of electron transfer from the titanium atom. For example, in a two component RIM polyurethane, the phosphito [X] group of KR 55 titanate will produce a gel time of 16 seconds while the pyrophosphato [X] group of KR 38S titanate will gel in 263 seconds as compared to a 27 second gel time for dibutyl tin dilaurate control. The [R] group will again provide compatibility while the [Y] group, containing amino functionality, will crosslink. For example, the amino titanate, KR 44 will gel in 6.3 seconds and provide an ultimate tensile strength of 1520 kg/cm^2 compared to 314 kg/cm^2 for a dibutyl tin dilaurate control at significantly lower cost. Titanate catalyst ability to react with both the polyol and the urethane during catalysis yields dense high molecular weight cross-link chains resulting in higher glass transition temperatures. For example, Tg, °C for KR 44U is 242 compared to 116 for tin. Furthermore, qualitative indications are that Tg increases with increased titanate catalyst as opposed to decreased Tg experienced normally with increased tin catalysts levels. See Table XIV.

Table XIV. Evaluation of Various Catalysts in Unfilled Two Component RIM Polyurethane

Formulation in PBW:

Side A (Polyol NIAX PPG2050 Union Carbide – 70.25
(Polyol NIAX 50-1180 Union Carbide – 29.7
(Catalyst (as shown) – 500 ppm

Side B (Isonate 143L Upjohn – 100

Feed Ratio 1:1, Feed Mix Temperature 40°C, Feed Rate 8 Kg./min.

Equipment: Cincinnati Milicron Model HT RIM RRIM Molding Machine

EXPERIMENTAL RESULTS

Catalyst	Shore D Hardness	% Elongation at Break	Ultimate Tensile 10^2 KG/cm^2	Die C Tear 10^2 kg/cm	Flex.Mod. @ 25°C 10^3 Kg	Tg°C	Gel Time Sec.	Demold Time Sec.
T12 (M&T)	58	190	3.14	1.27	26	116	27	253
T1 (M&T)	59	175	2.82	0.91	29	124	13	218
?OCA(Dupont)/Ti	60	170	3.63	0.94	31	138	15	209
KR TTS	57	210	3.71	1.43	24	127	38	214
KR 9S	58	205	2.82	1.28	22	112	149	227
KR 12	58	205	3.10	1.34	23	148	106	241
KR 38S	58	200	4.24	1.19	27	131	263	519
KR 44	62	220	12.4	2.63	36	216	8.5	158
KR 46B	59	210	4.80	1.69	29	117	13	290
KR 52S	60	210	4.91	1.46	28	148	11	236
KR 55	59	255	6.27	3.06	26	144	16	184
KR 62ES	60	190	4.24	1.28	23	109	71	271
KR 138S	57	195	1.31	1.31	24	121	57	286
KR 158FS	58	185	1.19	1.41	26	116	148	304
KR 212S	58	200	1.28	1.26	25	118	102	218
KR 238S	59	205	1.31	1.09	26	121	119	256
KR 262ES	56	205	1.14	1.16	24	104	151	271
KR 44U	64	235	15.2	2.74	39	242	6.3	147
Expt.Tit. PM 2	62	210	6.13	1.49	31	187	11	193
Expt.Tit. PM 22	61	200	5.89	1.51	31	164	10	189
KR 55/44U 1/1	63	240	7.84	2.08	28	212	11.5	165
KR 55/44U 2/1	62	255	13.9	4.51	29	209	11	142
KR 55/44U 5/1	62	220	10.2	2.71	28	201	13	197

A roller manufacturer, Aerofoam, Dunswart, South Africa, compression molded a urethane compound (see Table XV) around a steel core on Cincinnati Milacron equipment. Only 0.7 mm of surface wear was found on conveyor idler rollers (see Figures 9) after 13 months of use in a mining operation, as compared to the average 2 week life span reported for the steel-shell rollers they replaced. Durability along with the antistatic properties is essential in a product for mine use. A coupler with built-in steric hindrance, KR 12, and the other a coupler/catalyst, KR 44U, surmounted the problem of unacceptable trade-offs. Specifically, the problems involved in producing a void-free/conductive/minimal-shrinkage polyurethane elastomer (PUR) were: (1) Coupling agents, indispensable for bonding filler and polymer, tend to release water and other low-molecular-weight components trapped by additives incorporated exclusively to inhibit them. Freed, these components foam the PUR,; (2) Catalyst, typically reacts with metallic fillers necessary for low resistivity, reducing their effectiveness.

In the formulation, a zeolite paste is used to occlude water. Most couplers negate a portion of the additive's effectiveness. Because durability is key in the roller application, the uncontrolled formation of voids was unacceptable. KR 12, organo titanate has a steric hindrance function that prevents it from reacting with the zeolite. KR 44U amino titanate, substituted successfully for a prob-

Table XV.   Aerofoam Urethane Roller Formulation
S. African Patent, Polyeth Mktg.

Order of Addition	Substance	Parts By Weight	Maker(s)
Polyol			
1.	Voranol EP 1900	60	Dow Chemical
2.	Voranol CP 450	20	Dow Chemical
3.	1,4 Butandiol	20	Various
4.	Fyrol 6	10	Stauffer
5.	Baylith L Paste	10	Bayer
	(Store 24 hours with continual mixing.)		
6.	Ken-React KR 12	0.27	Kenrich
7.	Ken-React KR 44U*	0.4	Kenrich
8.	Silicone B 1048	1.0	Th. Goldschmidt
9.	Aluminum Powder	12 / 133.67	Various
Isocyanate			
10.	Diphenyl Methylene Di-Isocyanate	86.91	Various

*Note: Replaces Catalyst, Thorcat 535

Results

Gel Time - Less than one (1) minute
T_g - 200°C
Performance - 13 months on mining belt time with minimal wear.

Figure 9.   Aluminum powder filled polyurethane idler roller con-
            taining KR 44 and KR 11 as made by Aerofoam, Dunswart,
            South Africa. New Roller (top) and roller of similar
            composition after 13 month use in a gold mine belt
            application (bottom).

lematic tin catalyst. The roller formulation uses aluminum powder
for molded-in conductivity (eliminating the need for auxiliary antistatic brushes on the conveyor line), as well as for stabilizing part
size. The tin catalyst was counteracting the low resistivity imparted
by the aluminum by converting a percentage of it into nonconductive
aluminum oxide, according to a S. African patent. This reaction does
not occur with the amino titanate. In addition, the amino reagent's
adhesion to metal is reportedly superior; the result being enhanced
conductivity through better dispersion of metallic particles. It
appears that any foaming triggered by the amino catalyst/coupler is
effectively held in check by the KR 12. A decrease in gel time (less
than 1 min.) and a high glass transition temperature (above 200°C)
is also noted.[55]

     Damusis[56] reported that calcium metasilicate (22 micron, 1:d
15:1), when treated with KR TTS, yields significantly higher impact
and tear strength than equivalent treatment with A-1100 in RRIM urethane. Glass fibers are used in RRIM urethane to provide dimensional
stability and good physical properties - particularly impact strength.
However, 1/16" milled glass fibers have the disadvantages of high
cost (approx. 82¢/lb.),anisotropic impact properties and difficult
handling and process characteristics. Acicular type fillers, such
as Wollastonite, provide tempting substitution possibilities because
of lower price, isotropic impact properties and inherent easier handling. However, untreated Wollastonite grades provided inferior prop-

erties when compared to silane sized 1/16" milled fiberglass. Silane
treated Wollastonite, G Wollastokup 1100 0.5, has shown promise, but
falls short in the important area of impact.[57]

Table XVI is extracted from a paper[54] containing 14 tables and
7 figures. Apparently, Wollastokup KR TTS 0.5 (22¢/lb.) provides
equivalent impact while elongation, tear strength, tensile strength
and flexural modulus are higher, process viscosity 16% lower and
offering a 60¢/lb. savings when compared to silane sized 1/16" milled
fiberglass at the 40% loading level (24.1% on total composite) in
hydroxyl component.

Conductivity

Dispersion of composite ingredients with titanate coupling agents

Table XVI.   Comparison of Properties in RRIM Urethane

RRIM formulation with 40% reinforcing filler in the OH component only. Finished composite contains 24.1% total reinforcement.

OH Component	Equiv. Wt.	Grams	Equiv.
75% Niax 31-28	2070	60	.029
25% Niax 50-1180	47.5	20	.421
			.45
T-12 Catalyst	–	1 drop	–
Reinforcement		40	–
as shown	–		
Freon 11-B	–	1.7	– 100.00G

NCO Component			
Isonate	139.5	62.78	.45
		62.78 x 1.05	65.92G

$$\text{\% Composite Reinforcement (OH + NCO)} = \frac{40 + 0}{100 + 65.92} = 24.1\%$$

RRIM Paper

Composite No.	C-2	E-4	E-2
Properties	Silane Treated Milled 1/16" Glass Fibers	Wollastokup 1100 0.5	Wollastokup KR TTS 0.5
Tensile Strength, psi	5,212	5,005	5,126
Elongation, %	21	25	27
Flexural Modulus, psi			
A) 25°C	100,900	114,000	112,000
B) 0°C	184,274	211,472	228,985
C) 70°C	58,310	54,805	58,962
Tear Strength, pi	716	840	920
Izod Impact, in.-lb.	14	10	13.5
Shore D	80	80	84

is not always evidenced by viscosity reduction. Light transmission, semi-electron microscopy studies, electrical and other composite properties such as increased conductivity and magnetic intensity are often more indicative of interfacial coupling and dispersion. For example, 0.3% KR 238S titanate by weight of sub-micron dendritic nickel powder (INCO type 287) loaded to the 75% level in fluoroelastomer (Viton[R] E-430) reduced resistivity from 10,000 to 110 ohm/cm and in silicone elastomer (SWS B-124) reduced resistivity from 1.3 x 10^6 to 20 ohm/cm. Figure 10 shows field test flexibility and resistivity drop for 0.3% KR 238S, 70% nickel filled silicone elastomer. Monte and Sugerman[44] also reported a reduction in resistivity for

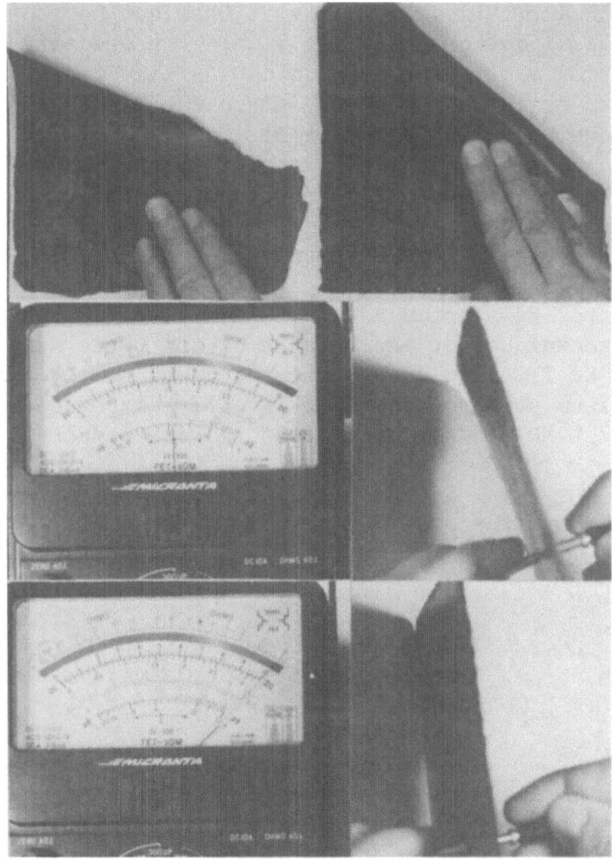

Figure 10. 70% nickel powder filled elastomer demonstrating titanate dispersion effects of flexibility and conductivity. (Top left) no titanate, (top right) 0.3% KR 238S titanate, (center) no titanate - high resistivity, (bottom) titanate - low resistivity.

conductive carbon (Ketjan) flexible PVC from 38 to 21 ohm/cm using
KR 138S at 1% by weight of carbon black. A patent[58] indicated an
increase of magnetic intensity in a magnetic iron oxide case vinyl
copolymer recording tape from 1210 Bm(gauss) for a raw soybean-leci-
thin prior art control to 1670 Bm(guass) using KR 38S. Figure 11
shows the "vacuum metalized-like" effect achieved by applying an
aluminum powder 0.35% KR 38 coating to the inside of a clear styrene-
acrylonitrile injection molded part effecting a systems composite
savings of from $0.06/part to $0.045/part. Light transmission examination of Figure 6 demonstrates uniformity of dispersion.

In an exploratory study[59] concerning rocket motors, uncured
nitrile elastomeric compositions were prepared on a water cooled two
roll mill consisting of 500 parts of Alcan MD101 aluminum powder with
and without 0.5% KR TTS and 100 parts of Chemigum N-612 (Goodyear).
As expected, the titanate containing sample, sheeted off the mill,
exhibited superior wet-out as evidenced by a significantly lesser
degree of stress cracking. However, there was little difference in
gloss or sheen between the KR TTS sample and the control. It was
then decided to cure the nitrile elastomer to make more durable
sheets so that conductivity and other property measurements could be
made. Four compositions, one of which was a control, were again prepared using 100 parts of the aforementioned polymer, 500 parts aluminum powder and 1.5 parts each of Agerite Staylite S, Sulfur, Altax
and Methyl Zimate and oven cured at 150°C for 45 minutes. Titanate
incorporation technique for the balance of the three nitrile rubber
samples were: KR TTS liquid was post added on the two roll mill
after the aluminum powder had been thoroughly dispersed into the nitrile; Drimix 72% KR TTS powder (a free flowing powder prepared by

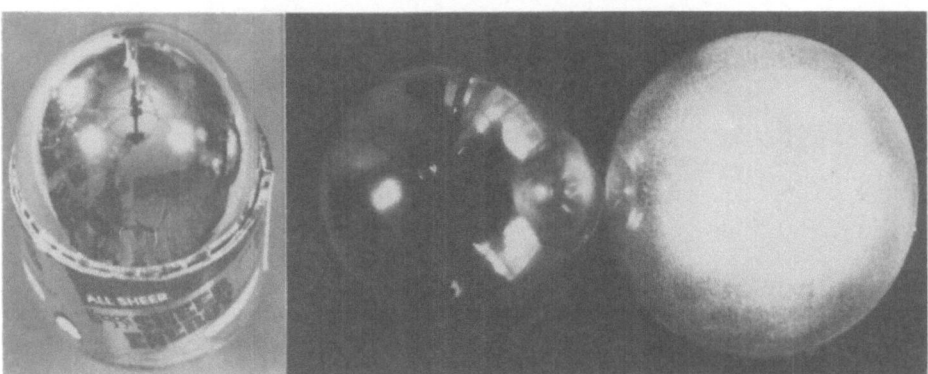

Figure 11. Aluminum powder coating spray applied to container
            simulating vacuum metalized appearance (left). Half
            of container blocked from light source behind it (center)
            and unblocked (right).

blending 72 parts of KR TTS liquid with 28 parts of Micro Cel E, Johns Manville) and KR 238S liquid were incorporated into the banded nitrile rubber followed by the dispersion of aluminum powder. After withdrawal of the four sheets from the oven, it was noted that two of the sheets containing the KR 238S liquid and Drimix 72% KR TTS (predispersed into the banded elastomer before addition of aluminum powder) had large bubble blisters while the control and post added KR TTS samples did not change. Also, the latter samples were similarly dull in appearance while the bubbled Drimix 72% KR TTS sample was brighter and glossier and the equally bubbled KR 238S samples had the brightest sheen of all and apparently was the most effectively dispersed sample. Three swatches, 1 inch in width, were then cut from the approximately 3/32" thick cured sheets and conductivity measurements were made as shwon in Table XVII. (We had decided not to measure the Drimix sample due to time limitation during testing.) Table XVII shows clearly that the pyrophosphato titanate (KR 238S) increased conductivity (decreased resistivity) by a factor of 118 when compared to the control. These observations appeared to us to indicated the following:

1. Dispersion of the titanate into the polymer before the incorporation of aluminum powder was best practice.
2. 0.5 titanate by weight of aluminum was too high a level and free titanate was apparently causing gassing to occur. The system was more sensitive to titanate level than say a $CaCO_3$ filled polyolefin thermoplastic composition wherein free titanate would have an ester plasticizer or beneficial lubricating effect. It was decided to make another sample using 0.35% titanate since this level had already proved optimum in a magnetic barium ferrite filled nylon and an aluminum powder/acrylic coating. Also, previous experiments in nickel filled coatings and elastomers showed that conductivity peaked at about 0.3% and began to drop off with higher titanate levels. Figure 12 shows a significant decrease in gassing at the 0.35% level as compared to 0.5% level.
3. The pyrophosphato titanate, KR 238S dispersed best. Again, experience in nickel conductive systems showed that the longer chain octyl pyrophosphato ligands, typical of KR 238S and KR 38S, gave somewhat lower initial levels of conductivity when compared to shorter chain butyl, methyl pyrophosphato ligands, typical of KR 62ES, but, upon aging, remained stable and exhibited the highest long-term conductive performance.

The potential effect proposed by the authors is smoother burn, more power and trajectory accuracy of the rocket motor composite when KR 238S and KR 38S are used at a level of 0.35% by weight of aluminum powder and ammonium perchlorate, respectively.

Table XVII. Conductivity Measurements of Granular Aluminum Powder
with Titanate in Cured Nitrile Elastomer

Item	Resistivity[a], K ohms
Al/Chemigum N-612 - 500/100	
Control	$701.600 \pm 20.6 \times 10^0$
KR TTS, post added to Al/nitrile	$713.900 \pm 6.7 \times 10^0$
KR 238S, preblended with nitrile	$5.968 \pm 2.7 \times 10^3$

a. Measured on Keithly Picoammetric breadboard located at IBM Corp.,
E. Fishkill, NY @ 1000 volts D.C. using 5000 Kg clamp pressure.

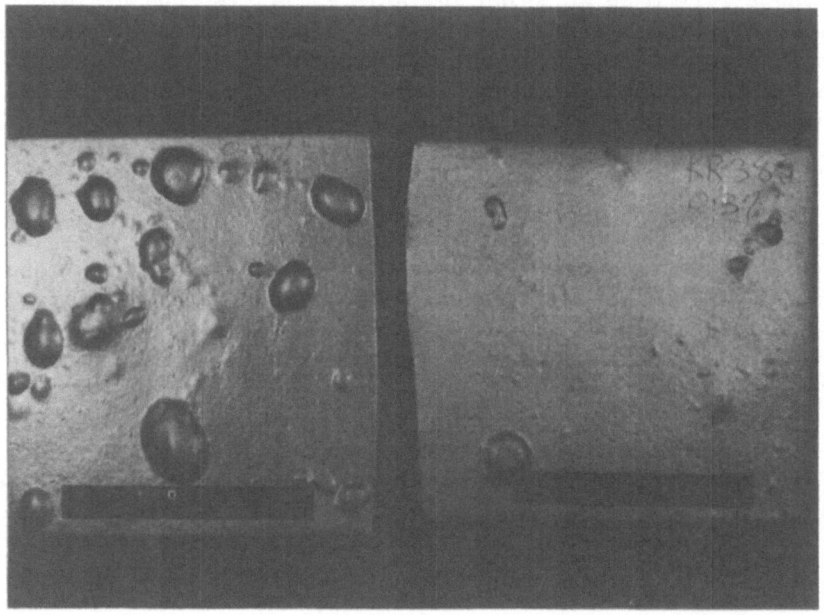

Figure 12. Aluminum powder filler with nitrile elastomer, 0.5% KR
238S titanate (left) and 0.35% KR 238S titanate (right).

Aging and Corrosion Resistance

It is our belief that the aging of filled polymers and corrosion
of polymer coated metals is caused by the continued presence of the
water of hydration on the surface of uncoupled fillers and metal substrates and by the "locked in" air voids present in undispersed filler
agglomerates or incompletely wetted metal substrate in the supposedly
"finished" thermoplastic or thermoset composite or coating.

Kummer[60] reported significant maintenance of elongation and
tensile strength with KR TTS titanate upon aging in a 60% $CaCO_3$/PP

composite and results are shown in Table XVIII. Table XIX shows 33%
HiSil 233, 1.0% KR 212S TPR (Kraton 1107) exhibits aged modulus, tensile and elongation properties significantly higher than the virgin
polymer. Table XX shows significant maintenance of aged properties
in a flexible PVC wire and cable compound. A paper[61] discussed ti-

Table XVIII.   Original and Aged Physical Properties of KR TTS
               Titanate-treated $CaCO_3$ in Polypropylene Homopolymer
               Injection-Molding Compound

	Formulation, parts by weight			
PP	40.0	PP	40.0
$CaCO_3$	60.0	$CaCO_3$	60.0
Titanate[a]	0.6	Titanate	0.0
Mineral Oil	0.6	Mineral Oil	0.6

Properties		
Unaged		
Tensile strength at yield, psi	3,042	3,659
Elongation, %	22	1
Unnotched Izod, ft-lb/in.	4.3	1.6
Notched Izod, ft-lb/in.	0.43	0.22
Gardner impact, ft-lb/in.	75	3
Flexural modulus, psi	452,000	570,000
Aged in water 6 days @ 93.3° (200°F)		
Tensile strength at yield, psi	2,952	2,678
Elongation, %	37.5	15
Gardner impact, ft-lb/in.	95.5[b]	80[c]

[a]KR TTS, isopropyl, triisostearoyl titanate

[b]Drew apart, did not shatter

[c]Shattered

Table XIX.   Aging Effects of KR 212S Titanate in Silica-
             Filled Styrene Butadiene Block Copolymer

Compound No.	1	2	3
Kraton 1107	100	100	100
Silica, HiSil 233	-	50	50
Stearic acid	0.25	0.25	0.25
KR 212S[a]	-	-	0.5
Properties, oven aged 24 hr. at 100°C			
100% Modulus, psi	260	240	360
200% Modulus, psi	300	-	400
300% Modulus, psi	310	-	410
Tensile Strength, psi	320	270	440
Ultimate elongation, %	200	170	310

[a]KR 212S, di(dioctylphosphato) ethylene titanate

Table XX. Effect of Various Chelated Pyrophosphato
and Monoalkoxy Amino Titanates of Original
and Aged Physical Properties of Vinyl

Formulation: PVC Resin (Geon 30) - 100; Calcium Carbonate (Supermite,
              lu) - 80; Plasticizers (DINP, 41.6; Chlorowax 40, 6.4);
              Titanate (% by weight of $CaCO_3$) as shown; Stabilizer
              (Dyphos) - 4.

Item	100% Modulus psi Orig.	Aged	Tensile Strength psi Orig.	Aged	% Elongation at Break Orig.	Aged
Control	2,200	-	2,400	2,400	150	80
0.50% KR 138S	2,200	2,100	2,400	2,100	160	100
1.00% KR 138S	2,000	1,900	2,300	2,000	180	100
0.50% KR 238S	2,200	2,400	2,400	2,400	180	120
0.50% KR 158FS	2,500	2,600	2,600	2,600	140	130
0.50% KR 44U	2,400	2,600	2,500	2,600	130	130

Mixing
Procedure: Resin and $CaCO_3$ were added to a Henschel Mixer and externally
            heated to 190 to 200°F. Titanates were dissolved in the com-
            bined plasticizers and added to the fluidized resin/filler mix
            during 5 minute period of mixing. Stabilizer added and mixed
            3 minutes. The resultant powder was fused on and mill mixed
            at 280°F then press molded @320°F prior to die cutting and
            evaluation.

Heat Aging: Die cut specimens were open oven aged for one week at 212°F
             prior to testing

tanate efficacy in clay, alumina trihydrate and $CaCO_3$ filled latex.
Table XXI indicates aged properties with $CaCO_3$ KR 138D, 1.5% at the
150 phr level are equal to 100 phr prior art loading in SBR latex.

A report[62] stated:

"KR 138D was diluted to 10% with water and applied by wiping
two polished mild cold rolled steel panels. The panels were
then allowed to dry at ambient for 15 minutes and 16 hours,
respectively. The treated panels along with an untreated
control were then placed in a humidity cabinet for 3 hours
at 50°C and 100% relative humidity. The untreated control
underwent severe rusting and pitting while the 15 minute
dry treated panel exhibited slight rusting and pitting and
the 16 hour dry panel was uneffected."

A patent[63] stated:

"Thus, 100 parts Esrex BM_2 (polyvinyl butyral) contg. Epikote
828 [25068-38-6] 40, I 40, and iso-Pr triisostearoyl titanate (II) [61417-49-0] 6 parts was mixed with talc 30, yellow
iron oxide 15, carbon black 2, phthalocyanine blue 3, pptn.
preventer 10, BuOH 200, iso-PrOH 200, and PhMe 200 parts, the

Table XXI. Original and Aged Property Effect on KR 138S Quat in CaCO₃ (Atomite) Filled SBR 2000 Latex (43.4%)

	Control				Titanate			
	1		2		3		4	
	Dry	Wet	Dry	Wet	Dry	Wet	Dry	Wet
43.4% SBR 2000 latex	100	230.5	–	–	–	–	–	–
10% Ammonium caseinate	1	10	–	–	–	–	–	–
60% Zinc oxide dispersion	5	8.55	–	–	–	–	–	–
68% Sulfur dispersion	2	–	–	–	–	–	–	–
65% Vanox 102 emulsion	1	–	–	–	–	–	–	–
50% Ethyl zimate slurry	1.5	–	–	–	–	–	–	–
			Varies below the line					
60% (Atomite) CaCO₃	100	167	150	250	100	167	150	250
KR 138S Quat (KR 138S:TEA 2:1)	–	–	–	–	1.5	1.5	2.25	2.25

Physical Properties–Unaged Spread Films

[(S) Stress at 300%, psi (T) Tensile, psi (E) Elongation at Break]

	1			2			3			4		
	S	T	E	S	T	E	S	T	E	S	T	E
Hot air cures at 104°C (220°F): 15 minutes	525	900	500	407	560	480	540	890	490	500	800	500
Films aged 24 hrs. at 121°C (250°F) in a circulating warm air oven: 15 minutes	800	850	320	–	560	260	450	515	350	500	755	490

KR 138S, titanium di(dioctylpyrophosphato) oxyacetate

mixt. sprayed on mold steel, and the steel coated with an
acrylic resin and baked at 170°C for 20 min. to form a
coating having better corrosion resistance than a coating
on an undercoat contg. no II."

A study[64] was made with the objective to determine the corrosion
resistance of an epoxy-polyamide coating in which KR 38S was reacted
in situ with silica and used in place of barium metaborate (Busan
11-M1). A control anti-corrosive primer formulation based upon barium
metaborate in an epoxy polyamide binder was prepared. Two additional
formulations were then prepared, one containing 1/2% KR 38S and another containing 1% KR 38S, based upon the weight of the silica. In
both latter formulations, the barium metaborate was replaced with
silica on an equal volume basis. The ingredients were added in the
order as shown in Table XXII and properly dispersed. The primers
were then applied to cold rolled steel at a dry film thickness of
2.0 mils and dried for 7 days at 25°C and 50% relative humidity prior
to testing. Based on the test results shown in Table XXII, the corrosion resistance of the KR 38S plus silica epoxy-polyamide primers
are essentially equivalent to the barium metaborate coating. Increasing the KR 38S from 1/2 to 1% had no beneficial effect on corrosion resistance. The primer containing the 1/2% KR 38S is more
economical exhibiting a cost saving of almost 70¢ per gallon of
primer.

A study[62] was made with the objective to determine the corrosion
resistance of an epoxy-polyamide coating in which KR 38S plus silica
is used in place of basic lead silico chromate, BLSC M-50. An anti-
corrosive primer formulation based upon BLSC M-50 in an epoxy polyamide binder was prepared. Two additional formulations were then
prepared. One containing silica with no BLSC M-50 or KR 38S and another containing 1/2% KR 38S, based upon the weight of the silica.
In both latter formulations, the BLSC M-50 was replaced with silica
or silica plus KR 38S to an equivalent systems viscosity. Because
of the higher oil absorption of the silica, the silica control formulation having no KR 38S resulted in a product having a lower PVC and
a reduced paint yield. However, the silica plus KR 38S formulation
allowed a higher level of silica to be used resulting in a primer
with a PVC and paint yield higher than that of the BLSC M-50 control.
The ingredients were added in the order shown in Table XXIII. The
primers were applied to cold rolled steel panels at a dry film thickness of 1.5 - 2.0 mils and dried for 7 days at 25°C and 50% relative
humidity. They were then exposed (without top coats) to the following environments: Salt Fog (5) - 1,000 hours and humidity (100%) -
500 hours. The tests indicated that the titanate must be added to
the formulation prior to the incorporation of any pigment in order
to maximize its effectiveness. The test results were positive and
are shown in Table XXIII. In as much as some observations are necessarily subjective, they have been reported using the following scor-
ing scheme:

Table XXII. Evaluation of the Corrosion Resistant Properties of
Silica Treated with 0.5 and 1.0% KR 38S and Barium
Metaborate (Busan 11-M1) in an Epoxy/Polyamide Primer

FORMULATIONS

	KR 38S		Busan
	0.5%	1.0%	11-M1
Part A			
Polyamide 815X70	249.0	249.0	249.0
Soya Lecithin	-	-	2.0
Ken-React KR 38S	2.2	4.4	-
Imsil A-10	80.3	80.3	-
Busan 11-M1	-	-	100.0
Desert Talc No. 80	150.0	150.0	150.0
RF-30 TiO_2	200.0	200.0	200.0
Bentone 34	2.0	2.0	2.0
Ethyl alcohol	1.0	1.0	1.0
Xylol	134.0	134.0	134.0

Disperse using high speed impeller

Total	Lbs.	818.5	820.7	838.7

Part B				
Araldite 471X75		301.0	301.0	301.0
MIBK		77.5	77.5	77.5
Total	Lbs.	378.5	378.5	378.5
Total Pt A&B	Lbs.	1197.0	1199.2	1216.5
Total Pt A&B	Gal.	109	109	109

Raw Material Cost ¢/gal			
KR 38S + Silica	12.9	19.3	-
Busan 11-M1	-	-	81.8

TEST RESULTS

	KR 38S		Busan
	0.5%	1.0%	11-M1
Viscosity KU			
Part A	86	82	92
Part B	72	72	72
Mixed	80	79	82
Fineness of Grind HEG	5	5	5
Salt Fog Resistance - 600 hrs. (ASTM B-117)			
Blistering (ASTM D-714)	2M	2M	2F
Rusting	Slight	Slight	V. Slight
Discoloration	Moderate	Moderate	Moderate

	KR 38S		Busan
	0.5%	1.0%	11-M1
Humidity Resistance - 1,000 hrs.			
Blistering (ASTM D-714)	8D	8D	4D
Rusting	None	None	None
Color Change	Moderate	Moderate	Slight
Sea Water Immersion - 60 days (ASTM D-1141)			
Blistering (ASTM D-714)	None	None	None
Rusting	None	None	None
Color Change	Slight	Slight	Slight

Table XXIII. Evaluation of the Corrosion Resistant Properties of Silica Treated with 0.5% KR 38S and Basic Lead Silico Chromate (BLSC M-50) in an Epoxy/Polyamide Primer

FORMULATIONS

	BLSC M-50 Control	Silica Control	Silica + 0.5% KR 38S
Part A			
Araldite 571 CX-80	210.0	210.0	210.0
BLSC M-50	480.0	–	–
KR 38S	–	–	3.3
Ti Pure R-900	30.0	30.0	30.0
R-2200 Red Oxide	15.0	15.0	15.0
DeGussa R-974 Silica	6.4	6.4	6.4
Asbestine 3x	235.0	235.0	235.0
Imsil A-108		260.0	400.0
Xylol	192.7	192.7	192.7
Diacetone Alcohol	96.2	96.2	96.2
Beetle 216-8	10.5	10.5	10.5
Part B			
Araldite 820	105.0	-----	-----
Xylol	24.0	-----	-----
Butanol	12.0	-----	-----
Total Weight Lbs.			
Part A	1275.8	1055.8	1199.1
Part B	141.0	141.0	141.0
	1416.8	1196.8	1340.1

	BLSC M-50 Control	Silica Control	Silica + 0.5% KR 38S
Total Yield Gals.			
Part A	91.8	89.2	95.5
Part B	18.8	18.8	18.8
	110.6	108.0	114.3
RMC $/Gal.	5.67	3.55	3.45

Note: KR 38S must be added to the mill base before the pigments are added.

TEST RESULTS

		BLSC Control	Silica Control	Silica + 0.5% KR 38S
Viscosity				
Initial	KU	100	99	102
Two weeks @ 25°C		96	99	93
Two weeks @ 49°C		100	97	100
Four months @ 25°C		121	109	108
Max change		21	10	9
Salt Fog Exposure				
500 Hours				
Blistering	ASTM D-714	10	10	10
Rusting overall	Score	10	9+	9+
Rusting at "X"	"	7	6	6
Peeling	"	10	10	10
1000 Hours				
Blistering	ASTM D-714	10	10	10
Rusting overall	Score	10	6	10
Rusting at "X"	"	5	2	4
Peeling at "X"	"	10	6	10
Humidity Resistance - 500 hours @ 49°C - 100% R.H.				
Blistering	ASTM D-714	10	6D	10
Rusting	Score	10	10	10
Appearance -	"			
Stripped		10	2	9+

Score	Performance	or	Effect
10	Excellent		No change
9			Trace
8	Very good		Very slight
6	Good		Slight
4	Fair		Moderate
2	Poor		Considerable
0	Very poor		Severe

Similar positive effects were noted[65] in a baked, epoxy ester spray, zinc rich primer composite.

SUMMARY AND CONCLUSION

We hope this brief overview of the use of titanate coupling agents for improved properties and aging of plastic composites and coatings inspires further investigation of the chemistry of the interface to achieve better understanding of the art.

REFERENCES

1. A. N. Gent, Adhesives Age, Vol. 25, (No. 2), pp. 27-31 (1982).
2. Reference Manual, Bulletin No. KR-0278-7, Rev. #2, Kenrich Petrochemicals, Inc.
3. G. R. Kritchevsky, "Polymer-Ceramic Interfaces," Massachusetts Institute of Technology, Cambridge, MA, June, 1977.
4. S. J. Monte and G. Sugerman, "A New Generation of Age- and Water-Resistant Reinforced Plastics," Polym.-Plast. Technol. Eng., 13(2), 115-135 (1979).
5. Modern Plastics, July, 1977, "New Developments in Coupling Agents."
6. U. S. Patent 4,207,226, 10 June 1980, "Ceramic Composition Suited to be Injection Molded and Sintered," Roger S. Storm, The Carborundum Co., Niagara Falls, NY.
7. U. S. Patent 4,269,756, 26 May 1981, "Use of Organotitanate in the Encapsulation of Electrical Components," Tsung-Yuan Su, Union Carbide Corp.
8. C. D. Han, H. L. Luo and J. Mijovic, "Effects of Coupling Agents on the Rheological Behavior and Mechanical Properties of Filled Nylon 6," pp. 82-83, May, 1982, SPE ANTEC Proceedings.
9. T. Bent, Adell Plastics, Baltimore, MD (Nov. 1980), personal communication.
10. C. D. Han, T. Van Den Weghe, P. Shete and J. R. Haw," Effects of Coupling Agents on the Rheological Properties, Process-

ability and Mechanical Properties of Filled Polypropylene,"
pp. 196-204, Polymer Engineering and Science, Vol. 21 (No.
4), (March, 1981).

11. S. J. Monte and P. F. Bruins, "New Coupling Agent for Filled
Polyethylene," pp. 68-72, Modern Plastics, Dec., 1974.

12. A. R. Mersberg, Briner Paint Mfg. Co., Inc., Corpus Christi,
TX, "Energy Conservation Utilizing Coupling Agents," Oct.,
1977, FSCT Proceedings.

13. C. D. Han, C. Sanford and H. J. Yoo, "Effects of Titanate
Coupling Agents on the Rheological and Mechanical Properties
of Filled Polyolefins," Polymer Engineering and Science,
Vol. 18, (No. 11), (August, 1978).

14. S. J. Monte, G. Sugerman and D. J. Seeman, "Titanate Coupling
Agents - Current Applications (V)," May 5, 1977, Paper No.
40, ACS Rubber Div Mtg., Chicago, IL.

15. S. J. Monte and G. Sugerman, "The Effect of Titanate Coupling
Agents on: Non-Black Reinforcing Agents; A New Vulcanizing
Agent System for Polyepichlorohydrin Elastomers; Azodicarbon-
amide Sponged Polyphosphazene Homopolymers; and Property
Maintenance of Lower Mooney Rubber Compounds," Paper #43,
ACS Rubber Division, October, 1979, Cleveland, OH.

16. T. J. Reinhart, U. S. Air Force Wright Aeronautical Laboratories,
(22 January 1980), personal communication.

17. T. J. Reinhart, U. S. Air Force Wright Aeronautical Laboratories,
(12 March 1980), personal communication.

18. G. Hutchins and R. Spell, LNP Corp., Malvern, PA, (Sept. 1981),
personal communication.

19. T. Skidmore, Siecor Corp., Hickory, NC, (Feb. 1981), personal
communication.

20. T. Baker, Raytheon Corp., Sudbury, MA, (March 1982), personal
communication.

21. M. Griggs, Colebrand Ltd., Rossendale, U.K., (March 1981) personal communication.

22. P. Westerman, BP Research, Ltd., Middlesex, U.K., (April 1981),
personal communication.

23. N. Eickman, Celanese Research, Summit, NJ, (May 1982), personal
communication.

24. J. Gabbert, Monsanto Co., St. Louis, MO, (March 1982), personal
communication.

25. K. Geddes, University of Lowell, Lowell, MA, (June 1981), personal communication.

26. A. Chattopadhyay, DeSoto, Inc., Des Plaines, IL, (July 1981),
personal communication.

27. M. Crystal, Uniplex Corp., Maplewood, NJ, (May 1982), personal
communication.

28. N. H. Sung, Department of Chemical Engineering, Tufts University,
Medford, MA 02155, "Role of Organo Silanes and Organo Tita-
nates in Promotion of Adhesion Strength of Aluminum Oxide-
Polyethylene Joint," 36th Annual Conference, Reinforced Plas-

tics/Composites Institute, SPI, February 16-20, 1981.

29. Jpn. Appl. 79/38,344, Furukawa Electric Co.

30. Jpn. Appl. 77/124,221, Furukawa Electric Co.

31. Jpn. Appl. 77/124,222, Furukawa Electric Co.

32. Jpn. Appl. 78/95,921, Furukawa Electric Co.

33. Jpn. Appl. 78/17,536, Furukawa Electric Co.

34. Modern Plastics, July, 1982.

35. Modern Plastics, Idea Exchange, August, 1981.

36. S. J. Monte and G. Sugerman, "Titanate Coupling Agents in Filler
    Reinforced Thermoplastics," SPI RP/C Instit., 1978, Paper 2B,
    Table 3.

37. U. S. Patent 4,083,820, 11 April 1978, "Low Smoke Polyphosphazene
    Compositions," R. L. Dieck, Armstrong Cork Co.

38. Jpn. Appl. 77/92,936, Tokuyama Soda Co.

39. S. J. Monte and G. Sugerman, "United States and Japanese Appli-
    cations of Titanate Coupling Agents in Polyolefins," Poly.-
    Plast. Technol. Eng., 17(1), 95-112 (1981).

40. K. Hayashida, T. Kamei and E. Kagoshima, "Influence of Coupling
    Agent upon the Flow Properties of Low Density Polyethylene
    melts Mixed with Calcium Carbonate or Barium Sulfate," Kyoto
    University of Industrial Arts and Textile Fibers, Vol. 9,
    No. 2, 1980.

41. Y. N. Sharma, R. D. Patel, I. H. Dhimmar and I. S. Bhardwaj,
    "Studies of the Effect of Titanate Coupling Agent on the
    Performance of Polypropylene-Calcium Carbonate Composite
    (IPCL Comm. No. 39), J. Appl. Polym. Sci., Vol. 27, 97-104
    (1982).

42. S. Walter, Saytech, Inc., New Brunswick, NJ, "Filled polypro-
    pylene can be flexible-and flame retardant, too," Plastics
    Engineering, Vol. 37, No. 6, June, 1981.

43. S. J. Monte and G. Sugerman, "Titanate Coupling Agents in PVC,"
    3rd Int'l. Symp. on PVC, Aug. 10-15, 1980, Case Western Re-
    serve University.

44. S. J. Monte and G. Sugerman, "Titanate Coupling Agents - Devel-
    opments of 1981," SPI RP/C Instit., Jan. 11-15, 1982, Paper
    22-B.

45. S. J. Monte and G. Sugerman, "Current Application of Titanate
    Coupling Agents in Filled Thermoplastics and Thermosets,"
    SPI RP/C Instit., Feb. 16-20, 1981, Paper 18-D.

46. J. J. Jakubowski is with the Dow Chemical Co., Midland, MI
    48640 and Prof. R. U. Subramanian, to whom inquiries should
    be made in the first instance, is with the Department of
    Materials Science and Engineering, Washington State Univ.,
    Pullman, WA 99164, USA.

47. Jpn. Kokai Tokkyo Koho 80,120,649, 17 Sept. 1980, Hitachi Chem-
    ical Co., Ltd.

48. Jpn. Kokai Tokkyo Koho 81,28,222, 19 Mar. 1981, Shin-Kobe Elec-
    tric Machinery Co., Ltd.

49. Cleveland Society for Coatings Techn., Tech. Comm., "Study of
    Organic Titanates as Adhesion Promoters," J. of Ctgs. Tech., *

        Vol. 51, (No. 655), (August 1979).

50. Jpn. Kokai Tokkyo Koho 81,999,266, 10 Aug. 1981, Dainippon Toryo Co., Ltd.

51. U. S. Patent 4,308,298, 29 Dec. 1981, "Upgrading of Cellulosic Boards," Yang-Hsien Chen, International Paper Co., New York, NY.

52. U. S. Patent 4,110,135, 29 Aug. 1978, "Control of Cure Rate of Polyurethane Resin Based Propellants," W. H. Graham, Thiokol Corp., Newtown, PA.

53. U. S. Patent 4,098,626, 4 July 1978, "Hydroxy Terminated Poly-butadiene Based Polyurethane Bound Propellant Grains," W. H. Graham, Thiokol Corp., Newtown, PA.

54. SPI Urethane Division, November, 1981, "Application of Titanate Coupling Agents in Mineral and Glass Fiber Filled RIM Urethane Systems," S. J. Monte and G. Sugerman, Ph.D., KPI, and A. Damusis, Ph.D. and P. Patel, Polymer Institute, University of Detroit.

55. Modern Plastics, Design & Application News, pp. 36, May, 1982.

56. A. Damusis and P. Patel, Polymer Institute, University of Detroit, (27 July 1981), personal communication.

57. SPI Reinforced Plastics/Composites Institute, Paper 14-C, 1979, "Silane Effects and Machine Processing in Reinforced High Modulus RIM Urethane Composites," E. G. Schwarz, F. E. Critchfield, L. P. Tackett and P. M. Tarin, Union Carbide Co.

58. German Patent No. 3038646, 13 Oct. 1980, Sony Corp.

59. S. J. Monte and G. Sugerman, "The Potential of Titanate Coupling Agents in Solid Rocket Fuel Systems," 1 June 1982, Joint Symposium on Compatibility of Plastics/Materials with Explosives, Propellants and Pyrotechnics and Processing of Explosives, Propellants and Ingredients, Phoenix, Arizona, sponsored by the American Defense Preparedness Association.

60. G. Crowe and P. E. Kummer, "Extending Resins with Calcium Carbonate," Plastics Compounding, Sept./Oct. 1978.

61. S. J. Monte and G. Sugerman, "Application of Titanate Coupling Agents in Latex and the Water Phase," ACS, Rubber Div., 113th Mtg., May 2-5, 1978, Montreal, Canada.

62. S. Spindel and S. B. Levinson, DL Laboratories, NYC, "Evaluation of the Corrosion Resistant Properties of Ken-React[R] KR 38S in an Epoxy/Polyamide Primer," Report DL-2871E, (November 1978).

63. Jpn. Kokai Tokkyo Koho 80,152,757, 28 Nov. 1980, Dainippon Toryo Co., Ltd.

64. J. H. Wilner, S. Spindel and S. B. Levinson, DL Laboratories, NYC, "Evaluation of the Corrosion Resistant Properties of Ken-React[R] KR 38S and Busan 11M-1 in an Epoxy/Polyamide Primer," Report DL-3299, 8 Sept. 1981.

65. S. J. Monte and G. Sugerman, "Adhesion Promotion and Polymer Composite Performance with Titanate Coupling Agents," Abstract No. 139, The Electrochemical Society, Inc., Symposium on

Adhesion Aspects of Polymeric Coatings, Minneapolis, Minne-
sota, 14 May 1981.

COMPOUNDING OF ADDITIVES AND FILLERS

Kurt Eise

Process Engineering
Werner & Pfleiderer Corporation
Ramsey, NJ 07446

INTRODUCTION

Almost all polymers today are modified by the addition of one or more additives to accomplish a variety of purposes. They are normally used in small quantities in the final product and a uniform distribution in the polymer matrix is necessary to achieve a well stabilized product.

Additives are most effective at their optimum concentrations. An insufficient amount can result in premature failure; however, an increase in the amount of additive will not produce a proportionate increase in performance. Incorporation of other fillers such as glass fibers, talc, $CaCO_3$ can minimize the loss of properties.

When properly compounded, multi-purpose concentrates can be used with a basic raw material to produce many different characteristics. Common multi-purpose concentrates available today include the following additives: slip, antiblock, antioxidant, stabilizer-processing aid, stabilizer and lubricant and many others which can be tailored to a specific end-use requirement.

DISCUSSION

The availability of improved processing equipment has had a major influence on the type and levels of additives used today. Multi-screw extruders by improving dispersion have allowed a reduction in the amount of additives required.

The equipment for compounding additives into a polymer system

must be capable of performing some of the following process tasks:
1. Incorporation and homogenization of additives without
    exceeding degradation temperatures;
2. Generation of high shear stresses for dispersion of
    non-reinforcing fillers or pigments;
3. Homogenization of two or more materials of differing
    melt viscosities without resulting in a stratified
    or layered final mix;
4. Provision of uniform shear stress at heat history to
    each particle;
5. Provision of precise control over the process to
    ensure narrow temperature distribution throughout the
    process and at discharge.

Good control over residence time distribution is an essential
feature of a continuous compounding system. The residence time distribution must be short and uniform to minimize heat history.

In addition to these general requirements, certain types of additives present special compounding problems. Slip, antiblock and
wax-free lubricants are low melting point additives which present
problems when compounded at master batch levels. They melt before
the polymer has become molten and act as excellent lubricants. This
presents conveying problems in conventional compounding equipment,
which can be overcome by the use of intermeshing twin-screw extruders
with their positive conveying characteristics. The viscosity differences present obstacles to effective melting and homogenizing.

To achieve effective homogenization, the components should have
nearly equal viscosities. This can best be achieved for liquid additives when the polymer is fully molten. By melting the additive
and injecting it downstream into the processing section after the
polymer melting has taken place, high levels can be incorporated and
homogenized in this type of compounder. Kneading elements effectively introduce high amounts of energy at specific locations within the
processing section. These kneading elements also, by changing their
shear direction, overcome the striation problem inherent in less sophisticated mixing equipment.

A case history involved incorporating a slip agent and antioxidant with 80°C melting point into polyethylene. In solid form,
a seven percent level was attained, but this was boosted to 16 percent by liquid injection downstream. Holding the level of liquid
slip and antioxidant at constant 15 percent, successful compounding
was achieved, with the same machine configuration and process section. The antiblock was fed into the first barrel.

Some engineering plastics are used in applications involving
very complex part shapes. A motor housing is a typical example
(Figure 1). The molder was encountering high reject rates caused

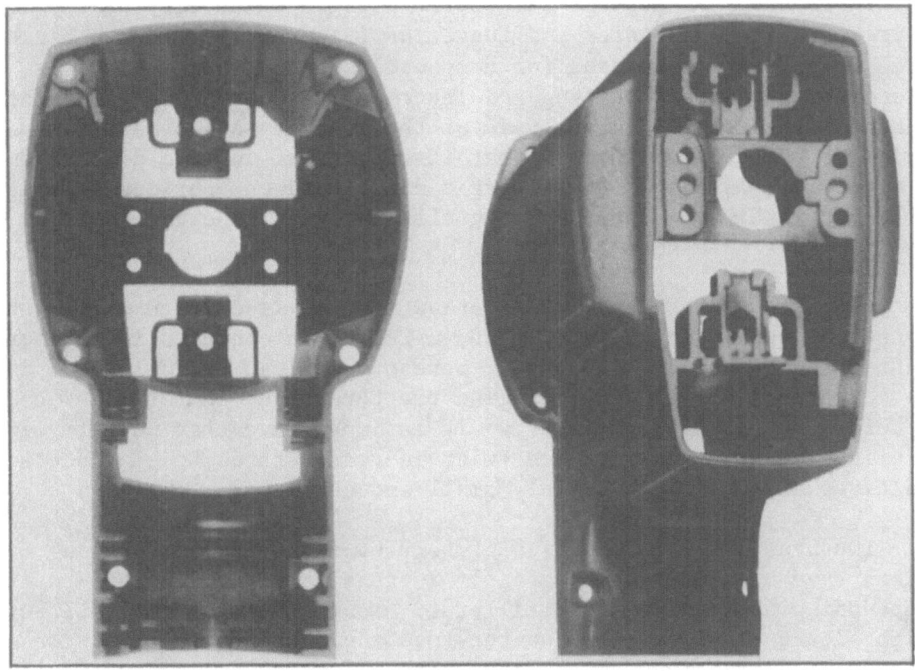

Figure 1. Motor housing with silicone concentrates.

by injection difficulties. Incorporation of silicone fluid concentrates improved the material flow and acted as a mold release agent
to overcome the high rate of rejection. These concentrates can be
made at 30 to 50 percent levels; which can subsequently be let down
to about one percent in the final material. Multiple injection is
required to compound such high levels of silicone in the concentrate
(Figure 2).

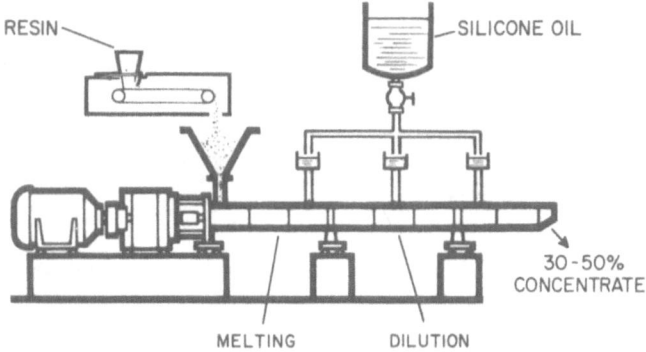

Figure 2. Compounding of silicone concentrate.

There are a variety of twin-screw designs which vary widely in their operating principles and functions (Figure 3). The two basic types which are widely used for compounding operations today are intermeshing counter-rotating and intermeshing co-rotating twin-screw machines (Figure 4). An analysis of the velocity and stress distribution of the two types shows that the degree of uniform dispersion is directly related to stress/strain distributions, particularly at low concentrations. The conveying efficiency is superior to that possible in single-screw machines.

The flow pattern in a co-rotating machine is different from that of the counter-rotating in that the material is conveyed in a figure-eight-shaped pattern. Zero shear stress point can be influenced in the co-rotating machine by changing the throughput and/or screw speed, as well as by changing geometry. It is imperative that uniform stress distribution be maintained even with increased throughput, since some additives are used in extremely small amounts.

Improvements in the design of both counter- and co-rotating machines have been made which greatly increase their efficiency in the compounding of additives and fillers. Among these are the interrupted flights/channels common in counter-rotating machines to break the material and re-orient the layers. The selection and arrangement of individual screw sections and kneading elements in co-rotating machines is carefully studied in order to further improve the unifor-

SCREW ENGAGEMENT		SYSTEM	COUNTER-ROTATING	CO-ROTATING
INTERMESHING	FULLY INTERMESHING	LENGTHWISE AND CROSSWISE CLOSED	1	THEORETICALLY NOT POSSIBLE 2
		LENGTHWISE OPEN AND CROSSWISE CLOSED	THEORETICALLY NOT POSSIBLE 3	SCREWS 4
		LENGTHWISE AND CROSSWISE OPEN	THEORETICALLY POSSIBLE BUT PRACTICALLY NOT REALIZED 5	KNEADING DISCS 6
	PARTIALLY INTERMESHING	LENGTHWISE OPEN AND CROSSWISE CLOSED	7	THEORETICALLY NOT POSSIBLE 8
		LENGTHWISE AND CROSSWISE OPEN	9 A	10 A
			9 B	10 B
NOT INTERMESHING	NOT INTERMESHING	LENGTHWISE AND CROSSWISE OPEN	11	12

Figure 3. Twin-screw designs and functions.

 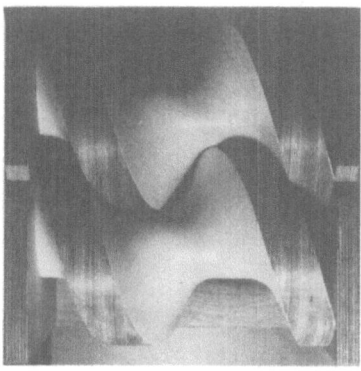

Figure 4. (a) Counter-rotating twin screw. (b) Co-rotating twin screw.

mity of stress distribution.

Two approaches to compounding of cross-linkable polyethylene have been successfully demonstrated in commercial operations using co-rotating twin-screw equipment. One involves feeding a preblend of solid peroxide fillers and additives in barrel one. The excellent materials conveying characteristics of this equipment overcome any feeding problems due to the lubricity of the peroxide. A separate feeder meters the polyethylene. The controlled high shear input of the kneading blocks enables good mixing, homogenization and dispersion of carbon black or fillers without generating a melt temperature that would initiate cross-linking. A variation of this requires preblending all of the components including polyethylene and feeding this preblend in barrel one. A second approach involves metering the liquid peroxide downstream after polyethylene melting has occurred. In this way, the problems of handling solid peroxide are eliminated, as the peroxide has a great tendency to block together. It may also be possible to eliminate the preblending operation and separately meter all of the solid components into barrel one.

Flame retardant additives, primarily used to meet certain requirements set by government or industry, cause deterioration of some polymer properties, such as impact strength, tensile strength, etc. With the objective of flame retardancy, it is imperative that all ingredients be thoroughly mixed. While the purpose of these ingredients is to retard flames, degradation can occur during the compounding operation. The result could, of course, be reduction of the flammability rating or an increase in the additive level. In a controlled processing technique, however, the reduction of flame retardant composition is possible without sacrifice of the flammability rating. This reduction can be 10 to 15 percent, which represents a considerable saving in costs, as well as of physical properties. Reinforcing fillers are usually added to offset property loss. In this case, compounding operations are accomplished in one step: high intensity controlled compounding of flame retardant and low intensity controlled reinforcement compounding (Figure 5).

SUMMARY

Each additive system can be most efficiently incorporated into the polymer by performing the mixing operations at conditions of most similar viscosities. To carry out the process objectives successfully, a compounding machine with excellent conveying characteristics is essential. A variable design machine configuration allows maximum flexibility in location of feed inputs. Without this flexibility it may be necessary to sacrifice material quality in order to compound at high levels, or lower levels of additive incorporation may have to be accepted. The ability to specify exactly where and how much shear input will be located within the processing section and the ability to control the degree of mixing intensity by using different screw hardware combinations are invaluable assets in the compounding of polymers with a variety of additives and fillers.

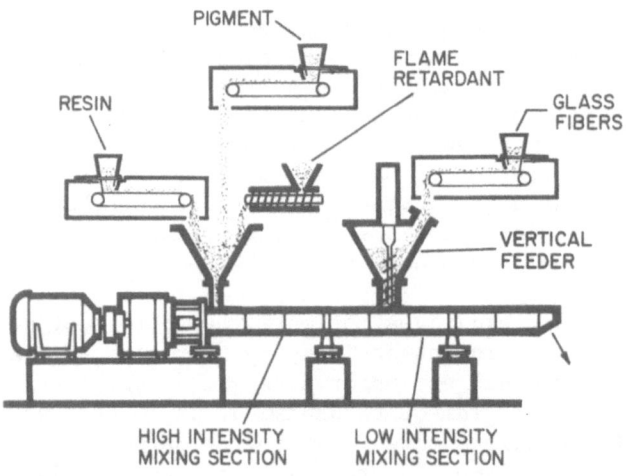

Figure 5. Compounding of flame retardant and glass fiber.

REFERENCES

1. A. A. Schoengood, <u>Additives</u>, SPE Journal, 1974.
2. Plastics Engineering Staff, <u>Additives: How and Why They Work</u>, Plastics Engineering, March, 1975.
3. J. R. Copeland and O. W. Rush, <u>Wollastonite: Short-Fiber Filler/ Reinforcement</u>, Plastics Compounding, November/December, 1978.
4. H. S. Katz, <u>Carbon Graphite Reinforcements Are Coming Of Age</u>, Plastics Compounding, March/April, 1979.
5. G. Crowe and P. E. Kummer, <u>Extending Resins with Calcium Carbonate</u>, Plastics Compounding, September/October, 1978.
6. T. D. Thompson, <u>Applications for Kaolin Fillers</u>, Plastics Compounding, May/June, 1979.
7. J. B. Marsden, <u>Functions, Applications and Advantages of Silane Coupling Agents</u>, Plastics Compounding, July/August, 1978.
8. J. G. Mohr, <u>Evaluating Glass-Fiber Reinforcements</u>, Plastics Compounding, July/August, 1978.
9. S. Jakopin, <u>Compounding of Additives</u>, SPE 37th ANTEC, New Orleans, LA.

THE EFFECT OF LUBRICANTS ON THE EXTRUSION

CHARACTERISTICS OF POLY VINYL CHLORIDE

E. A. Collins, T. E. Fahey and A. J. Hopfinger

Macromolecular Science Department
Case Western Reserve University
Cleveland, OH, 44106

INTRODUCTION

Poly vinyl chloride (PVC) has been one of the most widely used materials in the plastics industry for the last three decades. Consumption in the U. S. alone exceeded five billion pounds in 1980 and is expected to continue to grow. The extent of its usage results from the wide range of properties attainable with PVC, its low cost, and the relatively low energy requirements for its production. The achievement of a broad range of properties based on a common material arises from the almost limitless number of compounds which can be formulated from PVC through the use of a wide variety of additives.

The additives of greatest use in PVC may be broadly divided into two classes: those used to improve processability and those used to modify resultant properties. The first class of additives includes thermal stabilizers, lubricants, processing aids with plasticizers. The second class includes impact modifiers, pigments, gloss enhancers and fillers.

The need for additives to aid in the processing of PVC arises mainly from the inherently poor thermal stability of PVC in the temperature range used to process it. Although chemical stabilizers significantly improve PVC's thermal stability, lubricants are also necessary to process unplasticized PVC. The high shear fields encountered in most polymer processing equipment cause the development of local temperature increases which result in PVC degradation in spite of the presence of chemical stabilizers. Lubricants regulate energy transfer to the PVC matrix at PVC/metal interfaces while also reducing friction between PVC particles. At high temperatures, lubricants may also solubilize in the PVC, resulting in a viscosity reduction.

351

Numerous authors have attempted to classify the broad range of lubricants available for PVC.[1-6] Most of these studies attempt to rank lubricants as internal or external in nature. King and Noel[1] rated lubricants by their ability to depress the glass transition temperature of PVC. Illmann[2] developed a ranking of lubricants based on chemical structure. Polar lubricants were considered to yield internal lubricating effects, while nonpolar lubricants would migrate to a boundary layer, providing external lubrication.

Hartitz[3] used a Brabender Plasticorder to demonstrate two important points with regard to lubricants: a) they retard PVC "fusion" and b) they interact with each other. Two distinct classes of lubricants were delineated, in that significant retardation of PVC fusion could only be obtained by combination of an additive from each class. Hartitz's group A was composed primarily of metal soaps, while group B consisted of the less polar paraffin waxes, low molecular weight polyethylenes, and oils. Recently, work by Krzewki[5] has further demonstrated the importance of the interaction between these two classes of lubricants to the fusion of PVC.

These works demonstrate that lubricants, whether internal or external, reduce the rate of fusion of PVC. At low temperatures, this involves inhibition of diffusion of PVC molecules across particulate boundaries. As temperature increases, the rate of PVC diffusion will increase, causing a blurring of particle boundaries. When interparticle links become strong enough, an applied shear field will result in destruction of the particulate domains. By impeding the diffusion inherent to this process, lubricants may be expected to prolong the existence of the particulate structure at a given temperature, and to increase the temperature needed for complete conversion to a pseudo-melt state.

What remains unclear is the role of lubricants in the PVC melt once the particulate structure has been destroyed. Some works have suggested a decrease in viscosity due to lubricants,[4,7,8] although their effect generally decreases with increased temperature. It has been suggested that this is due to a solubilization of the lubricants in the PVC matrix at high temperatures.

Some works have also suggested that external lubricants can cause slip of the PVC melt at metal interfaces.[9-11] This effect has been disputed by others[8,12] and remains a subject of controversy. Clearly, at temperatures where a particulate flow dominates PVC flow behavior, slip will occur as a manifestation of particulate flow at metal surfaces. Wales[12] contends that this mode of behavior does not occur at temperatures above 180°C. Data of Collins and Nakajima[13] supports this conclusion. The work of Chauffoureaux[11] and others implies the existence of slip at higher temperatures. The discrepancies may result from the dependence of the slip-stick transition on formulation.

The present work attempts to define the role of lubricants during the extrusion of PVC by measurements of melt pressures, temperature and melt pool formation rate. Two common lubricants, calcium stearate (CaSt) and paraffin wax, are used for this work. Capillary rheometer data is also used to help define the mechanisms by which lubricants function during extrusion.

EXPERIMENTAL

Materials

All formulations were based on Diamond Shamrock PVC 430, a medium molecular weight suspension resin. To minimize stability considerations, all formulations also contained 1.5 phr of tin mercaptide stabilizer. The paraffin wax used was Boler 392A. All formulations were prepared by mixing in a Henschel high intensity mixer to a drop temperature of 220°F.

Extrusion Experiments

All extrusions were performed on a Modern Plastics Machinery 1-1/2" diameter extruder, with an L/D of 24:1. The machine is air cooled with four heating/cooling zones along the barrel and three for the adapter and die. For all experiments, the die was a slit die of dimensions 1" by .125". A standard breaker plate was used but no screen pack.

A single stage screw was used to facilitate measurements of melt pool growth rate. The screw is a square pitch screw with a compression ratio of 3:1.

Barrel temperature profiles were chosen such that a solid bed was maintained over most of the length of the screw. Screw cooling was used for the calcium stearate level series, with a neutral screw for the wax level series.

The extruder barrel is instrumented with six Dynisco PT422A pressure transducers and four melt thermocouples. Pressure and temperature signals are processed by Dynisco TCM601 and SCM600 signal conditioners and recorded on a Yokagawa six-channel chart recorder. The location of the pressure and temperature sensors, as well as the barrel heating/cooling zones, are shown in Figure 1.

The system used for measurement of melt temperatures was designed for this study. The thermocouples are 1/8" diameter, flush mounted and insulated from the barrel by polyimide sheaths. The

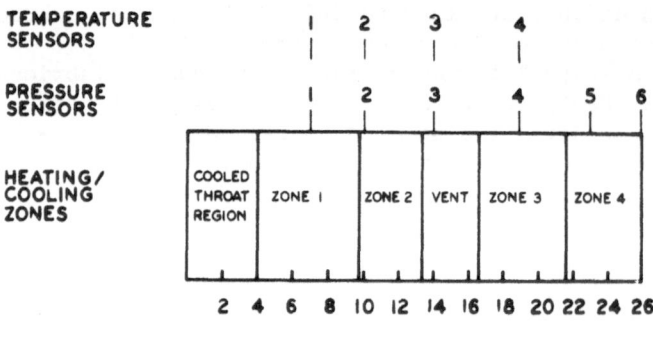

Figure 1.   Location of heat/cool zones and pressure and temperature
            sensors.

thermocouples are flush with the barrel surface while the extruder
is running. Measurements of temperatures within the screw channel
are made by stopping the screw and instantaneously injecting the
thermocouples into the melt. In this way, temperatures may be taken
at different positions within the channel both radially and axially,
depending on the orientation of the screw to the plane of the sensors
at the time the screw is stopped.

     For measurements of solid bed widths, the extruder was stopped
and cooled quickly (10-15 minutes). The filled screw was then ejec-
ted hydraulically and samples chipped off by flight. A small amount
of carbon black was added just prior to stopping the screw to facilitate differentiation of solid bed and melt pool.

Capillary Rheometry

     Viscosity vs. shear rate data was obtained with a Sieglaff-
McKelvey capillary rheometer. A .040" diameter capillary with an
L/D of 26 was used for all work. Measurements were taken at tempera-
tures of 190, 200, 210 and 220°C.

RESULTS AND DISCUSSION

Melt Pressure Measurements

     All pressure data is obtained as a range of pressure, the mag-
nitude of the fluctuation varying from 100 to 800 psi. This range
represents the maximum and minimum of a cyclic fluctuation in pressure, equivalent in frequency to the screw speed. This is an arti-
fact of the spike in pressure recorded each time a flight top passes
under the pressure sensor. The fluctuations subside to 100-200 psi
for sensors 4 and 5, located in the metering zone. The shallower

channel in this region results in a lower pressure differential between flight top and flight channel. Minima of the cyclic pressure
traces are used in all pressure graphs, assuming the maxima to be
related more to flight gap pressures.

Figure 2 illustrates melt pressure data from the extruder for
four CaSt levels. The pressures recorded by sensor #1 are essentially constant for all samples except the zero calcium stearate level.
This sensor is located at flight 7, near the end of the feed zone.
Screw ejection studies have shown that it is in this region that
pressure starts to build, resulting in the quick compaction of the
loosely conveyed solid present in flights 1-5. This densification
process was measured by a water displacement technique for samples
from the flights of the screw. The results, shown in Figure 3,
demonstrate that compaction occurs quickly in flights 6 through 10
where the melting process begins.

Figure 1 demonstrates that any differences in compaction rate
for the samples containing 1.0-3.0 parts calcium stearate are not
significant enough to affect the pressure. However, the low pressure generated by the sample with no calcium stearate suggests much
poorer compaction for this formulation. This conclusion has been
substantiated in a previous publication in a laboratory compaction
experiment.[14]

The sample with no calcium stearate displays a radically different behavior from the remainder of the series. While the other
formulations result in curves of similar shape with maxima at flight
18-19, the removal of all calcium stearate causes a much flatter
pressure profile peaking later in the extruder. This dramatic change

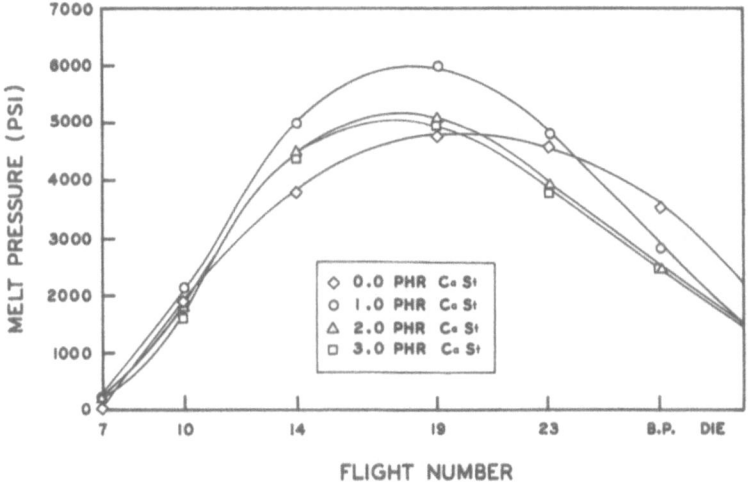

Figure 2. Pressure development vs. calcium stearate level - 60 rpm.

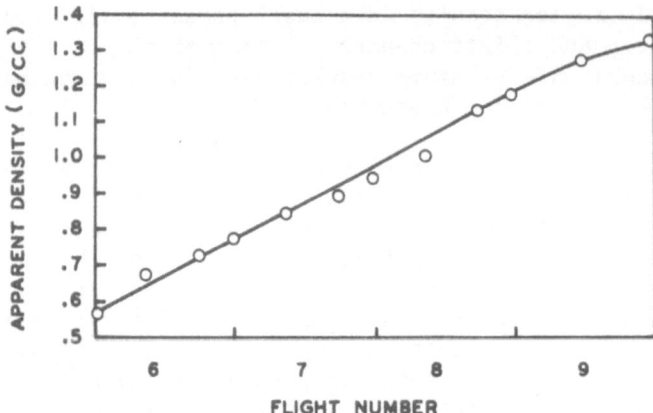

Figure 3.   Densification of solid bed at end of feed zone.

cannot be explained by simple viscosity differences.   It is thought
to result from an interaction between the calcium stearate and paraffin wax.   The complete removal of either material has been found to
result in vastly different fusion and lubrication properties.[3,5,14]

Figure 4 lists pressure profiles as a function of paraffin wax
level.   The data shows a general decrease in pressures with increasing wax level, although the trend is not as clear as with the calcium
stearate level series.   The 1.2 phr sample may have artificially
high readings due to the slightly higher flow rate.

Unlike the calcium stearate series, there is a clear drop in
pressure at sensor #1 with increasing wax level.   The data suggests

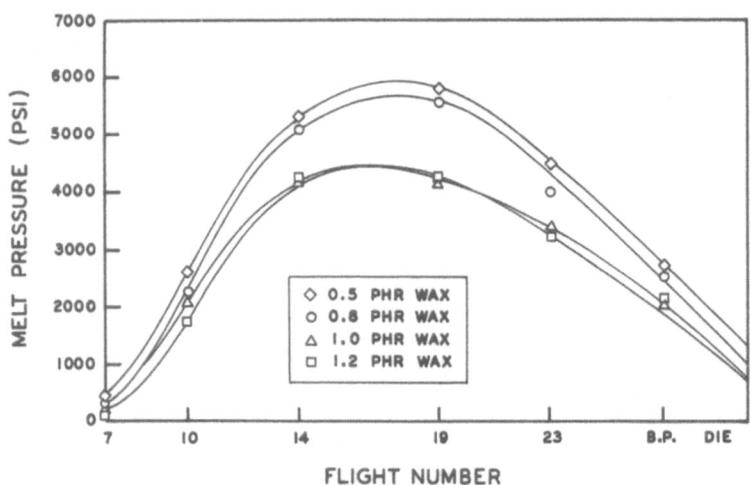

Figure 4.   Pressure development vs. paraffin wax level - 60 rpm.

a decrease in solids conveying rate or compaction rate with increasing wax level. This would imply a reduction in frictional coefficient at the barrel surface in the feed zone.

Melt Temperature Measurements

The melt temperature measurement system employed in this study is unique in that no previous measurements of radial temperature profiles within a screw channel have been reported in the literature. However, the data presented here is in general agreement with previous measurements of radial temperature profiles at the screw tip.[15,16] This correlation is supportive of the assumption that barrel temperature effects have been minimized by the polyimide sheaths, since the previous workers reported extensive precautions to avoid any such effects.

Figure 5 illustrates the effect of thermocouple depth within the screw channel on temperature readings. As shown in Figure 5, temperatures recorded at the barrel surface are often higher than the corresponding barrel set temperatures, implying that they reflect the temperatures of a hotter melt. Melt temperatures generally decrease as distance from the barrel surface increases.

Several general features of the temperature measurements are illustrated by Figure 5. Except for thermocouple #3, temperature increases as distance down the screw increases. Thermocouple #3, located at flight fourteen, reads temperatures considerably lower than would be expected from the other readings or the barrel temperature settings. The cause of this phenomenon is illustrated by reference to Figure 1. As shown, this thermocouple is located just prior to the unheated vent region. The vent causes a cold region along that section of the barrel, resulting in a lowering of melt temperatures. This conclusion is further substantiated by the melting rate data presented in the following section.

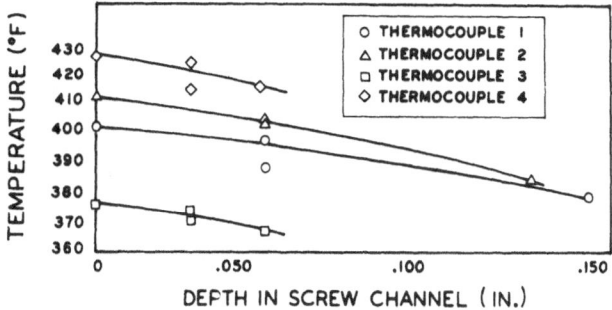

Figure 5. Effect of thermocouple on temperature readings.

Defining the relationship of axial position within the screw
channel to temperature is dependent on several factors. Except for
the depth vs. temperature data shown in Figure 5, all readings were
taken at the deepest settings. This was done to try to maximize
differences between solid bed and melt pool temperatures. Figure
6 illustrates the classic model of the melting mechanism in single
screw extruders. Clearly, in taking temperature measurements within
the channel, one would expect to see a temperature rise for measurements near the rear of the channel in the melt pool. Thus, measurements were made at several positions axially within the channel using
a pointer on the shank end of the screw to establish relative axial
position within the flight channels. Thermocouples were spaced such
that each thermocouple is located at the same axial position within
its flight channel.

The most interesting temperature data is that recorded by sensor
number two at flight 10. At this flight, a small melt pool is present at the rear of the channel in all experiments. Figures 7 and 8
illustrate the relationship of temperature to axial channel position
in flight 10. A significant rise in temperature at the back edge
of the channel is seen, due to the difference between solid bed and
melt pool temperatures.

Thus, the melt temperature data may be correlated with radial
and axial position within the screw channel as well as solid bed/melt
pool differences. However, no real correlation with formulation is
seen. This is probably not surprising in view of the complexity of
the temperature gradients within the channel.

Solid Bed Profile Measurement

The rate of melt pool formation, as assessed from screw ejection
experiments, is presented as the ratio of solid bed width to total
channel width (x/w). A sample of the data in detail is presented
in Figure 9. Melt pool formation begins around flights 8-9 for all

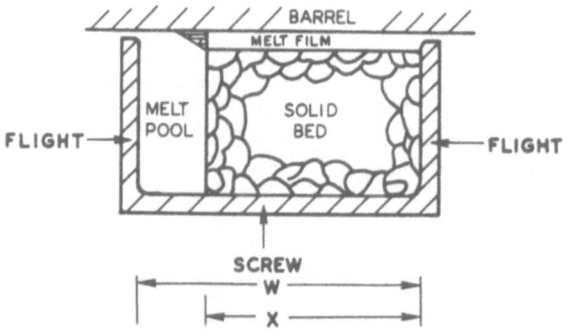

Figure 6. Classic solid bed/melt pool representation.

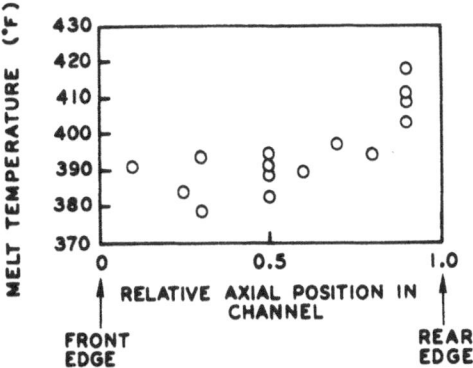

Figure 7.   Melt temperature vs. axial position in flight channel
            No. 10 for wax level series.

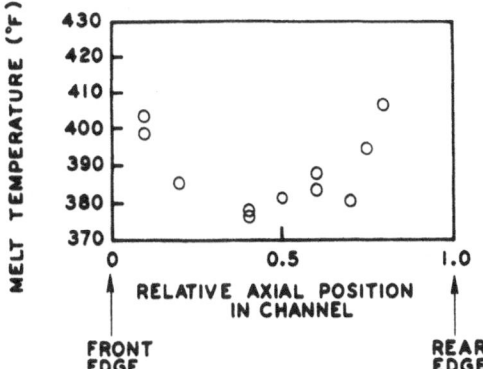

Figure 8.   Melt temperature vs. axial position in channel No. 10
            for calcium stearate level series.

samples. As shown in Figure 9, melt pool formation essentially
stops for the region from flight 14 to flight 18, corresponding to
the vent area. Fluctuations in melt pool width also occur in this
region and subsequent regions, due to flow patterns within the screw
channel. The apparent decrease in melt pool width around flights
16-19 is in part due to thickening of the melt film as the clear
delineation of melt pool and melt film becomes unsteady in the shallower channel.

     Although the use of a single stage screw in a vented extruder
results in the undesirable slowing of melting in the compression
zone, this screw was considered a more viable approach for differentiation of additive effects. Use of a two stage screw results
in fast rate of melt pool formation in the first stage of the screw,

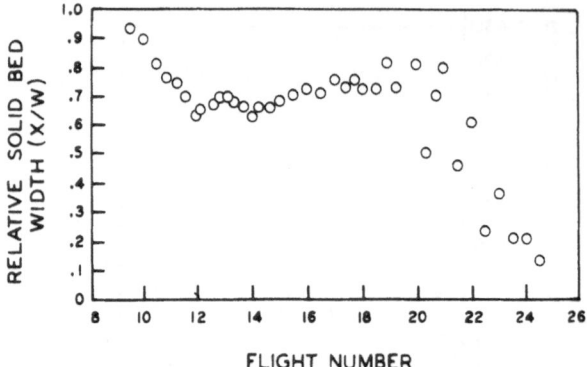

Figure 9.   Actual fluctuations in solid bed profile.

lessening differences between formulations. In addition, mixing of
the melt pool and solid bed at the decompression zone prohibits any
measurements in the second stage of the screw. For the purposes of
clearer data presentation, subsequent data is based on maximum melt
pool width for a given flight.

      Figure 10 illustrates solid bed profiles as a function of cal-
cium stearate level. For the 1.0-3.0 phr levels, melt pool formation
slows as stearate level is increased, due to the reduction in viscosity and, thus, viscous heat generation. As with the melt pressure data shown previously, the sample containing no calcium stearate
does not fit this trend. Up to flight 19, this sample melts slowest
of the four samples, but a sharp rise in melt pool formation rate
begins at flight 20, surpassing the melt pool growth of the 3.0 phr
sample. Clearly, this sample behaves completely differently from
those containing calcium stearate.

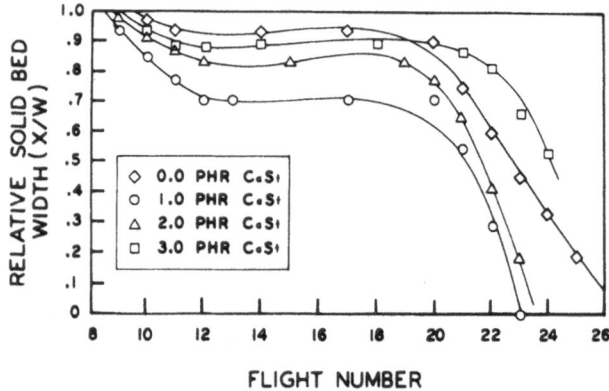

Figure 10.   Solid bed width as a function of calcium stearate level.

Figure 11 shows melt pool formation rate as a function of paraffin wax level. Melt pool formation rate decreases steadily as wax level is increased. As with the calcium stearate series, the effect appears to lessen at the higher levels, suggesting a non-linearity in the effect of wax level on melt formation rate.

Capillary Rheometer Measurements

Figures 12-14 show the effect of temperature on samples containing no paraffin wax and 1.0, 2.0, and 3.0 phr calcium stearate, respectively. At low shear rates, all three curves show a steady decrease in viscosity with increasing temperature as expected. However, each curve exhibits a break point at the lower temperatures,

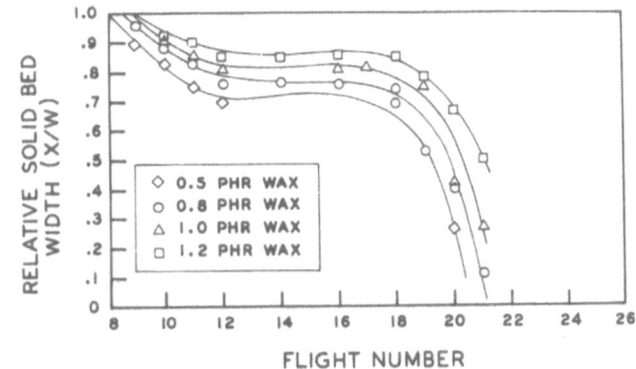

Figure 11. Solid bed width as a function of paraffin wax level.

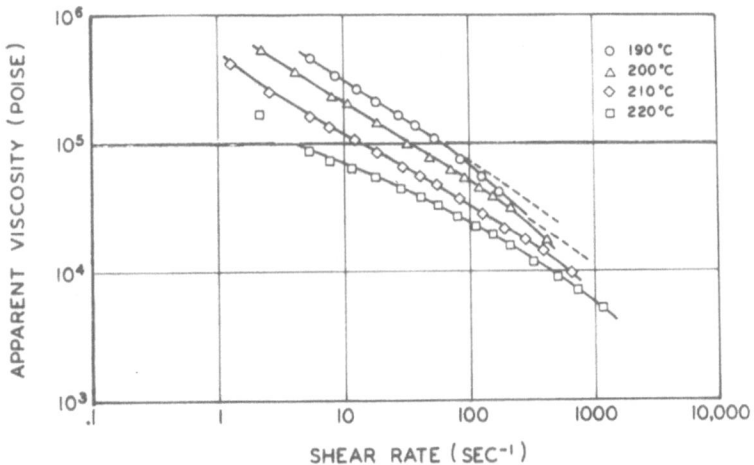

Figure 12. Effect of temperature on viscosity 1.0 phr, CaSt, no wax.

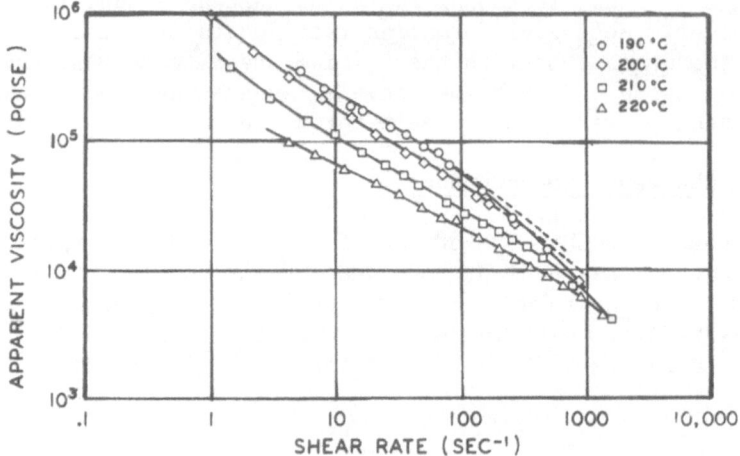

Figure 13. Effect of temperature on viscosity - 2.0 phr, CaSt,
            no wax.

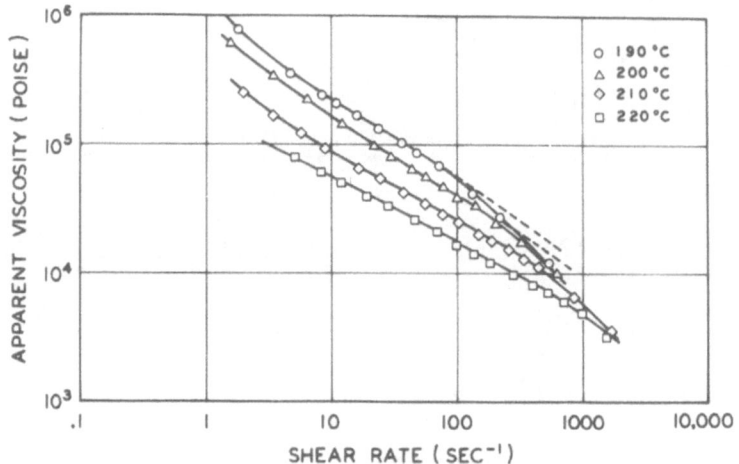

Figure 14. Effect of temperature on viscosity - 3.0 phr, CaSt,
            no wax.

above which viscosity drops off sharply, crossing the higher temperature curves. The shear rates at which the breaks occur correspond
to shear stresses of 5.0-5.5 x 10^6 dynes/cm. Chauffoureaux[11] has
postulated that slip should occur due to adhesive failure at the
capillary wall for shear stresses of 4-6 x 10^6 dynes/cm^2. The data
for samples with no wax correlates with this range, though the mechanism remains uncertain.

This effect is magnified by the addition of paraffin wax to the formulation. Figures 15-17 illustrate the effect of temperature for samples containing 0.5, 1.0, and 1.2 phr wax, respectively. At 0.5 phr, the 190°C data exhibits a discontinuity at approximately 70-80 sec^{-1}, above which the 190°C data falls below the 200°C data. At 1.0 phr wax, this break point occurs at 20-40 sec^{-1}, causing the 190°C data to eventually fall below both the 200°C and 210°C data. At 1.2 phr wax (Figure 17) the 190°C data falls below both the 200 and 210°C data over the entire range of shear rates. The 200°C data also yields viscosities below the 210°C data at shear rates above 100 sec^{-1}.

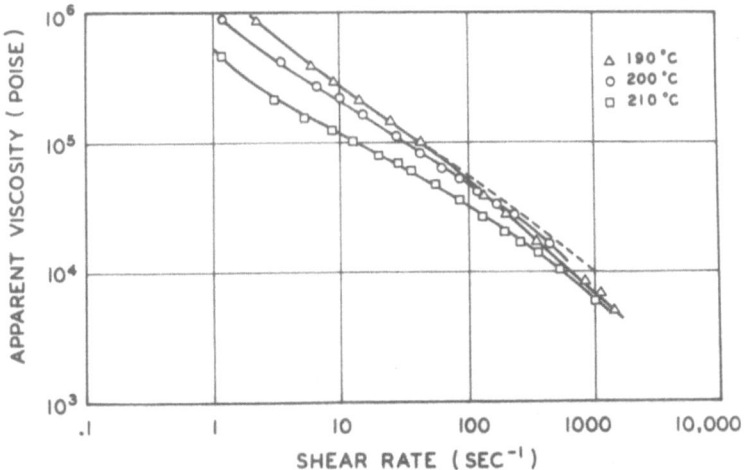

Figure 15. Effect of temperature - 0.5 phr wax.

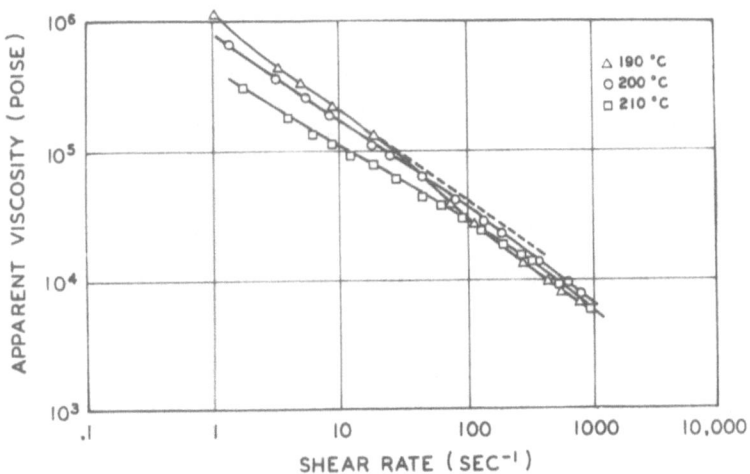

Figure 16. Effect of temperature - 1.0 phr wax.

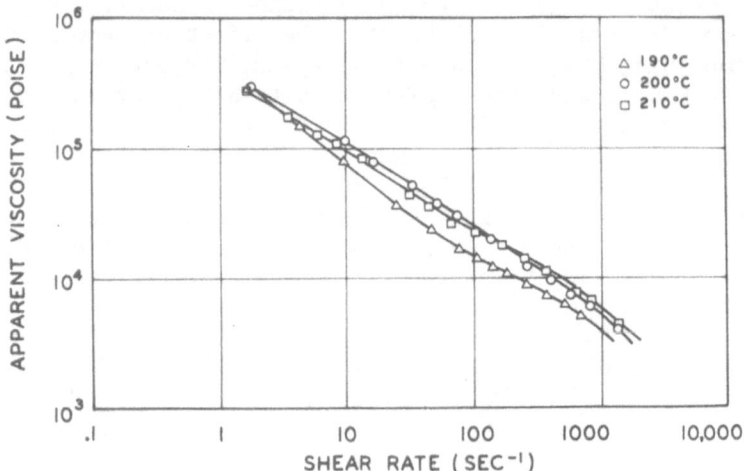

Figure 17. Effect of temperature - 1.2 phr wax.

The shear stress at which the discontinuity occurs appears to drop as wax level is increased. In all samples, with or without wax, it is clear that the phenomenon decreases in effect as temperature is increased. At 210°C or above, no discontinuities in viscosity vs. shear rate plots are observed.

The anomalous flow behavior at 190° and 200° strongly suggests the existence of particulate flow. PVC fusion will be severely impeded in the high wax level samples or the sample with no calcium stearate. As a result, particulate destruction and the resultant change in flow unit size will be delayed, resulting in lower apparent viscosities at 190° and 200°. The wax level data suggests that the transition in flow unit size is severely impeded by paraffin wax. Calcium stearate, in the absence of wax, appears to have less effect on the impedence of the fusion process. These results are in agreement with previous studies on the effect of lubricants on PVC fusion.[5,14]

The effect of wax level at 190°, 200°, and 210°C is shown in Figures 18-20, respectively. In comparing the three plots, it is clear that the effect of the wax on viscosity decreases with increasing temperature, and, in fact, at 220°C wax level has no effect on viscosity in the range of concentrations studied. Conversely, replotting the data from Figures 12-14 as a function of calcium stearate level in the absence of wax, it can be seen that the effect of calcium stearate on viscosity remains fairly constant over the entire temperature range (see Figures 21-24).

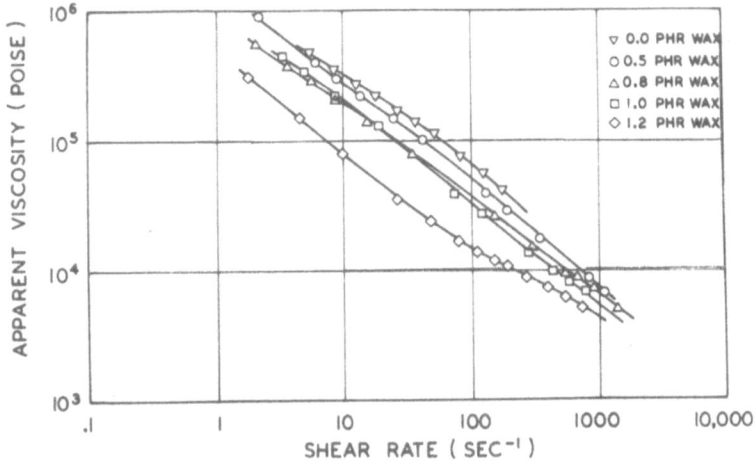

Figure 18.   Effect of wax level at 190°C.

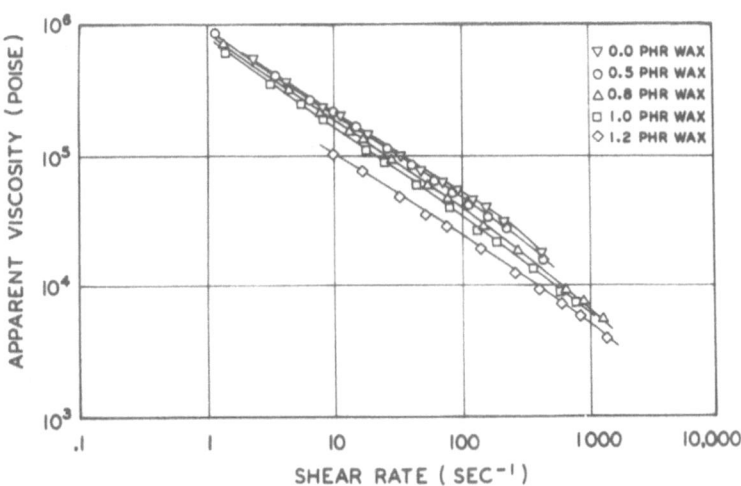

Figure 19.   Effect of wax level at 200°C.

Figures 25-27 illustrate the progressive effect of calcium stearate level in the presence of 1.0 phr wax. As might be expected,
the sample containing no calcium stearate yields very erratic behavior at low temperatures (Figure 25). In general, the inclusion
of paraffin wax blurs the effect of stearate level at 190° and 200°C,
minimizing the differences seen in the absence of wax. Only at 210°C
(see Figure 27) does the wax effect decrease enough to allow a clear
delineation by stearate level.

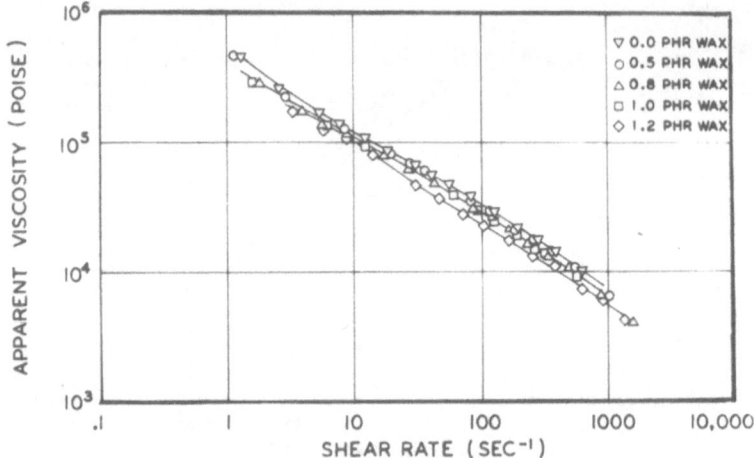

Figure 20.   Effect of wax level at 210°C.

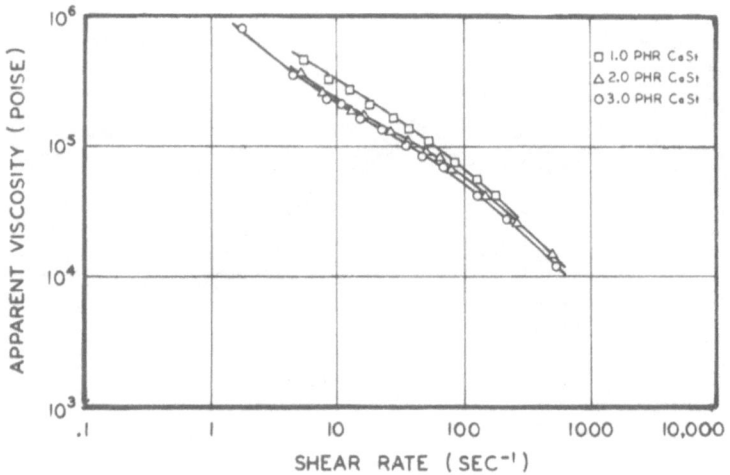

Figure 21.   Effect of calcium stearate level at 190°C in absence
              of wax.

CONCLUSIONS

      The effects of calcium stearate and paraffin wax on the single
screw extrusion of PVC have been examined. Both additives were found
to slow melt pool formation, as long as both additives are present
in the formulation. Complete removal of calcium stearate considerably lowers melting rate, implying an interaction between the two
additives. This is further substantiated by melt pressure data and
melt rheology.

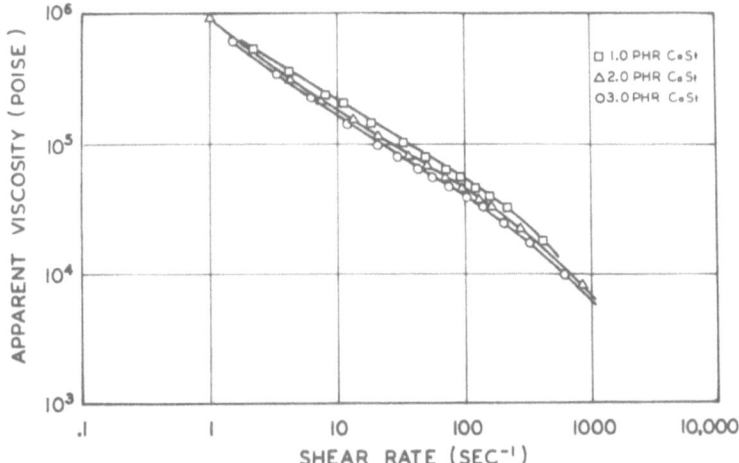

Figure 22.   Effect of calcium stearate level at 200°C in absence
of wax.

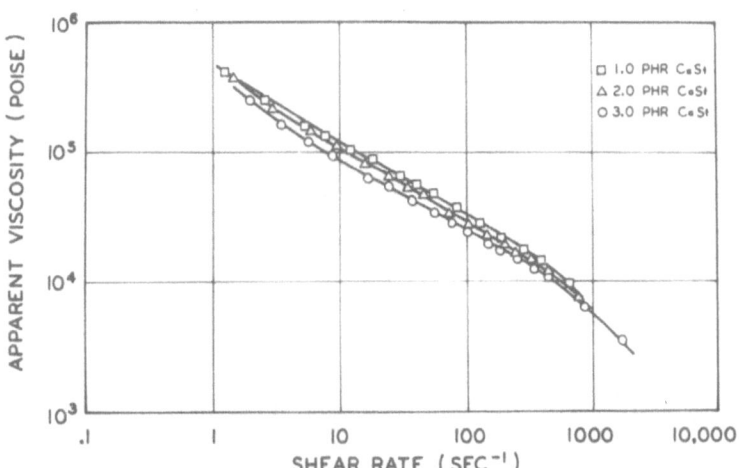

Figure 23.   Effect of calcium stearate level at 210°C in absence
of wax.

Calcium stearate was shown to reduce melt viscosity. This
result was supported by melt pressure data. Erratic rheological
data at 190° - 200°C was found for samples containing wax. The erratic low temperature rheology data suggests that the transition
from particulate to viscous flow is delayed by paraffin wax. This
observation is in agreement with the slower melt pool formation rates

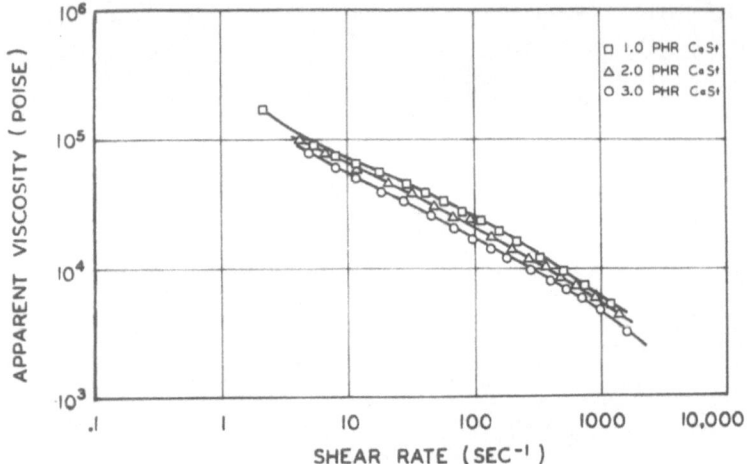

Figure 24.   Effect of calcium stearate level at 220°C in absence
             of wax.

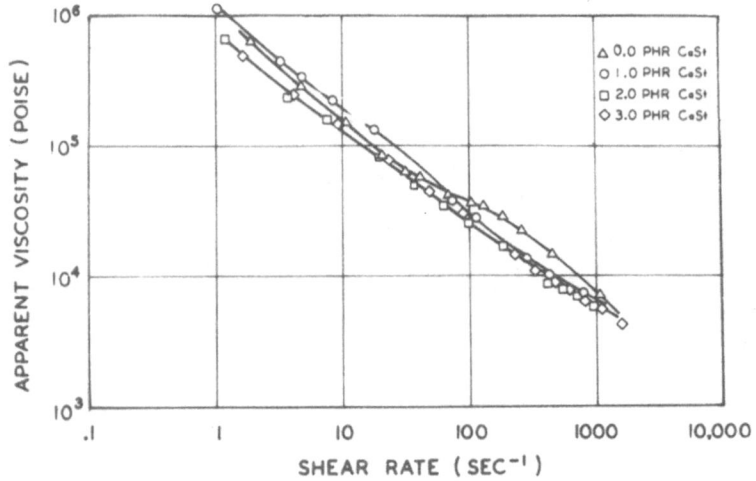

Figure 25.   Effect of calcium stearate level at 190°C with
             1.0 phr wax.

noted above. Reductions in feed zone pressure also imply a reduction in frictional coefficient by paraffin wax, further prolonging
the existence of the particulate structure.

    Melt temperature measurements within the screw channel demon-
strate both radial and axial temperature gradients. The results

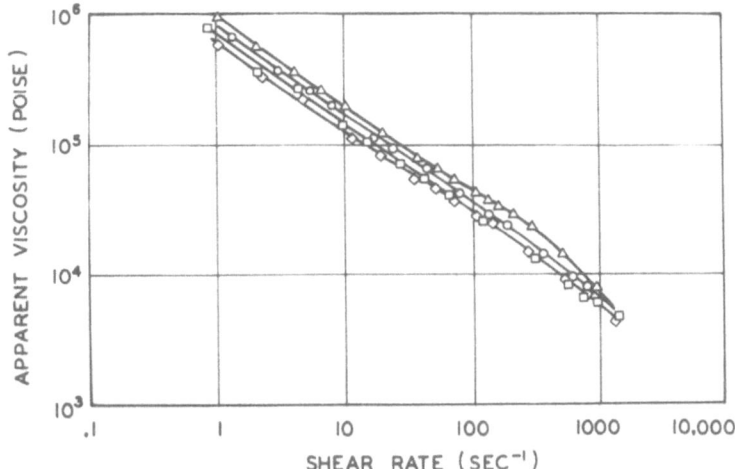

Figure 26.   Effect of calcium stearate level at 200°C with
              1.0 phr wax.

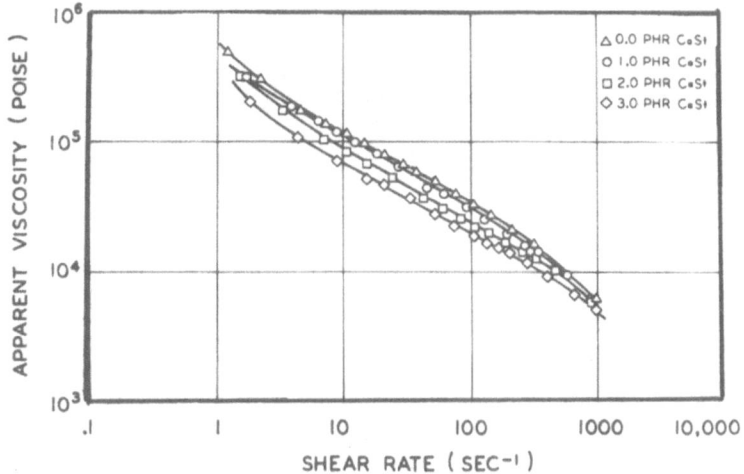

Figure 27.   Effect of calcium stearate level at 210°C with
              1.0 phr wax.

demonstrate significant differences in melt pool, melt film, and
solid bed temperatures.

ACKNOWLEDGEMENTS

The funding and support of Diamond Shamrock Corporation, where all experimental work was performed, is gratefully acknowledged. The assistance of Robert Shreve and Jan Zgodinski was also greatly appreciated.

REFERENCES

1. L. F. King and F. Noel, Polym. Engr. Sci., $\underline{12}$ (2), 112 (1972).
2. G. Illmann, SPE-J. $\underline{23}$, 71 (1967).
3. J. E. Hartitz, Polym. Engr. Sci., $\underline{14}$ (5), 392 (1974).
4. C. L. Sieglaff, Polym. Engr. Sci., $\underline{9}$ (2), 81 (1969).
5. R. J. Krzewki and E. A. Collins, J. Macromol. Sci., Phys., B20 (4), 465-478 (1981).
6. E. A. Collins and J. E. Hartitz, Plast. Des. Proc., $\underline{7}$, 14 (1978).
7. P. L. Shah, Polym. Engr. Sci., $\underline{14}$ (11), 773 (1974).
8. J. L. den Otter, Publication Central Laboratory TNO, Delft, 7-4-1970.
9. C. L. Sieglaff, SPE Trans., $\underline{4}$, 129 (1964).
10. W. C. Johnson, SPE ANTEC, $\underline{13}$, 514 (1967).
11. J. C. Chauffoureaux et al, J. Rheol., $\underline{23}$ (1), 1 (1979).
12. J. L. S. Wales, J. Polym. Sci., Symp. No. 50, 469 (1975).
13. E. A. Collins and N. Nakajima, J. Appl. Polym. Sci., $\underline{22}$, 2435 (1978).
14. T. E. Fahey, J. Macromol. Sci., Phys., B20 (3), 319-333 (1981).
15. E. A. Collins and H. T. Kim, Polym. Eng. Sci., $\underline{11}$ (2), 83 (1971).
16. H. T. Kim and J. P. Darby, SPE J., $\underline{26}$, 31 (1970).

PRECIPITATED CALCIUM CARBONATES AS ULTRAVIOLET STABILIZERS AND IMPACT MODIFIERS IN POLY (VINYL CHLORIDE) SIDING AND PROFILES

K. K. Mathur and D. B. Vanderheiden

Pfizer, Inc.
P. O. Box 548
Easton, PA

INTRODUCTION

Poly (vinyl chloride) is a low cost commodity polymer that is used in a wide variety of applications, such as pipe, tube, conduit, fittings and a host of others. PVC siding and profiles represent two of the more recent applications generating significant commercial interest. In order to successfully compete with aluminum siding, the PVC products must carry a similar long term performance warranty for periods up to 40 years. Special formulation considerations are required to meet these long term performance criteria. The key concern is polymer stability during long term weathering, with special emphasis on impact retention and color change.

Stability of poly (vinyl chloride) has been the subject of extensive research.[1] It has been well documented that PVC undergoes dehydrochlorination at its normal processing temperatures (180-200°C) or upon long term exposure to ultraviolet light.

$$-CH_2-CH-\overset{Cl}{\underset{}{|}}(CH_2-\overset{Cl}{\underset{}{|}}CH-)_n CH_2-\overset{Cl}{\underset{}{|}}CH- \longrightarrow$$

$$-CH_2-\overset{Cl}{\underset{}{|}}CH-(CH = CH-)_n CH_2-\overset{Cl}{\underset{}{|}}CH- + _n HCL \qquad (1)$$

Present literature[2] suggests that the dehydrochlorination can be initiated at several types of defect sites:
a) branch sites with tertiary chlorine;

b) allylic chlorine sites resulting from random unsaturation;
c) chain ends substituted with initiator residues;
d) chain end unsaturation;
e) head to head units; and
f) structures containing oxygen such as carbonyl and peroxides.

As indicated in reaction (1), evidence suggests that the dehydrochlorination proceeds like a chain reaction, producing long sections of conjugated polyenes. Both ionic and free-radical mechanisms have been proposed to explain the progressive hydrogen chloride "unzipping" process.[3,4] Though still debated, it is generally accepted that byproduct hydrogen chloride catalyzes further dehydrohalogenation of PVC. Mechanisms like reactions (2-4) are often proposed to account for this autocatalytic effect.[5]

$$
\begin{array}{c}
\underset{|}{Cl} \quad \underset{|}{Cl} \qquad\qquad\qquad\qquad + \quad \underset{|}{Cl} \\
-CH_2-CH-CH_2-CH- \; + \; HCl \;\rightarrow\; -CH_2-CH-CH_2-CH- \; + \; HCl_2^- \qquad (2)
\end{array}
$$

$$
\begin{array}{c}
+ \quad \underset{|}{Cl} \qquad\qquad\qquad \underset{|}{Cl} \\
-CH_2-CH-CH_2-CH- \;\rightarrow\; -CH_2-CH=CH-CH- \; + \; H^+ \qquad\qquad (3)
\end{array}
$$

$$
H^+ \; + \; HCl_2^- \;\rightarrow\; 2HCl \qquad\qquad\qquad\qquad\qquad (4)
$$

The overall mechanism for the degradation of PVC on exterior exposure would seem to involve two primary processes, dehydrochlorination and photooxidation. As a result of dehydrochlorination, PVC changes color from white to yellow to red to brown. These color changes are typical of systems containing long conjugated polyene sequences. While the color change, usually reported as yellowing, is aesthetically objectionable in siding and profiles, it generally appears long before any measurable loss in mechanical properties. Rapid oxidation of the polyene sequences, however, leads to the formation of carbonyl and hydroperoxide groups that are capable of promoting serious degradation and loss of physical properties via an autocatalytic free-radical mechanism similar to that generally accepted for thermal autoxidation.[6]

In order to achieve long term weatherability in PVC siding and profiles, the present trend is to use a high level of rutile grade TiO_2. This pigment is extremely efficient in scattering visible light and absorbing some ultraviolet light down to 300 nm. Sunlight in the 300-320 nm region is reported to be the most effective in initiating the dehydrochlorination of PVC.[7] About 10-12 phr of TiO_2 is being used for long term weathering protection. Impact strength of filled PVC is also retained because of the small TiO_2 particle

size. On the negative side, the use of high levels of TiO_2 makes
the formulation of dark shades impossible.

Based upon the review articles published by McKellar, Bentley,
Hawkins, Scott and others,[8] the overall photo-stabilization mechanisms
of polymers such as PVC can be classified into the following five
categories:
1. Light Scattering : Examples - TiO_2, ZnO, MgO, $CaCO_3$, Fe_2O_3, Cr_2O_3
2. Light Absorption : Examples - 2 Hydroxybenzophenones, 2-Hydroxy-
                                    benzotriazoles
3. Excited State Quenching : Examples - Ni(II) chelates, cycloocta-
                                    diene
4. Radical Scavenging : Examples - hindered phenols, secondary aryl
                                    amines, quinones, zinc dialkyl
                                    dithiocarbamates
5. HCl and Defect Chlorine Scavenging : Examples - Mixed metal fatty
                                    acid soaps (Ba-Cd, Ba-Zn, Ca-Zn);
                                    Organotin mercaptans and carboxyl-
                                    ates (dibutyltin tin dilaurate,
                                    dibutyl tin dimercaptide); cal-
                                    cium stearate.

Of the group of scattering pigments, calcium carbonate stood
out as a particularly attractive choice for PVC siding and profile
applications for several reasons. First, it is white in color with
a moderate scattering power of its own, in addition to being an excellent extender for TiO_2.[8] It can thereby improve the light scattering protection of the filled part in two ways. Second, calcium
carbonate, preferably with a calcium soap surface treatment, will
neutralize HCl and potentially assist in the removal of chlorine
defect sites in PVC, thereby contributing to enhanced polymer stability on exterior exposure. Third, calcium carbonates, especially
the precipitated products, can be tailored with regard to particle
shape, size, size distribution, surface chemistry and purity to retain
or improve the impact strength of the filled PVC product. Finally,
calcium carbonates are compatible with PVC, non-toxic, non-migrating,
non-staining, easy to store and handle, and are relatively inexpen-
sive.

Based on the suspected performance capabilities of calcium carbonate in PVC siding and profiles, the following study was undertaken
to identify the optimum calcium carbonate properties required to de-
liver improved PVC properties. Calcium carbonate variables studied
included (a) product origin (natural ground vs. precipitated), (b)
particle morphology, (c) particle size/surface area, and (d) particle
size distribution. The most efficient products were compounded in
siding and profile formulations and studied for their effects on key
mechanical properties and weathering stability.

EXPERIMENTAL

Materials

The calcium carbonate products used in this study were representative of the two general classes of products available today,
the ground natural products and the synthetic or precipitated calcium
carbonate (PCC) products. The ground natural products were made by
physically grinding limestone to a target particle size. Because of
the nature and practical limitations of the grinding process, the
ground natural products are generally larger in size and have a much
broader particle size distribution than the precipitated calcium carbonates. Both of these properties have been found to pose limitations
where light scattering, TiO_2 extending and impact properties in PVC
are concerned.

The precipitated calcium carbonate (PCC) products used in this
study were synthesized in controlled precipitation processes by the
reaction:

$$Ca(OH)_2 \ + \ CO_2 \ \rightarrow \ CaCO_3 \ + \ H_2O$$

As a mineral, calcium carbonate is distinguished by its existence in
three different crystalline polymorphs (calcite, aragonite, and vater-
ite) and literally dozens of crystal habits, the most common being
prismatic, tabular, cuboid, rhombohedral, scalenohedral and acicular.
Novel process modifications and controls were applied to produce the
various PCC sizes and morphologies evaluated in this work (see Table
I). In each case, the controlled precipitation processes yielded
high purity products of uniform size, shape, and surface chemistry.
Submicron particle sizes and narrow particle size distributions are
distinct advantages of the precipitated calcium carbonate products.

All of the calcium carbonate samples were surface treated with
a fatty acid. The levels of the fatty acid treatment were varied according to the surface area of the different samples to produce a
monomolecular surface coverage. Since the fatty acid reacts with
the calcium carbonate to form a calcium soap on the surface, the resultant filler becomes hydrophobic and easier to disperse in PVC.
This results in substantially improved impact strength in the filled
PVC. Solvent extraction and carbon X-ray photoelectron spectroscopy
analyses on the treated product confirm that the fatty acid had completely and irreversibly reacted with the calcium carbonate.

Since the filler dispersion is known to influence impact strength
to a significant degree, all of the calcium carbonate samples were
formulated and processed into PVC sidings and profiles which were
then characterized for dispersion and impact properties. The dispersion properties were evaluated in two ways. First, calcium elemental mapping was done on a KEVEX 5100-EDAX system tied into a Scan-

Table I. Morphology and Properties of PPTD, $CaCO_3$ Synthesized

	AVG. PSD (microns)	N_2 SURFACE AREA (m^2/g)	LIGHT SCATTERING COEFFICIENT
	0.02	55	405
	0.07	20	890
	0.50	6	898
	0.80	9	875
	5.0	5	463

ning Electron Microscope (SEM). A low temperature ashing technique was used in which an oxygen plasma at approximately 1.0 ton pressure and 13.56 M Hz was first applied to the PVC surface to erode away the organic material (see Figures 1 and 2). Second, cryogenically microtomed sections were examined via Transmission Electron Microscopy (TEM) to study the distribution of filler and other additives.[9] All of the calcium carbonates dispersed quite well with the single exception of the smallest particle size (0.02μ) sample.

The precipitated calcium carbonates found most suited for siding and profiles are commercial now and are sold by Pfizer Inc., MPM Division, under the tradename of Super-Pflex 200 and Ultra-Pflex. Other precipitated calcium carbonates used were synthesized to com-

plete the particle size and morphology grid and are not commercially
available at this time.

Figure 1.   CaCO$_3$ dispersion in extruded PVC siding by calcium
            mapping technique.

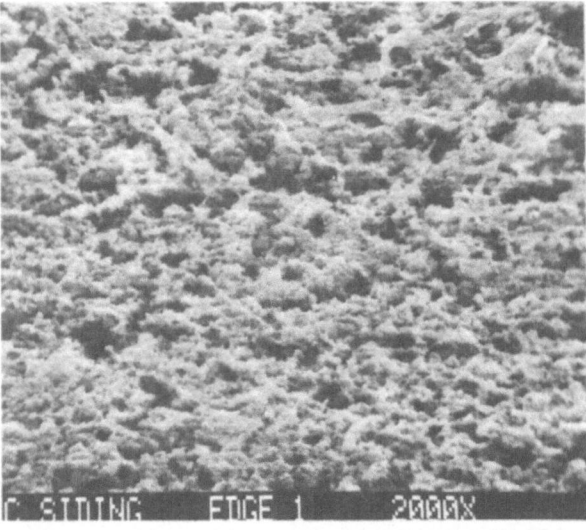

Figure 2.   CaCO$_3$ and TiO$_2$ dispersion in PVC siding by low tempera-
            ture plasma ashing and SEM/KEVEX analysis.

The poly(vinyl chloride) homopolymer used in this study was a
general purpose siding and profile grade homopolymer with a K value
of 66. The basic formulation contained butyl tin mercaptide stabilizer, process aid, processing lubricant, impact modifier, TiO_2, and
calcium carbonate as specified.

Dry Blending and Extrusion

Appropriate amounts of PVC powder and ingredients were blended
on a high intensity mixer using the standard blending procedure.
The final mix temperature was between 105-110°C. The dry blends were
cooled and extruded into 3" wide 40 mil thick tapes. A coat hanger
extrusion die was used to attain uniform extrusion flow. The extruded tapes were water cooled, dried, conditioned, and tested.

Testing - Physicals and Color

The impact strength of siding and profiles was measured by the
Gardner Variable Height Impact Tester throughout the study. The
Tinius-Olsen Dynatup 800 and Notched Izod Impact Testers were utilized
to relate the PVC impact property with $CaCO_3$ particle size. The sam-
ples were weathered on a QUV accelerated weatherometer for 1000 hours.
Additional testing was done by exposing the samples at 45° south in
Arizona and Florida for 18 months which equated to about 250,000
Langley hours. Color and impact measurements were made after 3, 6,
9, 12, and 18 month of exposure. The color was measured using a
Diano-Hardy Spectrophotometer and were computed as total color change,
yellowness, blueness, and lightness.

DISCUSSION

As shown in Table I, precipitated calcium carbonates of average
particle sizes 0.02, 0.07, 0.50, 0.80, and 5.0 microns and varying
morphologies were synthesized for the study. The surface area of
these $CaCO_3$'s varied from 55 m^2/g to 5 m^2/g. Natural ground coated
calcium carbonates in the particle size range of 0.80-5.0 microns
were used to complete the experimental matrix. As shown in this
table, calcium carbonates with average particle size between 0.07-
0.80 microns were optimum for light scattering.[10] This result differs from the published value of 2.2 microns.[11]

The 0.07 micron and 0.50 micron products showed very little
effect on the impact strength of poly(vinyl chloride) up to 10 phr
loading (Figure 4). At 15 phr loading, the 0.07 micron $CaCO_3$ was
most efficient for both drop weight and notched Izod impact.

The 10:1 aspect ratio 0.80 micron $CaCO_3$ primarily improved the
tensile properties and had little influence on the impact strength.
Other larger size precipitated calcium carbonates as well as natural

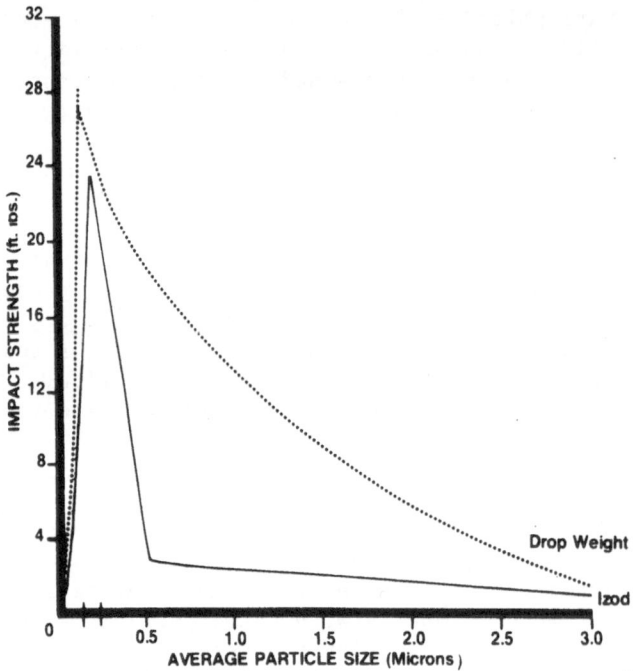

Figure 3. Effect of average intrinsic particle size of coated CaCO₃
fillers on impact properties of rigid PVC.

ground calcium carbonates lowered the initial impact strength sub-
stantially (Figure 3). Therefore, both the 0.07 and 0.50 micron
CaCO₃ were formulated in a siding formulation and were screened by
EMMAQUA and QUV weatherometer for approximate loading levels.

This study suggested that a loading of 5-10 phr of Coated Precipitated Calcium Carbonate (CPCC - 0.5 microns) and Coated Ultra-
fine Precipitated Calcium Carbonate (CUPCC - 0.07 microns) were most
effective for long term impact retention of poly(vinyl chloride)
siding. Simultaneously, a dramatic reduction in yellowing of poly
(vinyl chloride) was also seen by the use of these calcium carbonates.

In order to develop external weathering data, 5 phr and 10 phr
CUPCC (0.07 microns) and CPCC (0.50 microns) filled sidings were subjected to 45° South Arizona and Florida exposures and were tested for
color and impact properties after 3, 6, 9, 12, and 18 month exposures,
equating to a total of 250,000 Langley hours.

As shown in Figures 5 and 6, the 5 phr CUPCC (0.07 microns) and
CPCC (0.50 microns) filled sidings retained comparable impact strength
to the unfilled (referring to the Control containing no calcium car-

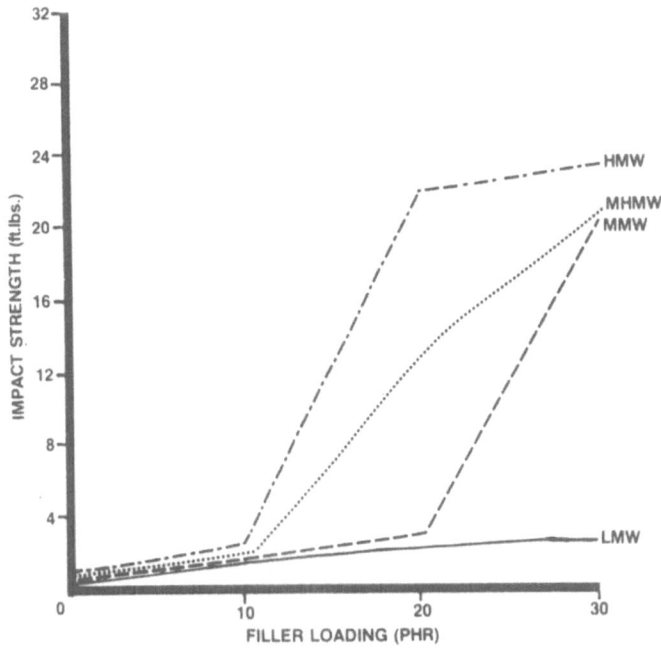

Figure 4. Effect of filler loading on the notched Izod impact of
different molecular weight PVC.

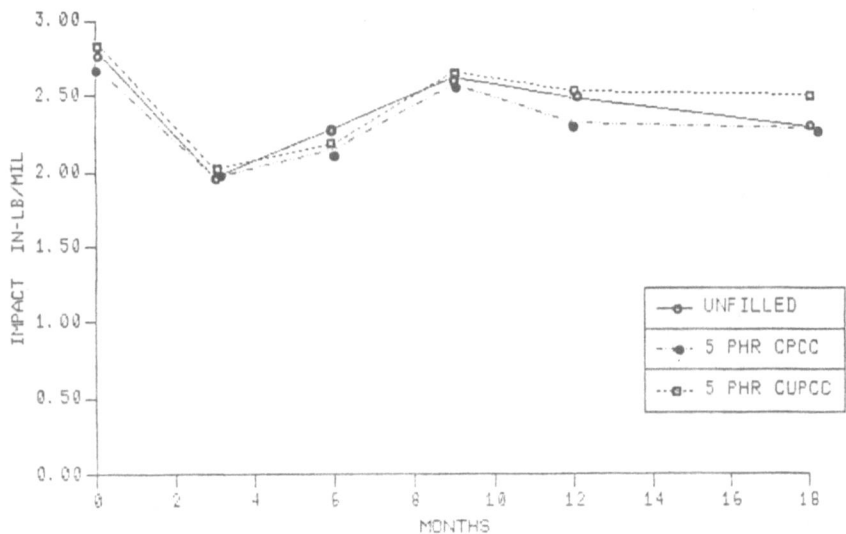

Figure 5. Gardner VHIT of 5 phr CPCC and CUPCC filled PVC siding -
45° S. Florida exposure.

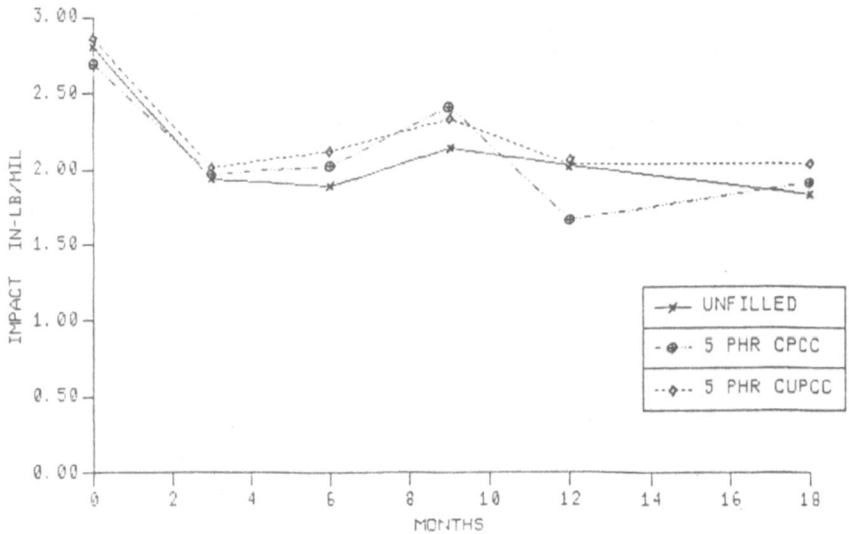

Figure 6.   Gardner VHIT of 5 phr CPCC and CUPCC filled PVC siding -
             45° S. Arizona exposure.

bonate but the standard 12 phr TiO_2) siding after 18 months exposure
in Florida. At 10 phr loading, CUPCC (0.07 microns) was practically
equal to the unfilled siding and slightly higher than the CPCC (0.50
microns) filled sidings (Figure 7).

     Similarly, both CPCC and CUPCC performed equivalently at 5 phr
loading; however, at 10 phr the CUPCC retained greater impact strength
after 18 months exposure in Arizona (Figure 8).

     As shown in Figures 9 and 10, the total color difference between
$CaCO_3$ filled and unfilled siding was much greater up to 9 months of
exposure in Florida, as well as Arizona. The unfilled siding turned
much yellower (Figures 11 and 12) on exposure. Between 12 and 18
months, the top oxidized layer chalked off leaving a fresh surface
with very little yellowing. The $CaCO_3$ filled siding followed a similar pattern, but imparted dramatically lowered initial yellowing
to the siding.

     Since the optimum loading for CPCC (0.50 microns) was between
5-10 phr, an additional accelerated study was conducted where both
CUPCC and CPCC were compared for performance at 7 phr level.

     As shown in Table II, the 7 phr addition of CPCC maintained
the impact strength after 1000 hours of QUV weathering. This also
permitted a reduction in TiO_2 from 10 phr to 6 phr level (Table II,
Figures 16, 17, 18). The total color difference, ΔE, yellowness,
(Δb) and lightness (ΔL) at reduced TiO_2 levels, also given in Figures

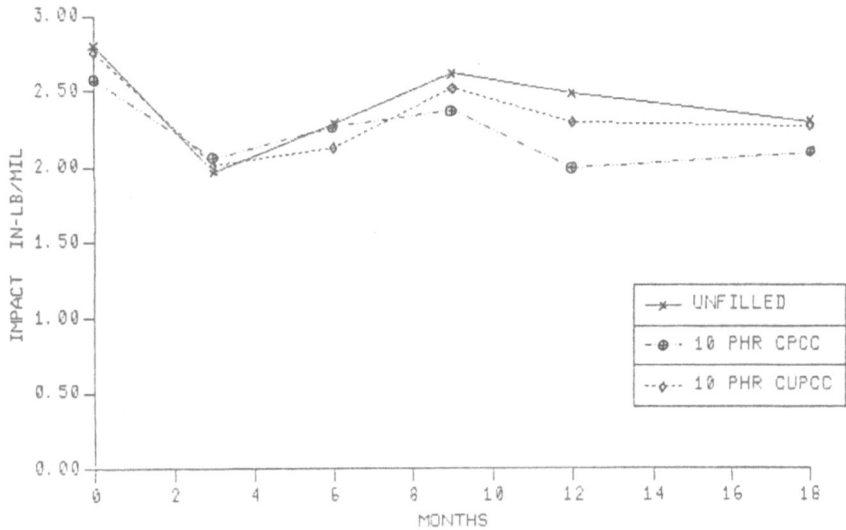

Figure 7.   Gardner VHIT of 10 phr CPCC and CUPCC filled PVC siding -
45° S. Florida exposure.

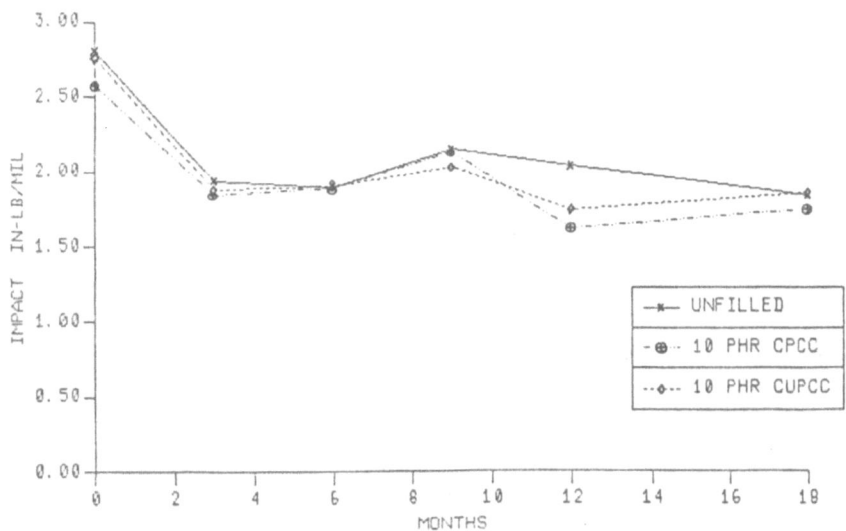

Figure 8.   Gardner VHIT of 10 phr CPCC and CUPCC filled PVC siding -
45° S. Arizona exposure.

16, 17 and 18, also shown that a reduction in TiO_2 would be possible
by the use of CPCC.

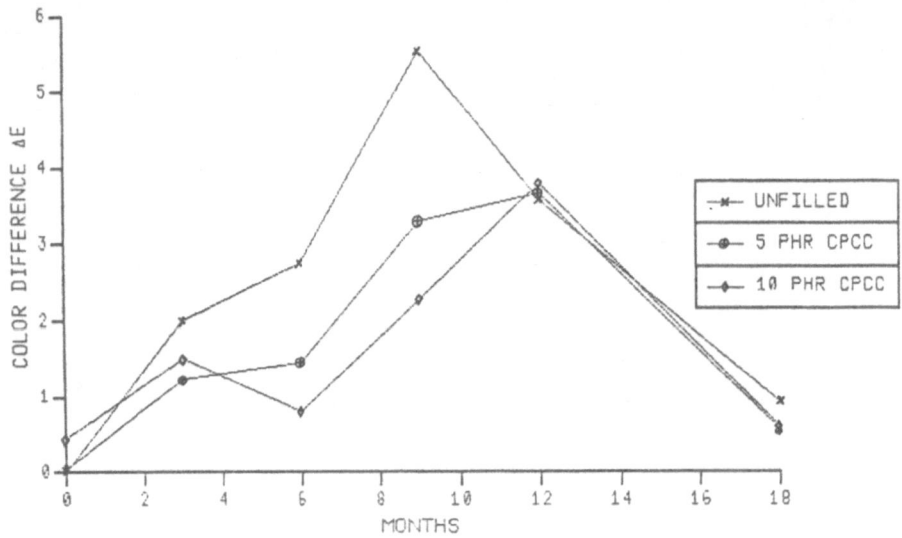

Figure 9.   Total color difference (ΔE) in PVC siding made with CPCC
            after 45° S. Florida exposure.

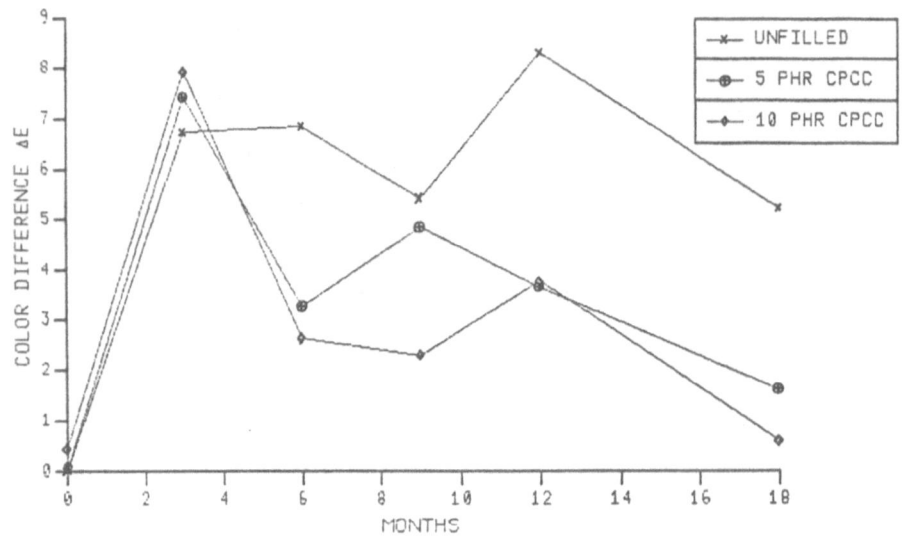

Figure 10. Total color difference (ΔE) in PVC siding made with CPCC
            after 45° S. Arizona exposure.

     Since the exterior poly (vinyl chloride) door and window pro-
files have less stringent impact requirements, the possibility of
using high loadings of CUPCC (0.07 microns) as a co-impact modifier
or the only impact modifier in the formulation was studied. As shown

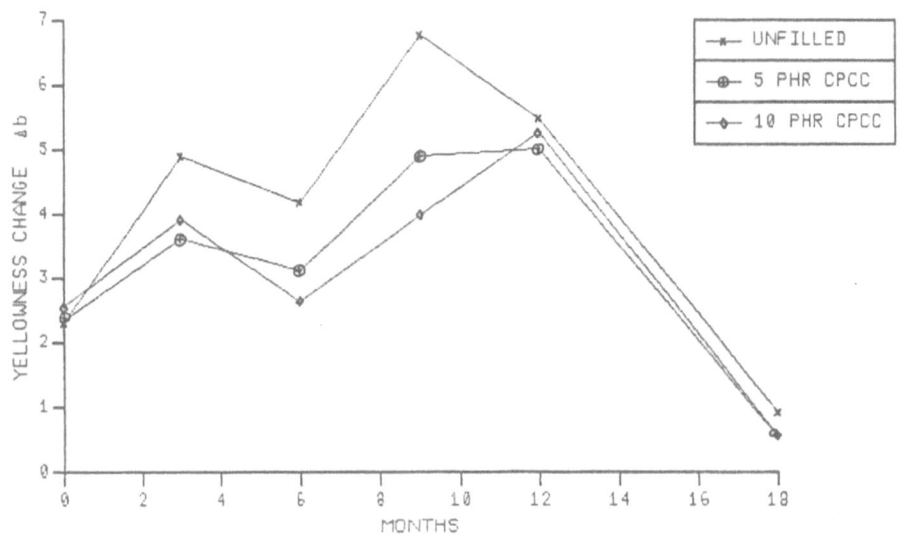

Figure 11.   Yellowness change (Δb) of PVC siding made with CPCC
             after 45° S. Florida exposure.

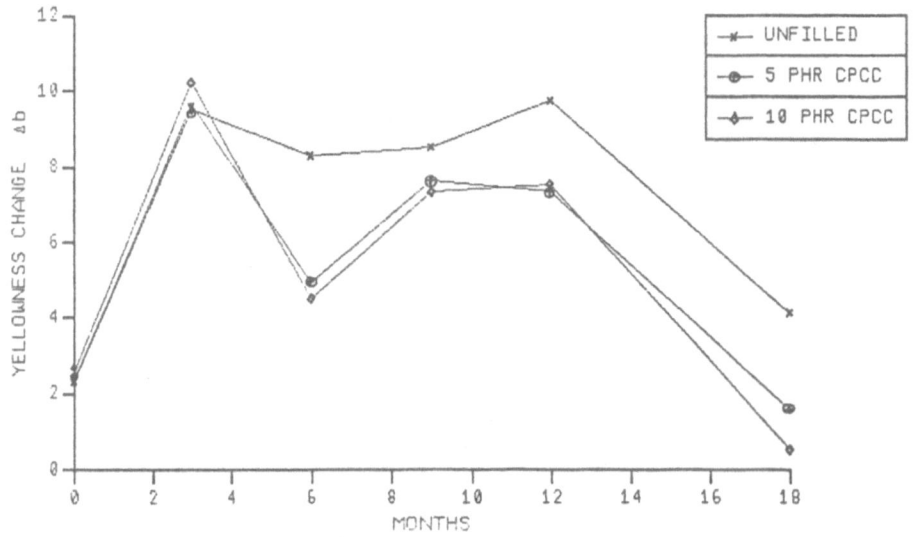

Figure 12.   Yellowness change (Δb) of PVC siding made with CPCC
             after 45° S. Arizona exposure.

in Table III, the 15 phr addition of CUPCC (0.07 microns) permitted
complete removal of acrylic impact modifiers and also reduced the
initial yellowing dramatically.

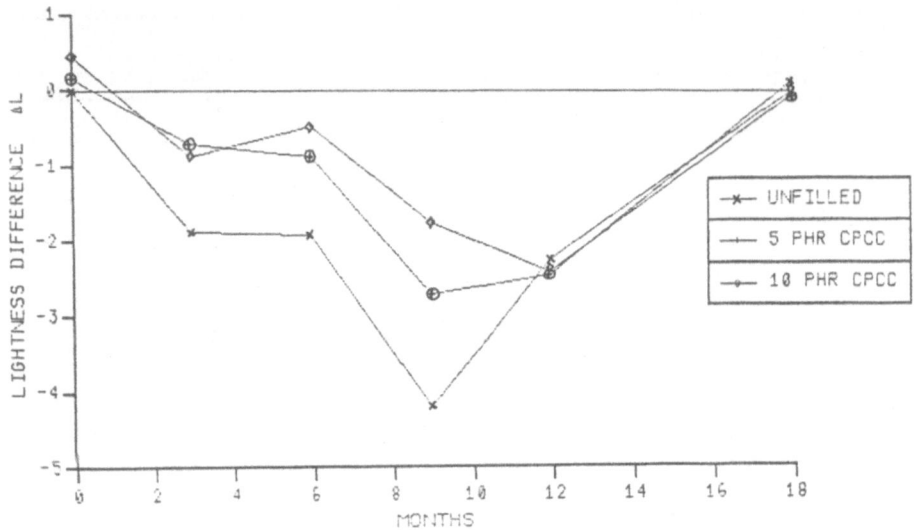

Figure 13.   Lightness difference (ΔL) in PVC sidings made with CPCC
             after 45° S. Florida exposure.

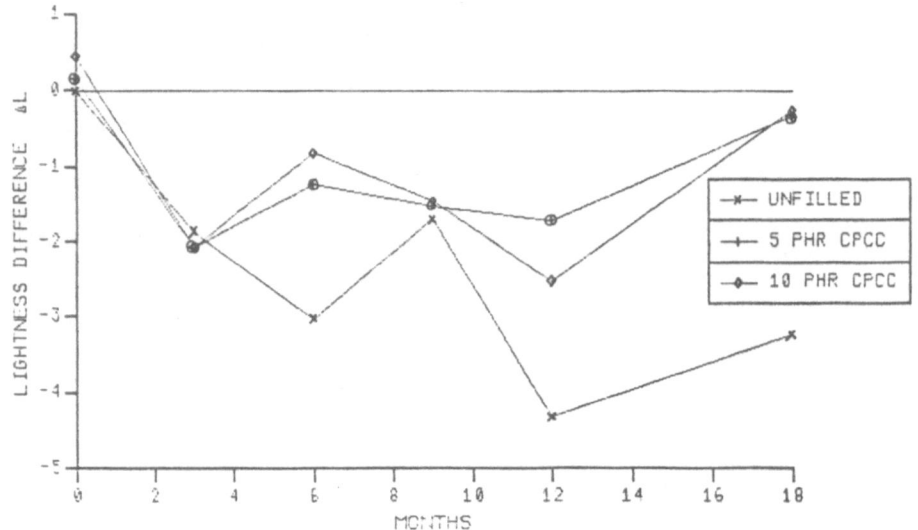

Figure 14.   Lightness difference (ΔL) in PVC sidings made with CPCC
             after 45° S. Arizona exposure.

        In an attempt to understand the ultraviolet stabilization mech-
anism by $CaCO_3$, the CPCC and CUPCC filled sidings were studied for
ultraviolet reflectance on the Cary Model 14 Spectrophotometer. The
scans made at 5°A intervals from 4000-2005°A gave the reflectance
density

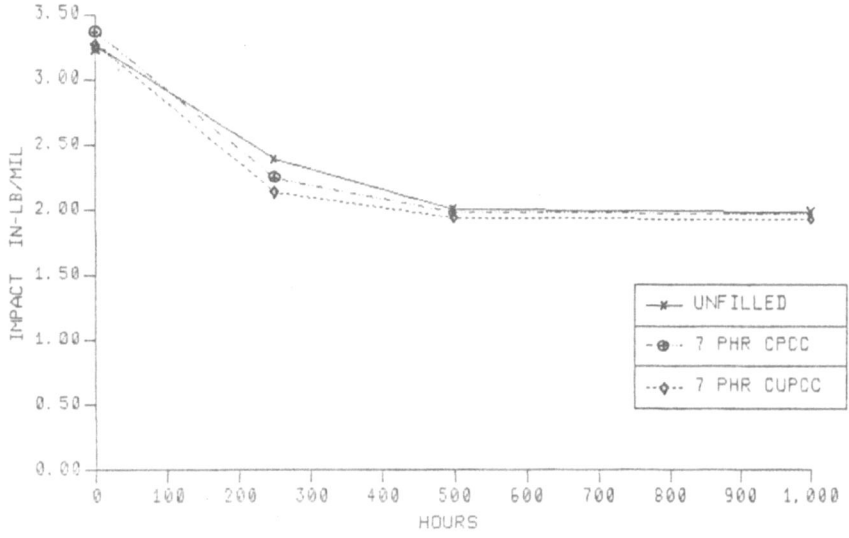

Figure 15.   Gardner VHIT of 7 phr CPCC and CUPCC filled PVC siding –
QUV weathering.

Table II.   Effect of CPCC on Color and Impact Retention
of PVC Siding by QUV Accelerated Weathering

Formulations	1	2	3	4
PVC - K66	100			
Butyl Tin Mercaptide	1.5			
Calcium Stearate	0.5			
Paraffin Wax	1.25			
Acrylic Process Aid	1.00			
Acrylic Impact Modifier	8.00			
TiO_2 (Rutile-Chalking)	10.00	10.00	8.00	6.00
CPCC (0.5 micron)		7.00	7.00	7.00
Results				
(A) Gardner Impact				
Initial "lb/mil	3.26	3.37	3.47	3.33
250 Hours "lb/mil	2.39	2.24	2.53	2.19
500 Hours "lb/mil	2.00	1.98	1.95	2.38
1000 Hours "lb/mil	1.98	1.97	2.28	2.33
(B) Color				
Yellowness (Δb)				
Initial	3.3	2.84	2.73	2.74
250 Hours	5.40	2.50	3.10	4.10
500 Hours	8.00	4.40	6.80	6.20
1000 Hours	10.40	7.90	9.29	9.96

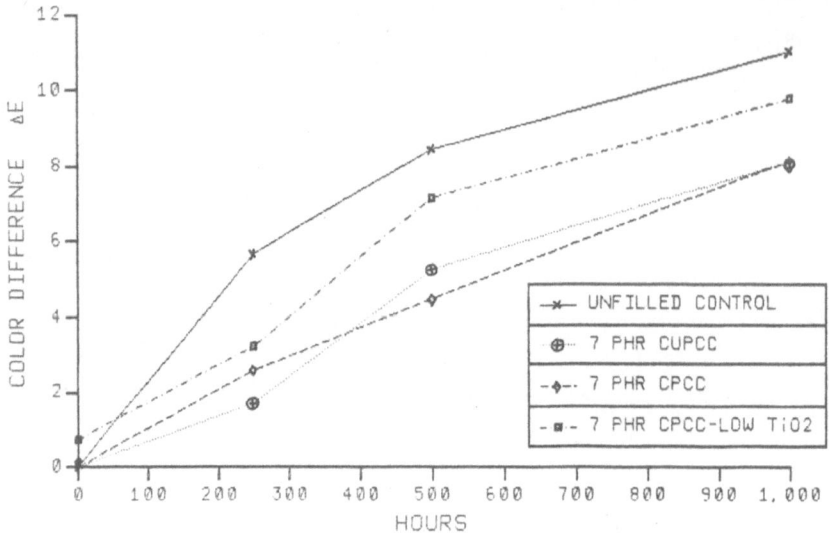

Figure 16.   Total color difference (ΔE) of PVC siding made with CPCC
             and CUPCC - QUV weathering.

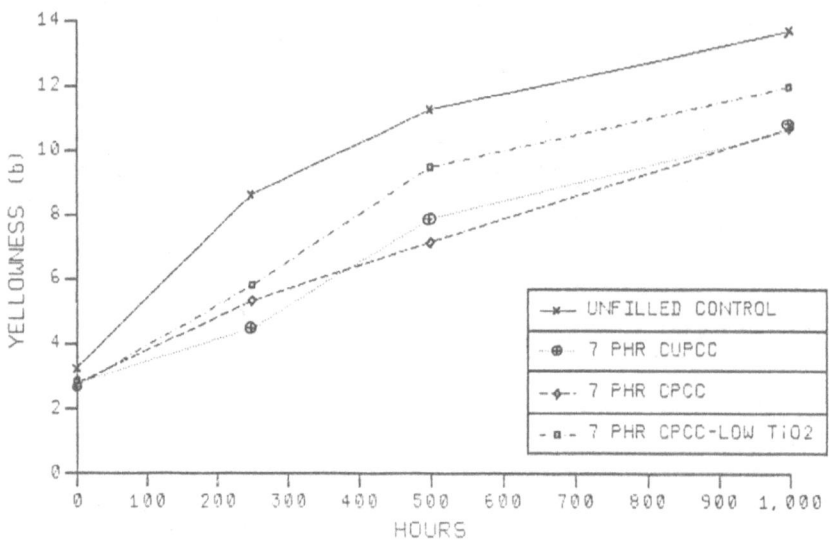

Figure 17.   Yellowness (b) of PVC siding made with CPCC and CUPCC -
             QUV weathering.

D(λ)   = $\log_{10}$ [1/R (λ)]
where D(λ) is the reflection density at wavelength λ and R is the
reflectance of wavelength λ expressed as a fraction. When converted
to absorptance and integrated for Spectral Energy, the difference
in $CaCO_3$ filled and unfilled sidings was very small, suggesting that

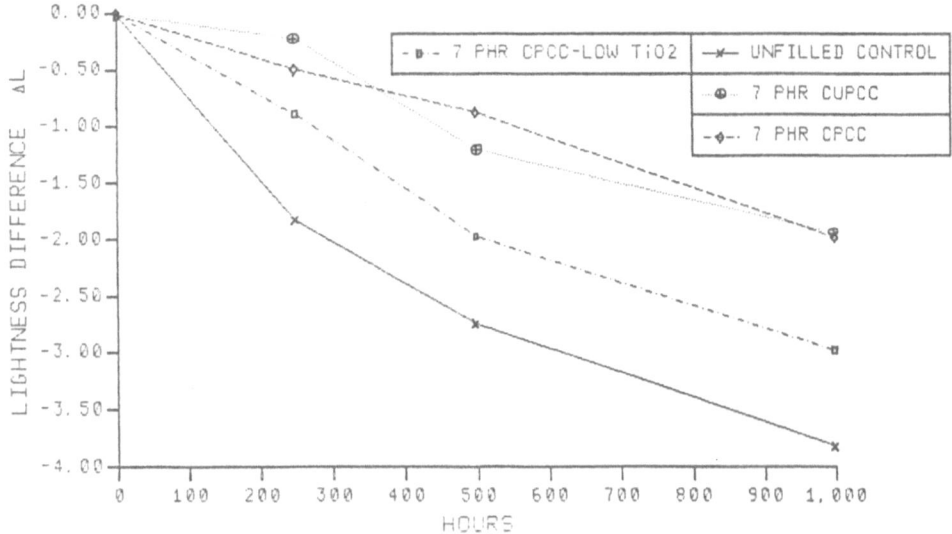

Figure 18. Lightness difference (ΔL) of PVC siding made with CPCC and CUPCC - QUV weathering.

Table III. Effect of CUPCC on Color and Impact Retention of PVC Profile by QUV Accelerated Weathering

Formulations	1	2	3	4
PVC - K66	100			→
Butyl Tin Mercaptide	1.5			→
Calcium Stearate	0.5			→
Paraffin Wax	1.25			→
Acrylic Process Aid	1.50			→
TiO₂ (Rutile-Chalking)	10.00			→
Acrylic Impact Modifier	6.00	6.00	4.5	0
CUPCC (0.07 micron)	-	15.0	15.0	15.0
Results				
(A) Gardner Impact				
Initial "lb/mil	3.51	3.31	3.22	3.42
250 Hours "lb/mil	2.12	2.12	2.37	2.18
500 Hours "lb/mil	2.09	2.03	2.20	2.32
1000 Hours "lb/mil	2.16	2.26	2.10	2.29
(B) Color Yellowness (Δb)				
Initial	2.96	3.09	3.66	2.60
250 Hours	4.71	2.58	2.15	3.20
500 Hours	7.08	6.88	3.70	5.00
1000 Hours	10.39	7.96	5.61	6.83

the increase in light scattering was not the dominant mechanism in the reduction of PVC yellowing by CPCC and CUPCC. It is hypothesized that the high surface area $CaCO_3$ pigments are extremely efficient in scavenging HCl, thereby slowing the dehydrochlorination of poly (vinyl chloride) and minimizing yellowing.

CONCLUSIONS

The present studies have shown that a precipitated calcium carbonate of optimum particle size can provide a unique and valuable combination of property improvements in PVC siding and profile applications.

PVC Siding

1. Precipitated calcium carbonates with a prismatic morphology, an average particle size in the range 0.07-0.50μ, and surface treated with a fatty acid were found to provide the best long term impact retention in PVC sidings.

2. The optimum loading level was found to be 5-10 phr.

3. At 5-10 phr, these precipitated calcium carbonates also inhibited the initial yellowing caused by exterior exposure and showed the potential for lowering the TiO_2 requirement by 30-40%, thereby lowering the cost of the siding product.

PVC Profiles

1. Addition of the 0.07μ coated precipitated calcium carbonate improved both Notched Izod and drop weight impact strength of PVC profiles.

2. The impact strength continued to improve with increased loading levels up to 15 phr.

3. The study also demonstrated the possibility of lowering or eliminating the impact modifier from the formulation containing this 0.07μ $CaCO_3$ product.

4. Similar to the PVC siding results, the yellowing of PVC profiles was also minimized by the use of the coated precipitated calcium carbonate products in the 0.07-0.50μ optimum size range.

REFERENCES

1. (a) L. I. Nass, ed., Encyclopedia of PVC, Vol. 1, Chap. 8 & 9, Marcel Dekker, Inc., New York and Basel, 1976; (b) W. L.

Hawkins, Ed., Polymer Stabilization, Wiley-Interscience, New York, 1972.

2. (a) G. Ayrey, B. Head and R. Poller, J. Polym. Sci., Macromol. Rev., 8, 1 (1974); (b) D. Braun, "Degradation and Stabilization of Polymers," G. Geuskens, Ed., Wiley, New York, 1975, p. 23.

3. A. Maccoll, Chem. Rev., 69, 33 (1969).

4. E. Arlman, J. Polym. Sci., 12, 547 (1954).

5. S. Van der Ven and W. deWit, Angew Makromol. Chem., 8, 143 (1969).

6. L. Bateman, Quarterly Review (London), 8, 147 (1954).

7. G. P. Mack, Modern Plastics, 31, 150 (1953).

8. J. McKellar, Rad. Res. Rev., 3, 141 (1975); P. Bentley, J. McKellar, G. Phillips, Rev. Progr. Color Relat. Tap, 5 (33), 1974; W. Hawkins, Polymer Stabilization, Wiley-Interscience, New York (1972); G. Scott, Eur. Polym. J. Suppl., 189 (1969), Br. Polym. J., 3, 24 (1972); Pure Appl. Chem., 30, 267 (1972).

9. F. G. Stieg, Official Digest of FSPT, 52-54 (Jan., 1959).

10. K. Mathur and S. Driscol - SPE ANTEC, Vol. XXVII (1981).

11. R. Zeller, TAPPI, 63, 13 (May, 1980).

12. P. F. Woener, "Synthetic Calcium Carbonate," Pigment Handbook, Vol. I, New York, John Wiley & Sons (1973).

IRON COMPOUNDS AS ADHESION PROMOTERS FOR BITUMEN

J. Shim-Ton, S. Varevorakul and R. T. Woodhams

Center for the Study of Materials
University of Toronto
Toronto, Canada, M5S 1A4

INTRODUCTION

Asphalt paving composites normally contain about 6 percent of a bituminous hydrocarbon binder and the remainder an inorganic aggregate comprising particles of sand, crushed rock, limestone or other minerals which may be indigenous to a particular region. Although it is recognized that prolonged exposure of asphalt pavements to moisture can cause a pronounced loss of strength and accelerate the deterioration of road surfaces, there does not appear to be any generally acceptable solution to the problem.[1,2]

There are, of course, numerous amino-type "anti-stripping" agents which are commercially available for the prevention of moisture damage but in several publications the effectiveness of these additives has been questioned since in a few cases the addition of anti-stripping agents, while initially aiding surface wetting, caused accelerated stripping in the later stages.[3] The tendency of amine functional groups to imbibe moisture and spontaneously form emulsified water droplets was responsible for this accelerated deterioration. Ciplijauskas et al[4] examined the flexural strength of asphalt concretes in which the ashpalt binder had been chemically modified. This study concluded that sulfonated or carboxylated asphalts, when used as binders, can produce composites which are virtually immune to moisture even after total immersion in water for several months. However, the results were not uniform and the retention of mechanical strength was dependent upon the choice of aggregate, most aggregates being of complex composition and containing various transition metals, particularly iron, which could have contributed to the observed effects. Subsequent contact angle and peel test studies by Shim-Ton[5]

revealed that polar hydrophilic substituents (sulfonic acids, car-
boxyls) were easily displaced by water when such acidic groups were
not insolubilized or complexed with transition metals such as iron
or chromium. Magnesium or calcium salts appear to have an intermediate adhesion retention and are not as effective for resisting moisture attack.

The use of silane coupling agents as adhesion promoters for
glass fiber reinforced plastics is well known, particularly as a
treatment to minimize the harmful effects of moisture. Morris and
Di Vito[6] were able to demonstrate that silane treatment greatly increased the moisture resistance of asphalt paving composites. Although silanes are certainly effective reagents for protecting asphalt
concretes from moisture, their high cost is likely to deter widespread
application.

In 1970, Fromm observed that a particular reddish-brown aggregate, which was used to pave a section of highway in northern Ontario,
was unusually resistant to moisture damage and more durable than
previous paving surfaces in that region. Subsequent laboratory investigation[3] revealed that iron naphthenate (or ferric complexes of
1,2-pentanedione) when added to asphalt in minor amounts (1%) were
particularly effective reagents for preventing the loss of adhesion
between asphalt and glass surfaces after immersion in water. The
following investigation was initiated to assess the performance of
iron naphthenate as an anti-stripping agent or coupling agent for
asphalt concrete. Borosilicate glass plates were selected to represent a typical inorganic substrate and the adhesion of bitumen coatings containing various additives was measured with an ASTM 90 degree
peel test procedure (after a suitable immersion period in distilled
water). The peel test measurements revealed changes taking place in
the cohesive properties of the bitumen during prolonged immersion and
also detected the eventual adhesive failure at any particular temperature.

PREDICTING ADHESION FAILURE

The equilibrium contact angle (θ) assumed by liquid bitumen (b)
on a silica surface (s) can be related to the surface energies by
Young's equation

$$\gamma_b \cos\theta = \gamma_s - \gamma_{sb} \qquad\qquad (1)$$

where γ_b = surface free energy of bitumen; γ_s = surface free energy
of silica; γ_{sb} = surface free energy of bitumen-silica interface.

The thermodynamic expression for the theoretical work of adhesion W_{adh} is given by

$$W_{adh} = \gamma_b (1 + \cos\theta) \tag{2}$$

In order to find out whether a bitumen-silica joint is stable towards moisture, the dispersion and polar force components of the bitumen, water and silica must be known. Table I provides the necessary data from previous publications. The assumption was made that the polar surface energy component of bitumen (γ^p) is negligible to the dispersion component (γ^d).

Using the Dupré relationship (Equation 1) it is possible to predict whether the bitumen-silica bond will fail when exposed to moisture. The condition for spontaneous separation is

$$\gamma_{bs} > \gamma_{bw} + \gamma_{sw} \tag{3}$$

where γ_{bs} = surface energy of a bitumen-silica interface (= 218 mJ/m^2); γ_{bw} = surface energy of a bitumen-water interface (= 51.5 mJ/m^2); γ_{sw} = surface energy of a silica-water interface (= 71.5 mJ/m^2).

Substitution of the values of Table I into the equation

$$W_{adh} = \gamma_{bw} + \gamma_{sw} - \gamma_{bs} \tag{4}$$

yields a value of -95 mJ/m^2 for the work of adhesion (i.e., spontaneous separation). It can be demonstrated that most hydrocarbons and polymeric resins will be displaced by water from inorganic surfaces unless special coupling reagents are employed.

MATERIALS

Shell Venezuelan asphalt, a paving grade bitumen used extensively throughout North America, was selected for testing. The properties of this asphalt are summarized in Table II.

The naphthenic acids are monocarboxylic acids of the naphthene series of hydrocarbons derived from petroleum.[9] These organic acids may be reacted with basic metal oxides to produce the corresponding metal carboxylates. Iron naphthenate, a black solution obtained from Nuodex Canada, Ltd., contains 6% iron by weight. Metal naphthenates are commonly sold as driers for coating enamels. The Pyrex brand borosilicate glass plates have a chemical composition as shown in Table III.

Portions of asphalt were maleated using a technique described by Ciplijauskas.[4] Sulfonations were accomplished using acetyl sulfate.[11] Sulfuric acid was added slowly to heated bitumen at 100°C in a stirred reaction kettle followed by the dropwise addition of a stoichiometric quantity of acetic anhydride. After 30 minutes

Table I. Dispersion and Polar Force Components of Surface Energies[8]

Substance	Surface Free Energies, mJ/m^2		
	γ^d	γ^p	γ
Water (w)	22.0	50.2	72.2
Bitumen (b)	34	~0	34
Silica (s)	78	209	287

Table II. Properties of Shell Venezuelan Asphalt

		Temperature °C	ASTM Test
Viscosity, Pa.s	1519	60	D-2170
Penetration	10	5	D-5
	96	25	
Penetration Index	-1.3		
Ductility	11.5	4	D-113
	150+	25	
Flash point, °C	299		D-92
Density, g/cm^3	1.011	25	D-70

the reaction was terminated and the mixture cooled to room temperature. Portions of the maleated and sulfonated bitumens were neutralized by the addition of calcium hydroxide to form the corresponding
calcium salts.

TESTING PROCEDURES

A weighed quantity of the additive (iron naphthenate) was mixed
into heated asphalt at 100°C and spread onto a clean glass plate.
A narrow strip of glass fiber cloth (satin weave) was impressed onto
the molten asphalt surface followed by a sheet of paper and another

Table III.   Composition of Borosilicate Glass[10]

	Percent
SiO_2	80.6
B_2O_3	11.9
Al_2O_3	2.0
Na_2O	4.4
Ca or Mg	1.1

glass plate. A pair of metal shims (0.60 cm thick) were placed
along each side of the cloth strip so that the final thickness of
the asphalt film could be exactly reproduced in each test specimen.
A schematic diagram (Figure 1) illustrates the construction of the
test assembly. Weights were placed on top of this assembly for several minutes until the asphalt had cooled and the desired film thickness was attained with the aid of shims. The sheet of paper simply
prevents the upper glass plate from adhering to the asphalt and does
not interfere with the subsequent test. These samples were conditioned in air or distilled water for 25 days prior to peel testing.
Before conditioning, two parallel longitudinal incisions were made
with a sharp utility knife producing a glass cloth strip 2.54 cm
wide ready for peel testing.

The peel tests were conducted at 90° using the ASTM B-533-70
technique which was originally developed to measure the peel strength
of metal films which had been plated onto a plastic surface. The
glass test plates were individually mounted in a sliding carriage
attached to the crosshead of an Instron Tester with one end of the
glass fiber tape attached to a load cell as shown in Figure 2. Most

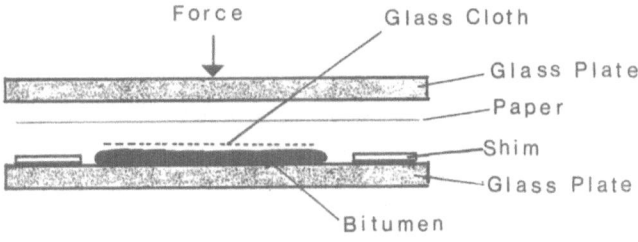

Figure 1.   The sample of heated bitumen is spread between two glass
            plates and a glass tape is impressed onto the surface.

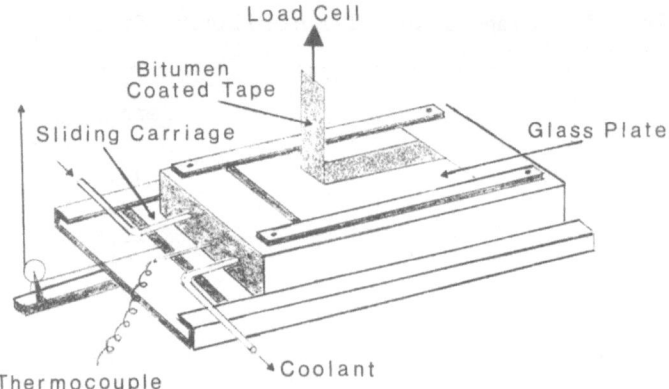

Figure 2.   The 90° peel test is a convenient technique for measuring
            the cohesive or adhesive properties of bituminous mix-
            tures in contact with various substrates over a range of
            temperatures (ASTM B 533-70).

measurements were performed at a constant temperature of 16.5°C, the
temperature being maintained by a refrigerant circulated through the
metal base plate of the holder. The peeling force was recorded at
a constant peeling rate of 1 cm/min for five independent samples in
order to obtain a reliable average value. The assembly could be
cooled to -20°C in order to measure the cohesive and adhesive properties of the bituminous layer at reduced temperatures. The glass
plates used in these experiments for the measurement of peel strengths
may be substituted by polished sections of mineral rocks, e.g., traprock, dolomite, in order to evaluate other substrates.

The glass plates were prepared by first washing them in laundry
detergent, then rinsing and placing in a warm (90-95°C) sodium hydroxide solution (1%) for 5 minutes. The chemically cleaned plates
were then immersed in warm (45-50°C) hydrochloric acid solution (5%)
for another 5 minutes before applying a final rinse with distilled
water. After cleaning, the plates were stored in distilled water
until ready for use.

Contact Angle Measurements

The technique described by Neumann and Good[12] was used for contact angle measurements. A small pin-head sized particle of the
bitumen mixture was obtained by pulverizing a sample previously cooled
to -20°C to render it brittle. The particle was placed on a clean
dry glass plate and heated to 100°C in an air oven. Care should be
exercised to prevent contamination of the exposed glass surface
inside the oven by placing the specimen plate inside a clean desic-

cator free from volatile substances. After 45 minutes the sample
could be removed from the oven and the equilibrium contact angle
measured with a telescope equipped with a goniometer. The test plate
was then immersed in water near the boiling temperature so that the
liquid bitumen droplets could reestablish their equilibrium contact
angles. After 30 minutes the heated vessel containing the plate was
removed and cooled to room temperature before remeasuring the contact
angles. The contact angles were measured for at least five droplets
and averaged. Further details of these procedures may be obtained
from the theses of Shim-Ton[5] and Varevorakul.[13]

Peel-Test Results

Three failure modes were observed during the peel test measurements which were characterized as cohesive, mixed cohesive-adhesive
(stick-slip type of response) or adhesive. The peel strengths were
dependent upon the thickness of the adhesive films, so that the use
of shims was necessary to obtain comparable results. The cooled
bitumen tended to slowly thicken at room temperature reaching a constant value after 12 days. Hence it was important to allow the
samples sufficient time to reach equilibrium before measuring the
peel strengths.

Under normal atmospheric conditions at 23°C, the failure mode
at the bitumen-glass interface was always cohesive. After a few days
immersion in distilled water the bitumen samples without additives
separated cleanly from the glass surface with negligible adhesion
force. Note in Figure 3 the complete loss of adhesion after 20 days
immersion time. The addition of ferric naphthenate solution produced
a dramatic increase in peel strength even at the smallest concentrations (0.1% Nuodex iron naphthenate). The slight decrease in cohesive strength was attributed to dilution by the solvent which was
present in the ferric naphthenate solution. Powdered ferric oxide
gave a similar response but was not as effective at the smallest
concentrations. The ferric oxide might have been more effective at
these small concentrations if the suspension had been heated for a
longer period to allow more of the ferric oxide to dissolve in the
bitumen. No attempt was made here to prereact the iron oxide.

The bitumen samples containing ferric naphthenate were most effective at concentrations less than 1% since no loss of peel strength
was evident even after 75 days immersion. The iron naphthenate also
appeared to inhibit emulsion formation. Ferric oxide behaved in an
almost identical fashion.

The reported strength is the average stress value over the entire
test strip whether cohesive or adhesive in nature. Such values are
usually sensitive to the methods used to clean the glass plates. For
example, treatment of the glass plates with chromic acid solution

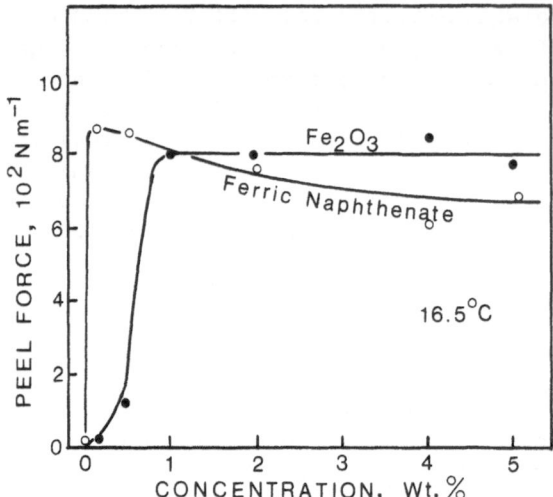

Figure 3. The addition of ferric oxide or ferric naphthenate to
bitumen prevents the loss of adhesion to a glass plate
after 20 days immersion in water. Without these additives the adhesive force is practically zero.

yielded surfaces which did not show any adhesion loss with bitumen
after water immersion. This suggests that residual chromate ions
may be responsible for the enhanced surface adhesion of glass in
this instance.

Contact Angle Results

The results of the contact angle experiments are summarized in
Table III. The contact angles of all the bitumen samples spontaneously increased after immersion of the glass plates in hot water.
The maleated bitumen tended to disperse so that no measurements could
be obtained. The sulfonated bitumen showed a very large increase in
the contact angle after immersion, even when neutralized with calcium
oxide.

DISCUSSION

It is apparent that a stable interfacial bond in the presence
of moisture is often difficult to achieve due to the powerful hydra-
ting effect of water molecules. Simple polar groups such as carboxylic or sulfonate groups are rapidly hydrated and are quickly displaced from most inorganic surfaces. Even when converted to their
relatively insoluble calcium or magnesium salts, these ionic salts
are gradually hydrated and displaced by water. Transition metal
salts form more stable chemical bonds with silicate surfaces, par-

ticularly ferric ions, since these ions can more easily fit into the tetrahedral coordination sites.[14] It is interesting to note that long chain fatty acids can be adsorbed on quartz if polyvalent cations (Cu, Mg, Ca, Ba, Zn, Pb, Fe^{3+}, Al^{3+}) are first adsorbed.[14]

The ferric naphthenate molecule probably forms a chemical bond with negatively charged silicate substrates, such behavior being analogous to that of a chromium complex (duPont Quilon or Volan bonding agents, for example). A proposed structure is shown in Figure 4 illustrating the multiple attachment sites of the ferric ion to the silicate substrate. The naphthenic acid molecule occupies one of the six coordination sites on each ferric ion.

These contact angle experiments may be employed to predict the wet strength behavior of asphalt composites. For example, a large contact angle (greater than 60°) predicts poor adhesion and loss of strength when such a binder is used in a composite. In this study, the sulfonated asphalt (calcium salt) gave a large contact angle (69.1°) in contact with glass. Earlier studies by Ciplijauskas[4] had shown that a sulfonated asphalt binder (magnesium salt) when used with a silica sand aggregate lost approximately 80% of its flexural strength after immersion in water. Hence contact angles can provide useful information concerning the resistance of asphalt composites to moisture attack.

This correlation has been used to anticipate the influence of other additives on bitumen adhesion with substrates other than glass (e.g., silica, limestone, traprock). It is interesting to note that iron naphthenate is effective with all the above substrates with respect to moisture resistance. These tests may be performed on small polished surfaces upon which a tiny particle of the bitumen

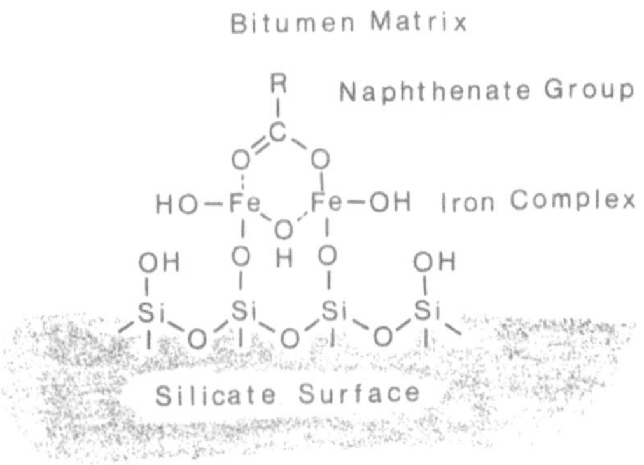

Figure 4. Schematic representation of the interfacial bond between ferric naphthenate and a silicate surface. The cohesive strength will be governed by van der Waals interactive forces between the naphthenate groups and the bitumen.

mixture is placed for subsequent examination under wet and dry conditions. The technique is likewise applicable to a wide variety of fillers and polymers and may have general utility for the study of interfacial adhesion.

The decrease in tensile strength of asphalt concrete paving surfaces due to ground moisture or a humid environment can occur during seasonal changes when rapid temperature changes induce large thermal shrinkage stresses, a common cause of cracking. These cracks can become quite numerous when the temperature falls below the brittle temperature of asphalt (near 0°C). The use of iron naphthenate coupling agents would be expected to reduce the number of stress-induced cracks and help to prolong the service life of roads.

Iron naphthenates are commonly used as driers (oxidation catalysts) in baking enamels so that some concern may exist concerning the long term oxidative stability of asphalt compositions containing iron. Most driers containing iron begin to show catalytic activity at temperatures greater than 66°C. Fromm[3] observed that cobalt naphthenate produced hard, brittle films when used as an additive, and attributed this effect to oxidation. Other polyvalent metal ions, such as aluminum, may be preferred if oxidation catalysis must be avoided.

The presence of acidic or basic functional groups in bitumen is quite likely to accelerate the loss of adhesion and adversely affect the mechanical strength of asphalt composites unless trace heavy elements in the aggregate are capable of forming strong water-resistant bonds. Such hydrophilic groups aid the formation of emulsified water in the bitumen, and reduce the cohesive strength of the bitumen unless rendered less hydrophobic by reaction with a heavy metal compound such as iron oxide.

The fixation of chemical bonds in the presence of water bears a similarity to the chemistry of flotation, flocculation or the use of mordants in dyeing, from which studies much useful information can be obtained.

CONCLUSION

Iron compounds, particularly ferric oxide and ferric naphthenate, are efficient adhesion promoters for bitumen, particularly in the presence of moisture. The presence of ferric iron also inhibits the tendency of bitumen to form emulsions with water. Since moisture damage is one of the principal factors contributing to the deterioration of asphalt paving surfaces, such additives should contribute to improved performance and reduced maintenance of asphalt roads. Contact angle measurement can be a useful diagnostic test for the prediction of adhesion retention in the presence of water. The techniques used in this investigation may also be applied to a systematic study of the interfacial wetting and adhesion characteristics of polymer materials containing fillers.

ACKNOWLEDGEMENTS

The authors wish to express their appreciation to H. J. Fromm of the Ontario Ministry of Transportation and Communications and to J. D. George, formerly with the Metro Toronto Roads and Traffic Department, for their advice and encouragement during these investigations. The research was supported by the National Science and Engineering Research Council of Canada (Grant No. A4873).

REFERENCES

1. J. A. N. Scott, "Adhesion and Disbonding Mechanisms of Asphalt ....," Association of Asphalt Paving Technologists, Conference Proceedings, February 13-15, 1978.
2. R. I. Hughes, D. R. Lamb and O. Pordes, "Adhesion in Bitumen Macadam," J. Appl. Chem., 10, 433-444 (1960).
3. H. J. Fromm, "The Mechanisms of Asphalt Stripping from Aggregate Surfaces," Proc. Assoc. Asphalt Paving Technol., 43, 191-206 (1974).
4. L. Ciplijauskas, M. R. Piggott and R. T. Woodhams., I and E. C. Product Research and Development, 18, 86-91 (1979).
5. J. Shim-Ton, "Adhesion of Asphalt to Glass," Department of Chemical Engineering and Applied Chemistry, University of Toronto, M.A.Sc. thesis, 1981.
6. J. A. Di Vito and G. R. Morris, "Silane Pretreatment of Mineral Aggregate to Prevent Stripping," Proceedings Transportation Research Board Meeting, January, 1982.
7. D. W. van Krevelyn and P. J. Hoftyzer, Properties of Polymers, Elsevier, pp. 168-171 (1976).
8. A. J. Kinloch, W. A. Dukes and R. A. Gledhill, "Durability of Adhesive Joints," Preprints ACS Coatings and Plastics Div., 35 (1), 546-559 (1975).
9. D. W. Kirk and D. F. Othmer, Encyclopedia of Chemical Technology, Interscience Publ., 13, 727-734 (1965).
10. L. Holland, "The Properties of Glass Surfaces," Chapman and Hall, London, 1966.
11. E. E. Gilbert, Sulfonation and Related Reactions, Interscience Publ., 1965.
12. A. W. Neumann and R. J. Good, "Techniques of Measuring Contact Angles," Plenum Press, pp. 31-91 (1979).
13. S. Varevorakul, "Adhesion Promoters for Bitumen," Department of Chemical Engineering and Applied Chemistry, University

of Toronto, B.A.Sc. thesis, 1981.
14. R. K. Iler, "The Colloid Chemistry of Silica and Silicates,"
     Cornell University Press, pp. 249-252 (1955).

INDEX

Page numbers indicate starting points of cited material. Often the pages immediately following will contain additional information.